MECHANICS OF

TRANSPORT

ADVANCED SERIES ON OCEAN ENGINEERING

Series Editor-in-Chief
Philip L- F Liu (*Cornell University*)

Vol. 1 The Applied Dynamics of Ocean Surface Waves
     by Chiang C Mei (MIT)
Vol. 2 Water Wave Mechanics for Engineers and Scientists
     by Robert G Dean (Univ. Florida) *and Robert A Dalrymple*
     (Univ. Delaware)
Vol. 3 Mechanics of Coastal Sediment Transport
     by Jørgen Fredsøe and Rolf Deigaard (Tech. Univ. Denmark)
Vol. 4 Coastal Bottom Boundary Layers and Sediment Transport
     by Peter Nielsen (Univ. Queensland)

Forthcoming titles:

Water Waves Propagation Over Uneven Bottoms
by Maarten W Dingemans (Delft Hydraulics)

Ocean Outfall Design
by Ian R Wood (Univ. Canterbury)

Tsunami Run-up
by Philip L- F Liu (Cornell Univ.), *Costas Synolakis* (Univ. Southern California),
Harry Yeh (Univ. Washington) *and Nobu Shuto* (Tohoku Univ.)

Physical Models and Laboratory Techniques in Coastal Engineering
by Steven A. Hughes (Coastal Engineering Research Center, USA)

Kalman Filter Method in the Analysis of Vibrations Due to Water Waves
by Piotr Wilde and Andrzej Kozakiewicz (Inst. Hydroengineering, Polish Academy of
Sciences)

Numerical Modelling of Ocean Dynamics
by Zygmunt Kowalik (Univ. Alaska) *and T S Murty* (Inst. Ocean Science, BC)

Beach Nourishment: Theory and Practice
by Robert G Dean (Univ. Florida)

Design and Construction of Maritime Structures for Protection Against Waves
by Miguel A Losada (Univ. da Cantabria) *and Nobuhisa Kobayashi* (Univ. Delaware)

Advanced Series on Ocean Engineering – Volume 3

MECHANICS OF COASTAL SEDIMENT TRANSPORT

Jørgen Fredsøe and Rolf Deigaard
Institute of Hydrodynamics and Hydraulic Engineering
Technical University of Denmark

 **World Scientific**
Singapore • New Jersey • London • Hong Kong

Published by

World Scientific Publishing Co. Pte. Ltd.

P O Box 128, Farrer Road, Singapore 912805

USA office: Suite 1B, 1060 Main Street, River Edge, NJ 07661

UK office: 73 Lynton Mead, Totteridge, London N20 8DH

Library of Congress Cataloging-in-Publication Data

Fredsøe, Jørgen.
     Mechanics of coastal sediment transport / Jørgen Fredsøe and Rolf
Deigaard.
          p.   cm. -- (Advanced series on ocean engineering ; v. 3)
     Includes bibliographical references and index.
     ISBN 9810208405 -- ISBN 9810208413 (pbk)
     1. Coast changes.   2. Sediment transport.   I. Deigaard, Rolf.
II. Title.   III. Series.
GB451.2.F74   1992
551.3'6--dc20   92-38727
                                                            CIP

First published 1992
First reprint 1993 (pbk)
Second reprint 1994
Third reprint 1995

10029718 60

For photocopying of material in this volume, please pay a copying fee through the Copyright Clearance Center, Inc., 222 Rosewood Drive, Danvers, MA01923, USA.

Printed in Singapore.

Contents

Introduction

Flow and sediment transport in the coastal zone are important in relation to several engineering topics like sedimentation and erosion around structures, backfilling of dredged channels, changes in near-shore morphology and long- and cross-shore sediment transport rates.

During the last decade the development in coastal sediment transport research has changed from simple phenomenological descriptions to sophisticated numerical models in which the flow as well as the resulting sediment transport are described in detail.

The longshore sediment transport calculations can be mentioned as an example. For many years these have been based on the so-called CERC formula (see p 329), which relates the longshore sediment rate to the longshore wave energy flux. Today it is recognized that such a simple model which does not include essential parameters like sediment grain size and bed morphology (for instance the presence of bars) does not give a full description of the complex problem of longshore sediment transport. On the other hand, more sophisticated methods like those described in this book often get very sensitive to variations in the input data, like the mean grain size. The cause might be that the cross-shore profile also varies with the sediment properties, and thus a longshore sediment transport calculation can not be made except with a fully three-dimensional model including grain sorting, a stage which until now has not been reached in the numerical modelling.

Another example of the development in coastal sediment transport modelling is the cross-shore profile modelling. Formerly, the average cross-shore profile was described by empirical rules (Bruun, Dean, see p 322), while the trend today is towards describing the on-offshore flow pattern as accurately as possible and then step by step describing the dynamic behaviour of the cross-shore profile by morphological computations. Like the longshore sediment transport calculation, this problem is not completely solved yet — particularly the resulting direction of sediment transport outside the surf zone and the detailed morphology of the bars which must be investigated much more thoroughly.

The purpose of the present book is to describe both the basic hydrodynamics and the basic sediment transport mechanisms which are of importance for the construction of a mathematical model for sediment transport in coastal areas.

The main effort has been to describe the hydrodynamics as is usual in sediment transport description. The first six Chapters deal only with the hydrodynamics of waves and current in and outside the surf zone.

The reader's background should be a basic course both in wave hydrodynamics and in fluid mechanics (laminar and turbulent flow).

The basic elements of wave hydrodynamics is consequently therefore only briefly listed in the introductory Chapter 1. Chapters 2 and 3 treat the turbulence made by non-breaking waves with and without the presence of a current. During the last decade significant progress has been made in this field, and today very accurate solutions can be obtained. However, these very accurate solutions are also very time-consuming and costly to compute, and it is not feasible today to introduce them in a numerical model of coastal sediment transport. Simpler as well as more advanced solutions to the wave boundary layer problem have therefore been described, from which the bed shear stress in combined wave-current motion can be obtained very accurately. Some of the simpler models are described quite in detail, so students should be able to work with these models on a PC.

Chapters 4–6 deal with aspects of surf zone hydrodynamics that are important for the sediment transport. Wave kinematics is described, including the strong production of turbulence in the surf zone which will have to be included in a detailed sediment transport model. The depth-integrated wave-driven currents are treated for more or less complicated situations on uniform coastal profiles, that can be represented by a simple computer model. Furthermore, the flow over complex three-dimensional topographies like rip channels is described. The vertical distribution of the forces from breaking and broken waves is described in detail. This has been applied to model the circulation current (undertow) in the surf zone which is of major importance for cross-shore sediment transport.

In Chapter 7 the sediment transport description starts with the basic concepts. The number of sediment transport formulae in current as well as in combined wave-current motion is very large today and rapidly growing. Many of these transport formulae are more or less empirical in their nature, and a simple deterministic description of sediment transport is still not available. In Chapter 7 an attempt is made to describe the construction of a simple transport formula for the bed load case, and some suggestions for the description of sediment transport at higher shear stresses are included as well. However, the sediment transport in the sheet flow layer as well as the bed concentration of suspended sediment are topics which still need to be investigated in more detail. Chapter 8 describes the behaviour of suspended sediment in waves. While the bed load even in this unsteady case usually can be calculated as the quasi-steady value from the instantaneous value of shear stress, the behaviour of suspended sediment gets quite complicated in the wave case due to the increase and decrease in turbulence intensity.

Chapters 9 and 10 concern bed waves, which are of major importance for the bed shear stress and for the transport of suspended sediment. The description again takes its starting point in the pure current case and extends these findings to the wave case as well.

Finally, the last two Chapters 11 and 12 describe the construction of models which can describe long- and cross-shore sediment transport.

It is not the intention of the book to give a broad review of the literature on this very wide topic. The structure of the book is more like a monograph where the subjects have been treated so much in detail that the book would be applicable as a student textbook.

As a help for further reading on the subject, a section with additional references has been included at the end of the book. The list is far from being complete but should provide students with a good basis for an in-depth study.

We would like to acknowledge the encouragement from Professor P.L.-F. Liu to make the book. Our Librarian Kirsten Djørup and Tom Foster, the Danish Hydraulic Institute, have tried to improve our written English. The book has been typewritten in TEX by Hildur Juncker and the drawings are produced by Eva Vermehren. Erik Asp Hansen, Danish Hydraulic Institute, has produced the figure for the front cover, and Julio Zyserman, Danish Hydraulic Institute, has been very helpful in constructing several figures in the book.

Finally we would like to thank the Danish Technical Council (STVF) for their large support to the field of Coastal Engineering in Denmark. This support has created a scientific environment which has made this book possible.

List of symbols

The main symbols used are listed below. Due to the large number of parameters, which stem from different disciplines of hydrodynamics, it has been decided to use double symbols. In most cases their use is restricted to a single chapter as stated in the symbol list. It should be easy to distinguish between these parameters from the context they are used in.

Main symbols

a	amplitude of near-bed wave orbital motion
a	migration velocity of bed forms (Chapter 8)
A	cross-sectional area of the surface rollers (Chapters 4, 5, 6)
b	level for the reference concentration (Chapter 8)
B	wave profile coefficient
c	wave celerity, phase speed
c	suspended sediment concentration
c_b	bed concentration
c_g	wave group velocity
c_2	turbulent energy dissipation coefficient
d	grain diameter
d_f	grain fall diameter
d_s	grain sieve diameter
d_v	grain spherical diameter
d_{50}, d_{16}, d_{84}	diameter corresponding to 50, 16 and 84 percent of the material being finer
D	mean water depth
$\mathcal{D}$	depth-integrated energy dissipation
D'	boundary layer thickness over the back of a dune
D_c	water depth at the bar crest

$\vec{e}$	unit vector in x-direction
E	wave energy
E	momentum exchange coefficient (Chapter 5)
E_f	wave energy flux
E_{fr}	energy flux due to the surface rollers
E_{fw}	energy flux due to the wave motion
E_{kin}	kinetic wave energy
E_{pot}	potential wave energy
f	friction factor
$f(\ )$	probability density
f'	skin friction factor
f_c	current friction factor
f_e	energy loss factor for the wave boundary layer
f_w	wave friction factor
f_w	wind friction factor (Chapters 5 and 12)
$F(\ )$	probability distribution (Chapter 1)
$F(\ )$	error function (Chapter 12)
$\mathcal{F}$	Froude number
F_m	momentum contribution to the radiation stress
F_p	pressure contribution to the radiation stress
F_r	roller contribution to the radiation stress
g	acceleration of gravity
h	local bed level
H	wave height
H_D	dune height
H_r	ripple height
H_{rms}	root mean square of the wave heights
H_s	significant wave height
H_x'	$\frac{\partial H}{\partial x}$
H_0	deep water wave height
H_0	H at $x = 0$ (Chapter 6)
I	gradient of the energy line
I'	energy gradient due to skin friction
I''	energy gradient due to the form resistance of the dunes
I_ℓ	submerged weight of the longshore sediment transport
k	turbulent kinetic energy
k	wave number
k_N	equivalent sand roughness of the bed
k_w	wave roughness
K	ratio between the wave height and the water depth
$K(\Delta\alpha)$	directional distribution of wave energy
K_c	factor in the CERC formula
ℓ	mixing length
ℓ_d	turbulent length scale

ℓ_{max}	turbulent length scale away from the bed
l_r	length of the surface roller
L	wave length
L_b	length of a longshore bar
L_D	dune length
L_r	ripple length
L_0	deep water wave length
M	Monin-Obukov parameter
M_{xy}	momentum exchange
n	porosity of the bed sediment
p	pressure
p	fraction of bed surface particles in motion
p^+	excess pressure above the mean hydrostatic pressure
p_r^+	pressure contribution from a surface roller
$P_{\ell s}$	longshore energy flux factor
PROD	production of turbulent energy
q	discharge per unit width of channel
q_b	bed load transport rate
q_{drift}	discharge due to wave drift
q_D	rate of sediment deposition at a dune front
q_ℓ	specific longshore sediment transport rate
q_{roller}	mean discharge due to surface rollers
q_s	suspended load transport rate
q_{sd}	suspended load transport rate due to the wave drift
q_{sx}	sediment transport rate in the x-direction
q_T	specific sediment discharge
q_0	sediment transport rate in the dune trough
Q	longshore discharge in the trough inshore of a bar
Q_ℓ	longshore sediment transport rate
$\mathcal{R}$	Reynold's number
RE	Reynold's number for the near-bed wave-orbital motion
s	relative density of the sediment
s	parameter in expression for directional wave spreading (Chapters 5 and 12)
S_{nn}	total wave momentum flux in the direction normal to the direction of wave propagation
S_{ss}	total wave momentum flux in the direction of wave propagation
S_{xx}	normal radiation stress in the x-direction
S_{xy}	shear component of the radiation stress
S_{yy}	normal radiation stress in the y-direction
S	mean water surface slope
t	time
T	wave period
T_s	significant wave period

u	velocity in the x-direction
u_f	friction velocity, time varying
u_l	Lagrangian wave drift velocity
u_s	streaming induced velocity
$u_{x'}$	velocity in the x'-direction
u_o	free stream wave-orbital velocity near the bed
U	wave period averaged velocity
U_B	velocity of grain in bed load transport
U_f	friction velocity, steady
U_f'	skin friction velocity
U_{fc}	current friction velocity
U_{fc}	critical friction velocity (Chapter 7)
$U_{f\,\mathrm{max}}$	maximum friction velocity during wave period
U_{fo}	friction velocity for the mean velocity profile in the wave boundary layer
U_δ	mean current velocity at the top of a wave boundary layer
U_{1m}	amplitude of near-bed wave-orbital velocity (1st order Stokes theory)
U_{2m}	amplitude of second harmonic of the near-bed wave orbital velocity
U_{10}	wind speed 10 m above the sea surface
v	velocity in the y-direction
$v_{y'}$	velocity in the y'-direction
V	depth mean velocity
V^*	local depth mean velocity
V_x	depth mean velocity in the x-direction
V_y	depth mean velocity in the y-direction
V'	mean velocity in the boundary layer over a dune
w	velocity in the z-direction
w^*	dimensionless settling velocity
w_s	terminal settling velocity of a sediment grain
w_∞	vertical velocity due to the displacement in a wave boundary layer
x	horizontal coordinate
x'	horizontal coordinate
y	horizontal coordinate
y'	horizontal coordinate
Y	position of the coastline
Y_b	width of a rip channel
z	vertical coordinate
z_c	height of the centroid of a suspended sediment concentration profile
z_0	zero level for velocity ($k_N/30$ at a rough bed)
α	wave direction
α_w	wind direction

α_0	deep water wave direction
α_0	slope of the lower boundary of a surface roller (Chapters 4 and 5)
β	bed slope angle of the coastal profile or transverse to the mean current direction
β	ratio between ε_s and ν_T (Chapter 8)
γ	specific gravity ρg
γ	angle of the bed slope in the mean current direction
γ	angle between the current and the wave propagation (Chapters 3 and 8)
γ_s	specific gravity of the sediment grains
δ	wave boundary layer thickness
δ^*	displacement thickness
δ_m	mean value of the wave boundary layer thickness
δ_s	sheet layer thickness
δ_v	thickness of the viscous sublayer
ΔD	wave set-up or set-down
ΔH	hydraulic head loss
$\Delta H''$	hydraulic head loss at dune front
Δt	time step
Δx	step in x
ΔY	step in Y
$\Delta \alpha$	deviation from mean wave direction
$\Delta \tau$	change in the shear stress over a wave boundary layer due to streaming
ε	dissipation of turbulent energy
ε'	apparent turbulent diffusion coefficient
ε_s	turbulent diffusion coefficient for sediment
ζ	hydraulic head loss coefficient
θ	Shields' parameter, dimensionless bed shear stress
θ'	dimensionless bed shear stress, skin friction
θ''	dimensionless form drag on dunes
θ^*	local value of θ at dune surface
θ_c	critical Shields' parameter
θ_0	dimensionless parameter ($\approx \frac{1}{2}\theta_c$)
θ_γ	Shields' parameter corrected for a bed slope
η	water surface elevation
η^+	surface roller thickness
κ	von Kármán's constant
λ	linear sediment concentration
μ	dynamic viscosity
μ_d	dynamic friction coefficient
μ_s	static friction coefficient
ν	kinematic viscosity
ν_T	eddy viscosity

ξ	surf similarity parameter
ρ	density of water
ρ_a	density of air
σ	normal stress
σ_g	geometric standard deviation
σ_G	normal dispersive stress
σ_0	Prandtl number for turbulent diffusion
τ	shear stress
τ'	skin friction
τ''	form friction
τ_b	bed shear stress
τ_c	critical bed shear stress
τ_F	shear stress carried by fluid
τ_G	shear stress carried by grain-grain interaction
τ_{max}	maximum bed shear stress during wave period
τ_s	near surface shear stress
τ_s	bed shear stress due to a water surface slope
τ_w	wind shear stress (Chapters 5 and 12)
τ_{wc}	instantaneous bed shear stress in waves and current
τ_{zx}	shear stress component
τ_{zy}	shear stress component
τ_δ	stationary part of the bed shear stress
ϕ	velocity potential
ϕ	friction angle
ϕ_d	dynamic friction angle
ϕ_s	static friction angle
Φ	dimensionless sediment transport rate
Φ	angle between the instantaneous bed shear stress and the mean current (Chapter 3)
Φ_B	dimensionless bed load transport
Φ_ℓ	dimensionless longshore sediment transport
Φ_S	dimensionless suspended load transport
ψ	flow direction
ψ_1	angle between the particle path and the drag force
ω	angular frequency
$-$	signifies a time-averaged quantity
$'$	denotes turbulent fluctuations (unless stated otherwise above)
$\sim$	the part of a quantity that varies with the wave period
$\rightarrow$	a vector

MECHANICS OF COASTAL SEDIMENT TRANSPORT

Chapter 1. Basic concepts of potential wave theory

This introductory chapter is mainly a summary of the basic concepts for wave mechanics which are necessary for the further developments presented later in this book. For this reason, nearly all derivations are omitted, and readers are instead referred to standard texts in wave hydrodynamics. (e.g. Dean and Dalrymple, 1990; Svendsen and Jonsson, 1976; Mei, 1990; Phillips, 1966; Wiegel, 1964).

1.1 Waves propagating over a horizontal bottom

Wave kinematics are usually described by potential theory requiring the fluid to be inviscid and irrotational. In this case, a potential ϕ can be introduced, which is related to the velocity field by

$$u = \frac{\partial \phi}{\partial x}$$

$$v = \frac{\partial \phi}{\partial y}$$

$$w = \frac{\partial \phi}{\partial z} \tag{1.1}$$

where x and y are horizontal coordinates, and z is the vertical coordinate (Fig. 1.1). Origin of the coordinate system is located on the seabed. u, v and w are the velocity components in the x-, y- and z-directions. By inserting Eq. 1.1 into the continuity equation

$$\frac{\partial u}{\partial x} + \frac{\partial v}{\partial y} + \frac{\partial w}{\partial z} = 0 \tag{1.2}$$

the Laplace equation is obtained

$$\frac{\partial^2 \phi}{\partial x^2} + \frac{\partial^2 \phi}{\partial y^2} + \frac{\partial^2 \phi}{\partial z^2} = \Delta\,\phi = 0 \tag{1.3}$$

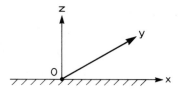

Figure 1.1 Definition of the coordinate-system.

The boundary conditions are

1. At the seabed, the flow velocity perpendicular to the bed is zero. For a plane horizontal bed this gives

$$w = \frac{\partial \phi}{\partial z} = 0 \qquad \text{for } z = 0 \tag{1.4}$$

2. A fluid particle located at the free surface must remain at the free surface giving

$$w = \frac{d\eta}{dt} \qquad \text{for } z = D + \eta \tag{1.5}$$

in which D is the mean water depth, and η is the surface elevation, see Fig. 1.2.

3. The pressure at the water surface must be equal to the atmospheric pressure and can be set to zero, whereby the Bernoulli equation gives

$$D + \eta + \frac{(u^2 + v^2 + w^2)}{2g} + \frac{1}{g}\frac{\partial \phi}{\partial t} = C(t) \qquad \text{for } z = D + \eta \tag{1.6}$$

in which C is a function of t only.

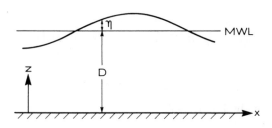

Figure 1.2 Definition sketch of D and η.

If the wave is two-dimensional and periodic with a wave period T and a direction of propagation in the x-direction, Eqs. 1.3 - 1.6 can be solved as a boundary value problem.

As the maximum surface elevation is assumed to be small compared to a typical dimension, for instance the wave length L, the problem can be linearized and solved analytically. As all higher order terms in $\eta_{\max}/L$ are neglected, the solution to the boundary value problem becomes

$$\phi = -\frac{H}{2}\frac{g}{\omega}\frac{\cosh(kz)}{\cosh(kD)}\sin(\omega t - kx) \tag{1.7}$$

$$\eta = \frac{H}{2}\cos(\omega t - kx) \tag{1.8}$$

$$c^2 = \left(\frac{\omega}{k}\right)^2 = \frac{g}{k}\tanh(kD) \tag{1.9}$$

$$u = \frac{\pi H}{T}\frac{\cosh(kz)}{\sinh(kD)}\cos(\omega t - kx) \tag{1.10}$$

$$w = -\frac{\pi H}{T}\frac{\sinh(kz)}{\sinh(kD)}\sin(\omega t - kx) \tag{1.11}$$

This solution is the linear wave solution, also called the Airy wave or a Stokes first order wave. In Eq. 1.8 H is the wave height (from trough to crest), k the wave number, and ω the cyclic frequency, k and ω being defined as

$$k = \frac{2\pi}{L}, \qquad \omega = \frac{2\pi}{T} \tag{1.12}$$

If second order terms in H/L are also included in the solution of Eqs. 1.3 - 1.6, then the Stokes second order solution is obtained, which is given by

$$\eta = \frac{H}{2}\cos(\omega t - kx) + \frac{1}{16}kH^2\left(3\coth^3(kD) - \coth(kD)\right)\cos\left[2(\omega t - kx)\right] \quad (1.13)$$

and

$$\phi = \phi^{(1)} + \phi^{(2)} \quad (1.14)$$

where $\phi^{(1)}$ is given by Eq. 1.7 and

$$\phi^{(2)} = -\frac{3}{32}ckD^2\frac{\cosh(2kz)}{\sinh^4(kD)}\sin\left[2(\omega t - kx)\right] + C_1 x - \frac{1}{16}c^2(kD)^2\frac{t}{\sinh^2(kD)} \quad (1.15)$$

where the constant C_1, which must be of the order $(H/L)^2$, represents a steady, uniform flow. If the mean value over a wave period of the mass flux through a vertical section is zero, then C_1 must have the value

$$C_1 = -\frac{1}{8}\frac{gH^2}{cD} \quad (1.16)$$

The last term in Eq. 1.15 does not affect the velocity profile, but appears from the Bernoulli equation 1.6.

The horizontal flow velocity in Stokes second order theory is found from Eqs. 1.1 and 1.14 to be

$$u = u^{(1)} + u^{(2)} \quad (1.17)$$

where $u^{(1)}$ is given by Eq. 1.10 and

$$u^{(2)} = \frac{3}{16}c\frac{(kD)^2\cosh(2kz)}{\sinh^4(kD)}\cos\left[2(\omega t - kx)\right] + C_1 \quad (1.18)$$

Similarly, the vertical velocity can be found to be

$$w = w^{(1)} + w^{(2)} \quad (1.19)$$

in which $w^{(1)}$ is given by Eq. 1.11 and

$$w^{(2)} = -\frac{3}{16}c\,(kD)^2\frac{\sinh 2kz}{\sinh^4(kD)}\sin\left[2(\omega t - kx)\right] \quad (1.20)$$

Example 1.1: Group velocity

Considering the superposition of two linear waves with same wave height but with slightly different wave numbers k and $k + \Delta k$:

$$\eta = \eta_1 + \eta_2 = \frac{H}{2}\sin\big(k(x-ct)\big) + \frac{H}{2}\sin\big((k+\Delta k)(x-(c+\Delta c)t)\big) \qquad (1.21)$$

or

$$\eta = H\cos\left[\Delta k\big(x - \big(c + k\frac{\Delta c}{\Delta k}\big)\,t\big)\right]\sin\big[k(ct-x)\big] \qquad (1.22)$$

This wave is shown in Fig. 1.3. For small values of Δk the envelope curve (shown dashed in Fig. 1.3) migrates in the x-direction with the velocity

$$c_g = c + k\frac{dc}{dk} = \frac{c}{2}\left[1 + \frac{2kD}{\sinh(2kD)}\right]$$

$$= \frac{c}{2}\left[1 + G\right] \qquad (1.23)$$

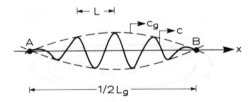

Figure 1.3 Wave group.

At the nodal points A and B, shown in Fig. 1.3, the surface elevation is always zero. At these points no energy transfer can take place, which means that the energy is transported with the group velocity c_g.

**Example 1.2:   Drift velocity (mass-transport velocity) in irrotational
flow**

In relation to sediment transport, it is essential to distinguish between the
mean velocity $\bar{u}$ measured at a fixed point and the drift velocity U, which is the
mean velocity of a fluid particle averaged over a wave period.

For two reasons the drift velocity U is always positive relative to the mean
velocity $\bar{u}$ (Longuet-Higgins, 1957):

  (1) A fluid particle will stay longer below the wave crest than below the wave
      trough, because the fluid velocity is positive below the crest and negative
      below the trough.
  (2) The particle path is elliptic in shape, with the particle travelling forward
      at the upper part of the orbit and backwards at the bottom of the orbit.
      At the top of the orbit, the velocity is slightly higher than at the bottom,
      resulting in a small positive contribution to the drift.

The drift velocity can be evaluated by Lagrangian considerations as follows
(Longuet-Higgins, 1957): consider the points P and Q, where P is a point on the
orbit of a particle, the mean position of which is Q, (Fig. 1.4.)

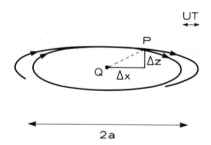

Figure 1.4 Drift velocity as seen from a Lagrangian point of view.

The difference between the instantaneous velocity at P and at Q is given
by

$$\Delta u = \frac{\partial u}{\partial x}\Delta x + \frac{\partial u}{\partial z}\Delta z \qquad (1.24)$$

where Δx and Δz are the horizontal and vertical displacements of P from Q. Δx and Δz are given by

$$\Delta x = \int u dt$$

$$\Delta z = \int w dt \qquad (1.25)$$

Inserting Eq. 1.22 into Eq. 1.21 and time-averaging over one wave period gives

$$U = \bar{u} + \overline{\frac{\partial u}{\partial x} \int u dt} + \overline{\frac{\partial u}{\partial z} \int w dt} \qquad (1.26)$$

The last two terms in Eq. 1.26 can easily be evaluated by applying the linear wave theory expressions Eqs. 1.10 and 1.11 for the orbital velocities giving

$$U = \bar{u} + \frac{1}{2c}\left(\frac{\pi H}{T}\right)^2 \left(\frac{\cosh^2(kz) + \sinh^2(kz)}{\sinh^2(kD)}\right) \qquad (1.27)$$

Close to the bed ($z \sim 0$), this expression becomes

$$U = \bar{u} + \frac{U_{1m}^2}{2c} \qquad (1.28)$$

in which U_{1m} is the maximum horizontal orbital velocity at the bed as predicted by the linear wave theory. It is seen that the mean drift velocity is a non-zero, second-order quantity.

The water discharge associated with the drift is found by integration over the water depth

$$q_{\text{drift}} = \int_0^D \frac{1}{2c}\left(\frac{\pi H}{T}\right)^2 \left(\frac{\cosh^2(kz) + \sinh^2(kz)}{\sinh^2(kD)}\right) dz =$$

$$\frac{1}{2c}\left(\frac{\pi H}{T}\right)^2 \frac{1}{\sinh^2(kD)} \frac{\sinh^2 kD}{2k} = \frac{\pi H^2}{4T} \frac{1}{\tanh(kD)} \qquad (1.29)$$

where

$$kc = \frac{2\pi c}{L} = \frac{2\pi c}{cT} = \frac{2\pi}{T} \qquad (1.30)$$

Wave drift, Eulerian analysis

In the preceding section the wave drift was analyzed by considering the paths of the water particles in the wave motion. In the following, the wave drift will be determined by considering the water motion from an Eulerian specification, i.e. at fixed positions in the $x - z$ coordinate system. Note that the Lagrangian

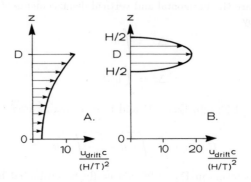

Figure 1.5 Example of the vertical distribution of the drift velocity. A: Calculated by Lagrangian analysis. B: Calculated by Eulerian analysis. $D = 10$ m, $T = 7$ sec, $H = 4$ m.

and the Eulerian analyses do not derive two different contributions but present two different approaches to net drift.

The time-averaged discharge of the water motion is calculated at a fixed position $(x = 0)$ by

$$q_{\text{drift}} = \frac{1}{T} \int_0^T \int_0^{D+\eta} u\,dz\,dt \tag{1.31}$$

where u and η are calculated from Eqs. 1.8 and 1.10. Below the wave trough level, the velocity u varies harmonically in time and does not contribute to q_{drift}. Eq. 1.31 can therefore be written

$$q_{\text{drift}} = \frac{1}{T} \int_0^T \int_{D-H/2}^{D+\eta} u\,dz\,dt \tag{1.32}$$

Only the lowest order solution in H/L is required, therefore u can be replaced with the horizontal orbital velocity at $z = D$

$$u = \frac{\pi H}{T} \frac{1}{\tanh(kD)} \cos(\omega t) = U_{1m} \cos(\omega t) \tag{1.33}$$

Inserting Eq. 1.33 into 1.32 gives

$$q_{\text{drift}} = \frac{1}{T} \int_0^T \left(\eta + \frac{H}{2} \right) U_{1m} \cos(\omega t)dt =$$

$$\frac{1}{T} \int_0^T \left(\frac{H}{2} \cos(\omega t) + \frac{H}{2} \right) U_{1m} \cos(\omega t)dt =$$

$$\frac{H U_{1m}}{4} = \frac{H^2 \pi}{4T} \frac{1}{\tanh(kD)} \tag{1.34}$$

which gives the same result as Eq. 1.29.

The Eulerian analysis thus gives the same discharge due to the wave drift as the Lagrangian analysis, but while the Lagrangian drift is distributed over the entire water depth, the Eulerian drift is located between the levels of the wave troughs and the wave crests. An example of the calculated vertical distribution of the drift velocities is given in Fig. 1.5.

1.2 Wave energy

Wave energy consists of two parts, potential and kinetic energy. The potential energy arises because water below the mean water level is moved above the mean water level, so for an Airy wave the potential energy per unit area is given by

$$E_{\text{pot}} = \frac{1}{L} \int_0^{L/2} \rho g \eta^2 dx = \frac{1}{16} \rho g H^2 \tag{1.35}$$

The kinetic energy is given by

$$E_{\text{kin}} = \frac{1}{2} \rho g \frac{1}{L} \int_0^L \left[\int_0^{D+\eta} (u^2 + w^2) dz \right] dx \tag{1.36}$$

which for a linear wave gives

$$E_{\text{kin}} = \frac{1}{16} \rho g H^2 \tag{1.37}$$

The energy is transported in the direction of wave propagation, at the group velocity. This can be seen either from Example 1.1 or from the following application of the momentum equation:

The work per unit time performed by the wave motion on the left side of a vertical line $A - A$ (Fig. 1.6) is given by

$$W = \int_0^{D+\eta} p u \, dz \tag{1.38}$$

The transport of kinetic and potential energy through the same vertical line $A - A$ is given by

$$Tr = \int_0^{D+\eta} \left[\rho g(z - D) + \frac{1}{2} \rho \left(u^2 + w^2 \right) \right] u \, dz \tag{1.39}$$

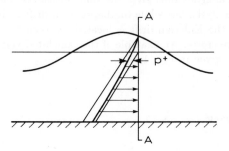

Figure 1.6 Pressure acting on a vertical line A–A.

The flux of energy E_f through $A - A$ from left to right is equal to the sum of the work per unit time given by Eq. 1.38, and the transport of energy given by Eq. 1.39. Hence, the flux of energy for small waves is given by

$$E_f = \int_0^D pu\,dz + \int_0^D \rho g(z - D)u\,dz \qquad (1.40)$$

The pressure p can be written as

$$p = p^+ - \rho g(D - z) \qquad (1.41)$$

in which p^+ is the pressure beyond the hydrostatic pressure from the mean water level. p^+ can be obtained for instance from the Bernoulli equation to be given by

$$p^+ = \rho\, g\, \frac{H}{2}\, \frac{\cosh(kz)}{\cosh(kD)} \sin(\omega t - kx) \qquad (1.42)$$

The energy flux Eq. 1.37 can now be written as

$$E_f = \int_0^D p^+ u\,dz = \frac{1}{8}\rho\, g\, H^2 c\left[1 + G\right] \cos^2(\omega t) \qquad (1.43)$$

where G is defined in Eq. 1.23.

The time-averaged value of the energy flux is obtained from Eq. 1.40

$$E_f = \frac{1}{16}\rho\, g\, H^2 c\left[1 + G\right] \qquad (1.44)$$

or

$$E_f = \left[E_{\text{kin}} + E_{\text{pot}}\right] c_g \qquad (1.45)$$

from which it is seen that the energy is transported with the group velocity c_g.

The energy transport can be explained by the following mechanism: close to the water surface p^+ is positive at the same time as u is directed in the direction of wave propagation (below the wave crest). Below the wave trough p^+ is negative, while the fluid is moving against the direction of wave propagation. This results in a work in the direction of wave propagation.

Example 1.3:   Shoaling

The change in wave height when water waves migrate from one water depth to another is usually calculated by the energy flux-concept. Neglecting energy loss (from dissipation in the wave boundary layer, see Chapter 2, or from wave-breaking, see Chapter 4) the energy flux can be taken as constant. Considering the case where waves with infinitely long wave crests propagate over a bottom with parallel bed contours, the wave orthogonals become parallel, so the energy equation reads

$$c_g E = \text{const.} \tag{1.46}$$

which, by use of Eqs. 1.9, 1.23, 1.35 and 1.36 gives

$$H^2 \tanh(kD)\big[1 + G\big] = H_0^2 \tag{1.47}$$

in which H_0 is the deep water wave height.

The wave table (Appendix I) provides a simple method for calculating linear wave shoaling as specified by Eq. 1.47.

1.3   Radiation stresses

The presence of waves will result in an excess flow of momentum, which is defined as the radiation stresses (Longuet-Higgins and Stewart, 1964). This flux of momentum is formed by two contributions: one due to the wave-induced velocities of the water particles and another one due to the pressure.

The first of the above mentioned contributions is evaluated as follows: the momentum per unit volume associated with a fluid particle is ρu, so its contribution to the flux of momentum across a vertical section normal to the s-axis (the direction of wave propagation) is ρu^2. Hence the total flux of momentum in the s-direction due to the wave-generated motion is

$$I_m = \int_0^{D+\eta} \rho u^2 dz \tag{1.48}$$

In the horizontal n-direction perpendicular to the s-axis there is no momentum flux due to the wave-generated motion, as the wave-generated velocity is zero in this direction.

The second contribution to the momentum flux originates from the pressure and is given by

$$I_p = \int_0^{D+\eta} p\,dz \qquad (1.49)$$

This value is independent of the orientation of the vertical section considered.

The principal component of the radiation stress in the direction of wave propagation is defined as the time-averaged total momentum flux due to the presence of waves, minus the mean flux in the absence of waves, i.e.

$$S_{ss} = \overline{\int_0^{D+\eta} (p + \rho u^2)\,dz} - \int_0^D p_0\,dz \qquad (1.50)$$

where p_0 is the hydrostatic pressure. The force per unit width caused by the presence of waves given by Eq. 1.44 is always directed towards the body of fluid considered, no matter whether the waves travel to the left or to the right.

As terms of third and higher order are disregarded, S_{ss} can be calculated from linear wave theory

$$S_{ss} = \frac{1}{8}\rho g H^2 \left(1 + \frac{2kD}{\sinh(2kD)}\right) = \frac{1}{2}E(1+G) \qquad (1.51)$$

Similarly, the radiation stress in the n-direction is given by

$$S_{nn} = \overline{\int_0^{D+\eta} p\,dz} - \int_0^D p_0\,dz \qquad (1.52)$$

which gives

$$S_{nn} = \frac{1}{8}\rho g H^2 \frac{2kD}{\sinh(2kD)} = \frac{1}{2}\,EG \qquad (1.53)$$

Hereby the radiation stress tensor becomes

$$[S] = \begin{bmatrix} S_{ss} & S_{sn} \\ S_{ns} & S_{nn} \end{bmatrix} = \frac{1}{2}E \begin{bmatrix} (1+2G) & 0 \\ 0 & G \end{bmatrix} \qquad (1.54)$$

In a coordinate system where the two horizontal axes x and y do not coincide with the s- and n-directions, the radiation stress tensor can easily be evaluated by considering the force balance on a small triangular vertical column indicated in Fig. 1.7. This gives the following radiation stress tensor

$$[S] = \begin{bmatrix} S_{xx} & S_{xy} \\ S_{yx} & S_{yy} \end{bmatrix} = \frac{1}{2}E \begin{bmatrix} (1+G)\cos^2\alpha + G & (1+G)\sin\alpha\cos\alpha \\ (1+G)\sin\alpha\cos\alpha & (1+G)\sin^2\alpha + G \end{bmatrix} \qquad (1.55)$$

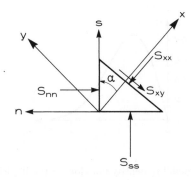

Figure 1.7 Radiation stresses acting on a small triangular element.

1.4 Irregular waves

The waves described by the preceding theories are assumed to be regular, i.e. to be periodic in time and space.

Natural, wind-generated waves will be irregular, with stochastically varying wave heights and periods. This book will not deal with the processes of irregular waves in any detail, but some of the later theories are extended to take irregular waves into account.

The most important information for the present treatment of the irregular waves is the statistical distribution of the wave heights. For wind waves, it has been found that the distribution of wave heights is often close to the Rayleigh distribution. This has been explained theoretically by Longuet-Higgins (1952) to be related to the spectrum - or Fourier decomposition - of the water surface elevation in natural waves. The probability density function for the Rayleigh distribution can be written

$$f(H) = \frac{2H}{H_{\mathrm{rms}}^2} \exp\left(-\left(H/H_{\mathrm{rms}}\right)^2\right) \tag{1.56}$$

where H_{rms} is the root-mean-square of the wave heights

$$H_{\mathrm{rms}} = \sqrt{\overline{H^2}} \tag{1.57}$$

H_{rms} is the wave height that represents the wave energy in the wave field, because the wave energy is proportional to H^2. The probability density function is shown in Fig. 1.8. Another wave height that is traditionally used to characterize a wave field is the significant wave height H_s or $H_{1/3}$, which can be defined from the

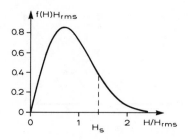

Figure 1.8 The probability density function for the Rayleigh distribution.

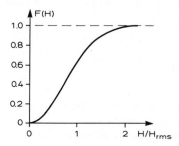

Figure 1.9 The probability distribution curve for the Rayleigh distribution.

wave height distribution as the average of the highest one third of the waves. The relation between $H_{1/3}$ and H_{rms} is

$$H_{1/3} \approx 1.4\, H_{rms} \qquad (1.58)$$

Traditionally, many coastal engineering formulae are based on $H_{1/3}$, but this tendency has weakened, among other things because of the difficulty in attributing a physical meaning to $H_{1/3}$.

The probability distribution curve for the Rayleigh distribution is written

$$F(H) = P(H' < H) = \int_0^H f(H')\, dH' = 1 - \exp\left(-\left(H/H_{rms}\right)^2\right) \qquad (1.59)$$

$F(H)$ is shown in Fig. 1.9.

Directional spreading

Both regular and irregular waves can be two-dimensional, which means that the wave fronts are very long, and that the wave-orbital motion occurs only in the $x - z$-plane. Natural, wind-generated waves are normally found to be three-dimensional in addition to being irregular.

Three-dimensional waves are normally described by wave spectra. The directional spreading is characterized by the directional distribution function $K(\Delta\alpha)$ where $\Delta\alpha$ is the angle between the direction considered and the x-axis. The function $K(\Delta\alpha)$ describes the distribution of the wave energy (or wave energy flux for a particular frequency) over the different directions. If the total wave energy is E, then the amount of wave energy associated with waves propagating in the interval from $\Delta\alpha - d\alpha/2$ to $\Delta\alpha + d\alpha/2$ will be given by

$$dE = EK(\Delta\alpha)d\alpha \tag{1.60}$$

Since the total wave energy is given, it follows that

$$\int_{-\pi}^{\pi} K(\Delta\alpha)d\alpha = 1 \tag{1.61}$$

A commonly used simple representation of the three-dimensional waves is made by describing $K(\Delta\alpha)$ as a cosine squared, Fig. 1.10.

$$K(\Delta\alpha) = \frac{2}{\pi}\cos^2(\Delta\alpha) \;,\; -\frac{\pi}{2} < \Delta\alpha < \frac{\pi}{2} \tag{1.62}$$

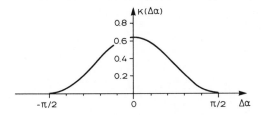

Figure 1.10 The directional spreading of the wave energy according to the cosine-squared, cf. Eq. 1.62.

A more elaborate distribution, where the variance in spreading is not fixed, is used to estimate the effect of directional spreading on wave-driven currents in Chapter 5.

REFERENCES

Dean, R.G. and Dalrymple, R.A. (1990): Water Wave Mechanics for Engineers and Scientists. Singapore, World Scientific, Advanced Series on Ocean Engineering, Vol. 2.

Longuet-Higgins, M.S. (1952): On the statistical distribution of the height of sea waves. Journal of Marine Research, 11:245-266.

Longuet-Higgins, M.S. (1957): The mechanics of the boundary-layer near the bottom in a progressive wave. Appendix to Russell, R.C.H. and Osorio, J.D.C.: An experimental investigation of drift profiles in a closed channel. Proc. 6th Int. Conf. Coast. Eng., Miami, Fl., pp. 184-193.

Longuet-Higgins, M.S. and Stewart, R.W. (1964): Radiation stress in water waves; a physical discussion with applications. Deep-sea Research, 11:529-562.

Mei, C.C. (1990): The Applied Dynamics of Ocean Surface Waves. Singapore, World Scientific, Advanced Series on Ocean Engineering, Vol. 1.

Phillips, O.M. (1966): The Dynamics of the Upper Ocean. Cambridge, At the University Press.

Svendsen, I.A. and Jonsson, I.G. (1976): Hydrodynamics of Coastal Regions. Lyngby, Den Private Ingeniørfond, Technical University of Denmark.

Wiegel, R.L. (1964): Oceanographical Engineering. Englewood Cliffs, NJ, Prentice-Hall.

Chapter 2. Wave boundary layers

2.1 Introduction

Bottom shear stress and turbulence in wave motion are key parameters for moving the sediment and keeping it in suspension. In this and the next chapter, we quantify these two parameters in the case of non-breaking waves.

In non-breaking waves without any resulting current, the turbulence is restricted to the thin boundary layer just above the seabed. Due to the no-slip requirement, the potential theory fails in this region because the shear stresses become important, and the full Navier-Stokes equation (with boundary layer approximations) must be solved here. This layer is of great importance for bringing sediment into suspension above the seabed, and further, this boundary layer is of importance for the vertical distribution of a co-present current.

This chapter starts by considering the pure oscillatory boundary layer, which is the boundary layer formed by a uniform, purely oscillatory outer flow, which for instance is formed in a u-tubed closed conduit, see Fig. 2.1. In this case the bottom boundary layer is the same at all x-values (except at the edge boundaries), while in real waves, Fig. 2.2, the boundary layer-characteristics change with x, as does the outer wave-induced flow.

The variation in the x-direction gives rise to the so-called streaming, which is analyzed later in this chapter. Except for this streaming, which is a term of second order, the convective terms are only of minor importance, and can be neglected

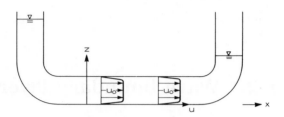

Figure 2.1 Purely oscillatory flow in a u-tube. The flow is uniform in the
x-direction.

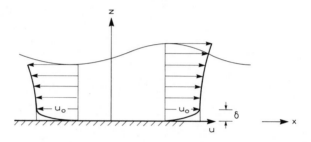

Figure 2.2 Flow including boundary layer in real waves. The potential
flow near the bed is indicated by a thin line.

when calculating bed shear stresses below waves. The general flow equations to
be solved in the boundary layer read

$$\rho \frac{du}{dt} = \rho \left(\frac{\partial u}{\partial t} + u \frac{\partial u}{\partial x} + w \frac{\partial u}{\partial z} \right) = -\frac{\partial p}{\partial x} + \frac{\partial \tau}{\partial z} \tag{2.1}$$

in the flow direction and

$$\frac{\partial p}{\partial z} = 0 \tag{2.2}$$

perpendicular to the flow direction. In the case of uniform flow, Eq. 2.1 is reduced
to

$$\rho \frac{\partial u}{\partial t} = -\frac{\partial p}{\partial x} + \frac{\partial \tau}{\partial z} \tag{2.3}$$

Outside the boundary layer, the shear stresses vanish, so Eq. 2.3 is reduced to

$$\rho \frac{\partial u_0}{\partial t} = -\frac{\partial p}{\partial x} \tag{2.4}$$

in which u_0 = the free stream velocity. The longitudinal pressure gradient $\partial p/\partial x$ remains constant through the boundary layer, as seen from Eq. 2.2, whereby Eq. 2.3 can be written as

$$\rho \frac{\partial}{\partial t}(u_0 - u) = -\frac{\partial \tau}{\partial z} \tag{2.5}$$

Example 2.1: The laminar boundary layer

In the case of laminar flow, τ is related to the velocity gradient by

$$\frac{\tau}{\rho} = \nu \frac{\partial u}{\partial z} \tag{2.6}$$

where ν is the kinematic viscosity. Eq. 2.5 now reads

$$\frac{\partial}{\partial t}(u_0 - u) = -\nu \frac{\partial^2 u}{\partial z^2} \tag{2.7}$$

Taking

$$u_0 = U_{1m} \sin(\omega t) \tag{2.8}$$

the solution to Eq. 2.7 (with the boundary conditions $u = 0$ for $z = 0$ and $u \to u_0$ for $z \to \infty$) is

$$u = U_{1m} \sin(\omega t) - U_{1m} \exp\left(-\frac{z}{\delta_1}\right) \sin\left(\omega t - \frac{z}{\delta_1}\right) \tag{2.9}$$

in which

$$\delta_1 = \sqrt{\frac{2\nu}{\omega}} \tag{2.10}$$

is the thickness of the laminar boundary layer, which is called the Stokes length. The bed shear stress is given by

$$\frac{\tau_b}{\rho} = \nu \frac{\partial u}{\partial z}\bigg|_{z=0} = \frac{\nu U_{1m}}{\delta_1}\left[\sin(\omega t) + \cos(\omega t)\right] \tag{2.11}$$

From Eqs. 2.8 and 2.11 it is seen that there is a phase shift of 45° between the shear stress and the outer flow velocity.

2.2 The momentum integral method for the turbulent wave boundary layer

Eq. 2.5 is valid in the laminar as well as in the turbulent case. In relation to sediment transport, the turbulent case is the most important. Several solution procedures exist for solving Eq. 2.5, ranging from simple approximate solutions to refined turbulent closures. Some of those are described in the following, starting with a simple integral approach.

The momentum equation, Eq. 2.12, is obtained by integrating Eq. 2.5 across the boundary layer thickness δ

$$- \rho \int_{z_0}^{\delta + z_0} \frac{\partial}{\partial t}(u_0 - u)dz = \int_{z_0}^{\delta + z_0} \frac{\partial \tau}{\partial z}dz = -\tau_b \qquad (2.12)$$

In the derivation of Eq. 2.12, the zero shear stress at the top of the boundary layer has been used. τ_b is the bed shear stress, and z_0 is the zero level of the velocity. The ordinary way of solving the boundary layer momentum equation is to assume a reasonable shape of the velocity profile, which fulfils the boundary conditions. In the case of a *hydraulically rough bed*, the velocity profile, as an approximation, can be taken to be logarithmic

$$u = \frac{U_f}{\kappa} \ln\left(\frac{z}{z_0}\right) \qquad (2.13)$$

where U_f = friction velocity ($= \sqrt{|\tau_b|/\rho}$), κ = the von Kármán's constant, which is around 0.40, and z_0 = the bed level, which following Nikuradse (1932) can be taken equal to $k_N/30$, in which k_N = bed roughness. As a first simple approximation, the development in the wave boundary layer can be found by inserting Eq. 2.13 into Eq. 2.12 with the boundary condition that at the top of the boundary layer u must be equal to the free stream velocity u_0.

Example 2.2:   Velocity profiles in turbulent steady channel flow and turbulent boundary layers

Eq. 2.13 is adopted from the velocity profiles in steady channel flow. If the case of uniform steady flow in a channel with flow depth D is considered initially, the bed shear stress is given by

$$\tau_b = \rho U_f^2 = \rho g D S$$

where S is the slope of the water surface.

The vertical distribution of τ is given by

$$\tau = \tau_b\left(1 - \frac{z}{D}\right) \tag{2.14}$$

In a turbulent flow, the shear stress is usually related to the velocity gradient by

$$\frac{\tau}{\rho} = \nu_T \frac{du}{dz} \tag{2.15}$$

in which ν_T is the eddy viscosity. In open channel flow, this varies as

$$\nu_T = \kappa U_f z\left(1 - \frac{z}{D}\right) \tag{2.16}$$

in which κ is the von Kármán's constant (~ 0.40). Eqs. 2.14, 2.15 and 2.16 give

$$\frac{du}{dz} = \frac{U_f}{\kappa}\frac{1}{z} \tag{2.17}$$

which is integrated to

$$\frac{u}{U_f} = \frac{1}{\kappa}\ln z + c_1 \tag{2.18}$$

The constant appearing in Eq. 2.18 must be determined experimentally (or by numerical solution of the full Navier-Stokes equation, Spalart (1988)). Nikuradse (1932, 1933) determined the constant to be

$$c_1 = 8.5 - \frac{1}{\kappa}\ln(k_N) \tag{2.19}$$

for the rough wall case (viscous sublayer $\delta_v < k_N$). Hereby Eq. 2.18 reads

$$\frac{u}{U_f} = \frac{1}{\kappa}\ln\left(z\Big/\frac{k_N}{30}\right) \tag{2.20}$$

If the wall is smooth, a viscous sublayer exists in the vicinity of the wall, the thickness being

$$\delta_v = 11.6\,\nu/U_f \tag{2.21}$$

In this case, Nikuradse found that the constant, appearing in Eq. 2.18, was determined by

$$c_1 = 5.7 - \frac{1}{\kappa}\ln\left(\frac{\nu}{U_f}\right) \tag{2.22}$$

so Eq. 2.18 becomes

$$\frac{u}{U_f} = 5.7 + \frac{1}{\kappa}\ln\left(\frac{zU_f}{\nu}\right) \tag{2.23}$$

In the smooth-wall case, the viscous terms dominate close to the bed. Here the flow equation reads

$$\frac{\tau}{\rho} \approx \frac{\tau_b}{\rho} = U_f^2 = \nu \frac{du}{dz} \qquad (2.24)$$

which gives

$$\frac{u}{U_f} = \frac{U_f}{\nu} z = z^+ \qquad (2.25)$$

where z^+ is a dimensionless near-wall coordinate. At the top of the viscous sublayer, given by Eq. 2.21, the velocity u is the same from the inner (Eq. 2.25) and the outer (Eq. 2.23) solutions. Usually the thickness of the viscous sublayer is only a few millimeters.

In steady turbulent boundary layers, the velocity profiles tend to be logarithmic. This can be deduced by assuming the shear stress to be a constant, which is equal to the bed shear stress in the lower part of the flow

$$\tau \approx \tau_b = \rho U_f^2 \qquad (2.26)$$

The eddy viscosity is assumed to increase linearly away from the bed

$$\nu_T \sim U_f z \qquad (2.27)$$

Eqs. 2.26 and 2.27 again give a logarithmic shape of the velocity profile through the use of Eq. 2.15. The logarithmic velocity profile can also be found from the more sophisticated turbulence models by expressing local equilibrium between production and dissipation of turbulent kinetic energy in the boundary layer, (see also Section 2.3 and Appendix I). Production of turbulent kinetic energy is given by

$$\text{Prod} = \tau \frac{du}{dz} \qquad (2.28)$$

while dissipation is given by

$$\text{Diss} = c_2 \frac{k^{3/2}}{l_d} \qquad (2.29)$$

where c_2 is a constant, k is the turbulent kinetic energy, which (Laufer, 1954) from measurements turns out to be

$$k \sim U_f^2 \qquad (2.30)$$

and l_d a length scale of turbulence

$$l_d \sim z \qquad (2.31)$$

Equating the production Eq. 2.28 and the dissipation from Eq. 2.29 gives again the well-known logarithmic shape, Eq. 2.18. In a boundary layer flow, the constants must, respectively, be matched with the flow near a rough and smooth wall, giving the same constants as those appearing in Eqs. 2.19 and 2.22.

In an oscillatory flow, the instantaneous production and dissipation are not necessarily the same, due to the unsteady behaviour. For this reason, slight deviations from the logarithmic shape are to be expected, especially for rapid oscillations (small wave periods).

Example 2.3:   Numerical solution of the momentum equation

By introduction of the dimensionless parameter

$$Z = \frac{u_0}{U_f}\kappa \tag{2.32}$$

the boundary condition at the top of the boundary layer reads (cf. Eq. 2.13)

$$Z = \ln\big[(\delta + k_N/30)/(k_N/30)\big]$$

or

$$\delta = \frac{k_N}{30}(e^Z - 1) \tag{2.33}$$

By inserting Eqs. 2.13 and 2.32 into the momentum equation Eq. 2.12, we obtain

$$-U_f^2 = -\delta\frac{du_0}{dt} + \frac{1}{\kappa}\frac{dU_f}{dt}\frac{k_N}{30}\big[e^Z(Z-1)+1\big] \tag{2.34}$$

This relation can be converted into a differential equation in the dimensionless parameter Z (from Eq. 2.32), utilizing

$$\frac{dZ}{dt} = \frac{Z}{u_0}\frac{du_0}{dt} - \frac{Z}{U_f}\frac{dU_f}{dt} \tag{2.35}$$

Eq. 2.34 can now be written

$$\frac{dZ}{dt} = \left[\frac{30}{k_N}\kappa^2 u_0 - Z\big[\exp(Z) - Z - 1\big]\frac{1}{u_0}\frac{du_0}{dt}\right]\Big/\big(e^Z(Z-1)+1\big) \tag{2.36}$$

If the flow just above the boundary layer is assumed to be described by sinusoidal wave theory, u_0 is given by

$$u_0 = U_{1m}\sin(\omega t) \tag{2.37}$$

and Eq. 2.36 becomes

$$\frac{dZ}{d(\omega t)} = \left[\beta\sin(\omega t) - Z\big[\exp(Z) - Z - 1\big]\frac{1}{u_0}\frac{du_0}{d(\omega t)}\right]\Big/\big(e^Z(Z-1)+1\big) \tag{2.38}$$

in which

$$\beta = \frac{30\kappa^2}{k_N}\frac{U_{1m}}{\omega} = 30\kappa^2\frac{a}{k_N} \tag{2.39}$$

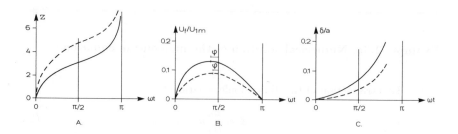

Figure 2.3 Variation in Z (Fig. A), U_f/U_{1m} (Fig. B), and δ/a (Fig. C) with time. —— : a/k = 10. -- : a/k = 100.

where a is the free stream particle amplitude $(= U_{1m}/\omega)$. From Eq. 2.20 it is seen that the variation in the dimensionless quantity Z with dimensionless time only depends on the parameter a/k_N. Eq. 2.38 is easily solved numerically. The boundary condition at $t = 0$, where $u_0 = $ zero is taken to be that the flow velocity in the wave boundary layer is zero. To be fully correct, the flow in the boundary layer is not totally zero when U_0 is zero, because of the flow inertia. This effect is especially pronounced for small values of a/k_N. In the main part of a wave period, however, the velocity profile is very near a logarithmic shape and is described by Eq. 2.13.

Fig. 2.3 shows two examples of the numerical solution of Eq. 2.38, namely for $a/k_N = 10^1$ and 10^2. In the numerical solution of Eq. 2.38 it is convenient to introduce a stretched value t^* of the dimensionless time ωt in the initial stage of the solution, because the variation in Z is very large as seen from Fig. 2.3. At small values of ωt, Eq. 2.38 can be written as

$$\frac{dZ}{d(\omega t)} = 2\beta \frac{\omega t}{Z^2} - \frac{Z}{\omega t} \tag{2.40}$$

which has the solution

$$Z = \sqrt[3]{\frac{6}{5}\beta}\,(\omega t)^{2/3} = t^* \tag{2.41}$$

The time ωt can be replaced by t^* for small values of ωt in order to account for the large initial variation in Z.

The wave friction factor f_w can be defined by

$$\tau_{\max} = \frac{1}{2}\rho f_w U_{1m}^2 \tag{2.42}$$

which is quite similar to ordinary open channel hydraulics. The theoretical solution for the wave friction factor over a rough bed, obtained by the momentum method, is depicted in Fig 2.4 together with the variation in the boundary layer thickness.

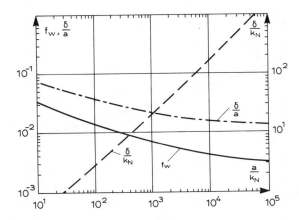

Figure 2.4 Variation in friction factor f_w and boundary layer thickness with a/k_N in the case of a rough bed.

As seen from Fig. 2.3C, the boundary layer thickness varies with time. The boundary layer thickness depicted in Fig. 2.4 is the one which occurs at $\omega t = \pi/2$, i.e. at the maximum outer flow velocity. At very small values of a/k_N the friction factor becomes higher than predicted, because the flow only moves a distance back and forth, equal to a few times the grain roughness. In this case, the flow around the individual grains must be modelled more accurately instead of adopting an average expression such as Eq. 2.13. The theoretical expression for the friction factor f_w can be approximated by

$$f_w = 0.04\left(\frac{a}{k_N}\right)^{-\frac{1}{4}} \quad , \frac{a}{k_N} > 50 \qquad (2.43)$$

Here a is the free stream amplitude, given by

$$a = \frac{U_{1m}T}{2\pi} \qquad (2.44)$$

Similarly, the following expression can be obtained for the wave boundary layer thickness

$$\frac{\delta}{k_N} = 0.09\left(\frac{a}{k_N}\right)^{0.82} \qquad (2.45)$$

In Fig. 2.5 a comparison between the theoretical expression and measured values of the wave friction factor is presented. The measurements are carried out over a bottom with natural sand fixed to the bed, except those by Jonsson and Carlsen (1976) where the bottom consists of two-dimensional triangular roughness

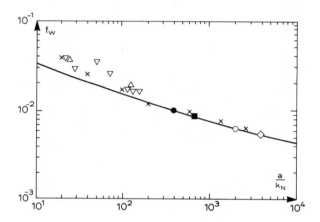

Figure 2.5 Wave friction factor versus a/k_N (rough wall) $\Diamond$, 0, ■, ●: Jensen et al. (1989), x: Kamphuis (1975), $\triangle$: Jonsson and Carlsen (1976), $\triangledown$: Sleath (1987). The solid line is the theoretical solution from Fig. 2.4.

elements. The friction factor is seen to be predicted quite well, except for small values of a/k_N, where the theory underpredicts the friction. As seen later, even more refined theories are not capable of predicting this variation. This is probably because the models do not model the flow around the individual grains in the bottom with enough details.

Kamphuis (1975) suggested that the data for small a/k_N- values could be approximated by

$$f_w = 0.4\left(\frac{a}{k_N}\right)^{-0.75} \quad , \frac{a}{k_N} < 50 \tag{2.46}$$

Example 2.4:   Boundary layer behaviour in broken waves

In non-linear waves, the outer velocity above the boundary layer can not be described by the simple harmonic variation Eq. 2.8. One of the most pronounced non-linear cases is broken waves, where the outer velocity variation u_0, due to the very asymmetric shape of the waves, can be described by a so-called "saw-tooth profile" (Schäffer and Svendsen, 1986), as sketched in Fig. 2.6.

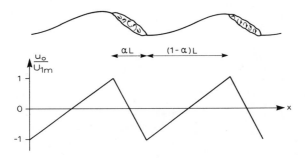

Figure 2.6 Velocity variation below a bore-like broken wave.

The time variation in the outer flow velocity here can be approximated by

$$\frac{u_0}{U_{1m}} = \begin{cases} \dfrac{1}{\alpha}\dfrac{2t}{T} & 0 < t < \alpha\dfrac{T}{2} \\[2ex] \dfrac{1 - 2\frac{t}{T}}{1 - \alpha} & \alpha\dfrac{T}{2} < t < \dfrac{T}{2} \end{cases} \tag{2.47}$$

where U_{1m} is the maximum velocity and α, which indicates the skewness of the wave, is defined in Fig. 2.6. Eq. 2.47 can easily be inserted in Eq. 2.36, which can then be solved numerically. Fig. 2.7 shows an example of the calculated time variation in the bed shear stress. For reasons of comparison, a more refined solution of the same problem, by means of a turbulence model, has been included in the figure.

The picture differs significantly from that obtained under harmonic waves (see Fig. 2.3B). It is interesting to note from Fig. 2.7 that the maximum positive τ - value is larger than the maximum negative value. Due to non-linear effects between flow and shear stress, the resulting bed shear stress, defined by

$$\bar{\tau} = \frac{1}{2\pi} \int_0^{2\pi} \tau \, d(\omega t) \tag{2.48}$$

differs from zero, and is directed in the direction of wave propagation. Fig. 2.8 shows the calculated variation in $\bar{\tau}$ with the skewness parameter α for different values of a/k_N. $\bar{\tau}$ is quite a small quantity, and much smaller than that originating from streaming, see Section 2.4.

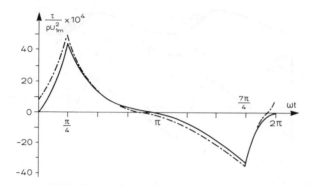

Figure 2.7 Temporal variation of bed shear stress under broken waves, $\alpha = 0.25$ and $a/k_N = 10^3$. ———— : Integrated momentum equation. — · — · — · : $k - \varepsilon$ model.

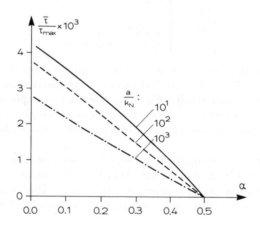

Figure 2.8 Resulting bed shear stress in the direction of wave progagation as a function of α.

Wave boundary layer over a smooth bed

If the bed instead of being rough is hydraulically smooth, the expression given by Eq. 2.13 for the rough case must be replaced by the following expression

$$u = \frac{U_f}{\kappa} \ln\left(\frac{9.8\, U_f z}{\nu} + 1\right) \tag{2.49}$$

in which ν = the kinematic viscosity (= $10^{-6}\mathrm{m}^2/\mathrm{s}$ for water at 20° centigrade). Like Eq. 2.13 Eq. 2.49 is taken from the wall law. To be fully correct, Eq. 2.41 should be modified in the thin viscous sublayer. However, because the momentum method is based on an integral formulation, this modification is of no significance. Now, Eq. 2.41 can be inserted in the momentum equation Eq. 2.13, in which z_0 for the smooth bed case is zero.

Example 2.5: Numerical solution for the smooth bed

In the smooth bed case the momentum equation can be solved in the same way as in the rough bed case, Example 2.2. Still defining the quantity Z by Eq. 2.32, the momentum equation now gives the following differential equation in Z (similar to Eq. 2.36 for the rough wall case)

$$\frac{dZ}{dt} = \frac{9.8\kappa^3}{\nu} \frac{u_0^2}{Z^2\left(\exp(Z) - 1\right)} \tag{2.50}$$

For sinusoidal waves (Eq. 2.37), Eq. 2.50 can be written as

$$\frac{dZ}{d(\omega t)} = 9.8\kappa^3 RE \frac{\sin^2(\omega t)}{Z^2\left(\exp(Z) - 1\right)} \tag{2.51}$$

in which RE is 'the amplitude Reynolds number' defined by

$$RE = \frac{U_{1m}a}{\nu} \tag{2.52}$$

At small values of t, Eq. 2.52 can be written as

$$\frac{dZ}{dt} = \frac{9.8}{\nu}\kappa^3 U_{1m}^2 \frac{\omega^2 t^2}{Z^3} \tag{2.53}$$

which has the solution

$$Z = \sqrt[4]{\frac{4}{3}9.8\kappa^3 RE(\omega t)^3} = t^* \tag{2.54}$$

By solving Eq. 2.52 (numerically with the analytical solution, Eq. 2.54 as the starting point), the variation in Z is now a function of dimensionless time ωt and the Reynolds number. Analogous to the rough wall case, Fig. 2.4, the variations in wave friction factor and boundary layer thickness, made dimensionless with the free stream particle amplitude a, are depicted in Fig. 2.9. Both quantities are now dependent on the Reynolds number.

For this smooth bed case, the variation in the friction factor can be approximated by

$$f_w = 0.035 RE^{-0.16} \tag{2.55}$$

while the boundary layer thickness is approximated by

$$\frac{\delta}{a} = 0.086 RE^{-0.11} \tag{2.56}$$

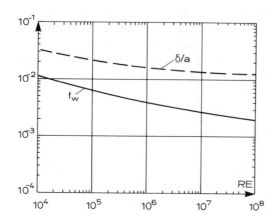

Figure 2.9 Variation in friction factor and boundary layer thickness with Reynolds number in the case of a smooth bed.

Example 2.6:   Transition from laminar to turbulent flow

The wave boundary layer is either laminar or turbulent depending on the value of the Reynolds number. Fig. 2.10 presents several experimental investigations on the variation in friction factor with RE-numbers.

It is seen that the transition from laminar to turbulent flow takes place for the RE-numbers in the interval between 2×10^5 and 6×10^5. However, even at higher RE-numbers, the flow is not fully turbulent during the entire wave period. Fig. 2.11 presents a plot, where the temporal value of the friction coefficient c_w is plotted against RE for various phases ωt. The coefficient c_w is defined by

$$c_w = \frac{2\frac{\tau_b}{\rho}\Big/ U_{1m}^2}{\cos(\omega t - \pi/4)} \qquad (2.57)$$

In a laminar flow, c_w should be a constant, and located on the curve

$$c_w = \frac{2}{\sqrt{RE}} \qquad (2.58)$$

as seen from Eqs. 2.10 and 2.11. From Fig. 2.11, it can be seen that the flow is fully laminar for RE less than 5×10^4, but even at very high Reynolds numbers, the flow might be partly laminar at the phases where the outer flow velocity is small.

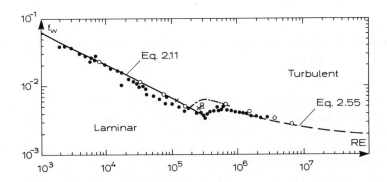

Figure 2.10 Measured variation in wave friction factor with RE-number in flow over a smooth bed. ●: Kamphuis (1975), × : Sleath (1987), o: Jensen et al. (1989), △: Hino et al. (1983).

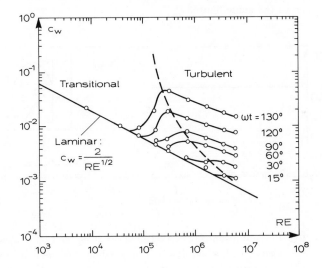

Figure 2.11 Normalized friction coefficient versus RE at different phase values. Smooth bed. From Jensen et al. (1989).

Example 2.7:   Phases between free stream velocity and bed shear stress

In an oscillatory flow, the bed shear stress is out of phase with the outer velocity. In the laminar case, see Example 2.1, the bed shear stress is exactly 45° in advance of the outer flow velocity. This phase is introduced because the pressure gradient, which is constant over the depth, can easily turn that part of the flow which is located close to the bed in the boundary layer (because the flow velocity is a minimum in this region). In the turbulent case, the near wall velocities are not decreased as much as in the laminar case, because of the vertical exchange of momentum by the eddies. For this reason, the bed shear stress in the turbulent case should not be as much ahead of the free stream velocity. Fig. 2.12 which shows the measured phase lag confirms this, as the phases drop from 45° to about 10° in the turbulent case. This phase is also predicted by the momentum integral method, as seen in the same figure.

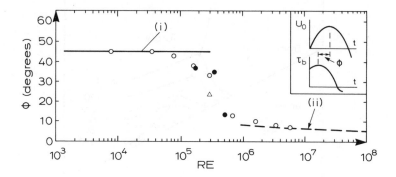

Figure 2.12 Phase difference between bed shear stress and outer flow velocity. (i): laminar flow. (ii): phase difference predicted by the momentum integral method.

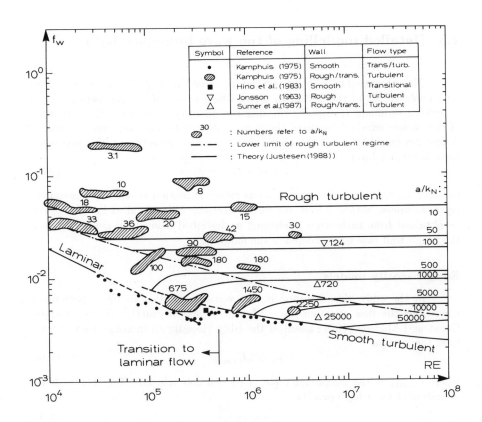

Figure 2.13 Friction factor diagram based on experiments and a one-equation turbulence model. (From Justesen, 1988a).

Transition from smooth to rough wall

In general, a sand bed exposed to waves will be hydraulically rough, when the waves are large enough for the sediment to move. Several experimental data exist for the friction factor in the rough, the smooth and the transitional regimes. Fig. 2.13 indicates the limit for the rough turbulent regime as a function of the Reynolds number and the parameter a/k_N. Also shown in this diagram is the theoretical findings of Justesen (1988a) by means of a one-equation turbulence model, see Section 2.3.

2.3 Detailed modelling of the wave boundary layer

The integrated momentum method, described in the preceding section, gives reasonable values of the bed shear stress which is one of the key parameters in describing sediment transport. Another key quantity is the vertical distribution of turbulence which determines the vertical distribution of suspended sediment. A first reasonable guess is to assume that the eddy viscosity ν_t varies across the wave boundary layer

$$\nu_t = \kappa U_f z (1 - z/\delta) \tag{2.59}$$

in which U_f and δ are the instantaneous values of friction velocity and boundary layer thickness, obtained from the former section. More detailed knowledge can be obtained from turbulence modelling, as described below. For simplicity, only the rough wall case is considered in the following.

Eddy viscosity models

In the eddy viscosity models, the vertical variation in ν_T is prescribed, whereafter the flow equation, Eq. 2.5, can be solved exactly. As an example, Grant and Madsen (1979) assumed the eddy viscosity to increase linearly away from the bottom

$$\nu_T = \kappa U_{f,\max} z \tag{2.60}$$

where $U_{f,\max}$ is the maximum value of the friction velocity. Together with the definition of the eddy viscosity

$$\frac{\tau}{\rho} = \nu_T \frac{\partial u}{\partial z} \tag{2.61}$$

Eq. 2.60 gives the following equation of motion

$$\frac{\partial}{\partial t}(u_d) = -\frac{\partial}{\partial z}\left(\kappa U_f z \frac{\partial u_d}{\partial z}\right) \tag{2.62}$$

in which u_d is the deficit velocity in the boundary layer, defined by

$$u_d = u - u_0 \tag{2.63}$$

The boundary conditions to u_d are

$$u_d = 0 \qquad for \quad z \to \infty \tag{2.64}$$

and, in the case of a rough bed

$$u_d = -u_0 \qquad for \quad z = k_N/30 \tag{2.65}$$

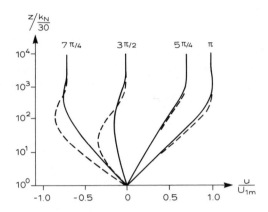

Figure 2.14 Velocity profiles in the boundary layer at different phases predicted by eddy viscosity models. Full line: steady eddy viscosity (Eq. 2.60). Dashed line: time-varying eddy viscosity (Eq.
2.67). Rough wall ($a/k_N = 159$). After Davies (1986).

Finally, the flow is periodic

$$u_d(t) = u_d(t + T) \tag{2.66}$$

Eq. 2.62 can be solved analytically, whereby the exact shape of the instantaneous velocity profile can be found, see Fig. 2.14.

It is seen that the profile close to the bottom follows the logarithmic shape,
while at the outer edge a small deviation from the logarithmic profile exists. Several other guesses on the shape of the eddy viscosity profile have been suggested
(e.g. Kajiura, 1968; Brevik, 1981; Myrhaug, 1982; Christoffersen and Jonsson,
1985). The calculated velocity profiles become nearly identical in all these calculations. In the above models, the eddy viscosity is taken to be time-invariant.
Trowbridge and Madsen (1984), and later Davies (1986), have incorporated the
effect of time variation into the eddy viscosity description. The dashed line in Fig.
2.14 shows the calculated velocity profiles for the following time variation in the
eddy viscosity

$$\nu_T = \kappa U_{f,\max} z \left[\frac{1 - \cos(2\omega t)}{2} \right] \tag{2.67}$$

Including the effect of the time variance has a certain effect on the velocity
profiles. However, as seen later, the time variation in ν_T is usually smaller than
that given by Eq. 2.67.

Mixing length theory

The mixing length hypothesis, introduced by Prandtl (1926), states that fluid is transported by turbulent fluctuations a certain distance l (the mixing length) across the flow before it is mixed with the surrounding water, and thereby adjusts its longitudinal flow velocity to that occurring at the new level. Prandtl found that the shear stress is now given by

$$\frac{\tau}{\rho} = l^2 \left|\frac{\partial u}{\partial z}\right| \frac{\partial u}{\partial z} \tag{2.68}$$

Inserting this expression into the equation of motion, Eq. 2.5, the following partial differential equation in u_d is obtained (defined in Eq. 2.63)

$$\frac{\partial u_d}{\partial t} = \frac{\partial}{\partial z}\left(l^2 \left|\frac{\partial u_d}{\partial z}\right| \frac{\partial u_d}{\partial z}\right) \tag{2.69}$$

In the mixing length theory, it is necessary to prescribe the variation of l. The simplest expression is

$$l = \kappa z \tag{2.70}$$

in which κ is the von Kármán's constant. Different variants of Eq. 2.70 exist without significantly changing the calculated velocity profiles. Bakker (1974) was the first to solve the mixing length model for the wave boundary layer by application of the finite difference method. Results of the mixing layer theory will be presented later in this section.

Example 2.8: Prandtl's mixing length theory

Consider Section A-A parallel with the bed in a uniform channel flow (see Fig. 2.15). Perpendicular to A-A the eddies in the turbulent flow cause an exchange of fluid and thereby an exchange of momentum. The eddies in the flow are assumed to have a characteristic size, called l. An eddy with a length scale l will be able to transport fluid from level I to level II, where the mutual distance between I and II is of the order l, see Fig. 2.15. For reasons of continuity the same amount of fluid must be transported downwards from II to I. Hence, a fluid discharge of ρq will occur in both directions, where q can be interpreted as a typical value of the turbulent fluctuations perpendicular to the wall.

The fluid particles transported from level I will keep their longitudinal velocities unchanged until they reach level II, where they will be mixed with the surrounding fluid and adapt to the mean flow velocity at level II. Thus the particles would have increased their velocity by

$$u_{II} - u_I \simeq l\frac{du}{dz} \tag{2.71}$$

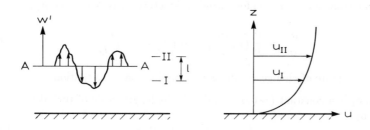

Figure 2.15 Exchange of momentum perpendicular to the flow direction.

The total rate of exchange of momentum per area unit through A-A is given by

$$\Delta B = \rho q l \frac{du}{dz} \qquad (2.72)$$

In order to formulate an equation of motion in which the turbulent fluctuations are neglected, the influence of these turbulent fluctuations must be substituted by an equivalent shear stress given by

$$\tau = \rho q l \frac{du}{dz} \qquad (2.73)$$

The magnitude of q is related to the vertical fluctuations, which, for reasons of continuity, again have the same magnitude as the horizontal velocity fluctuations. These fluctuations are of the same order as the mean velocity difference over a distance equal to the mixing length, whereby q can be related to this mean velocity difference

$$q \simeq l \frac{du}{dz} \qquad (2.74)$$

Inserting Eq. 2.74 into Eq. 2.73, the basic expression given by Eq. 2.68 is obtained.

One-equation models (k-model)

In a one-equation formulation, a transport equation for turbulent kinetic energy k is included in the flow description, this allows a more accurate description of the vertical distribution of turbulence possible. The turbulent kinetic energy k is defined by

$$k = \frac{1}{2}\left(\overline{u'^2} + \overline{v'^2} + \overline{w'^2}\right) \qquad (2.75)$$

in which u', v' and w' are the turbulent fluctuations in the x-, y- and z-direction.

The transport equation for turbulent kinetic energy reads (see Appendix II)

$$\frac{\partial k}{\partial t} = \frac{\partial}{\partial z}\left(\nu_T \frac{\partial k}{\partial z}\right) + \nu_T\left(\frac{\partial u}{\partial z}\right)^2 - c_2 \frac{k^{3/2}}{l_d} \tag{2.76}$$

(rate of change = diffusion + production − dissipation)

in which c_2 is a constant (~ 0.08), and l_d is the length scale of turbulence, which is given by

$$l_d = \kappa \sqrt[4]{c_2}\, z \simeq 0.213\, z \tag{2.77}$$

In the one-equation model, the eddy viscosity is related to the turbulent kinetic energy through the Kolmogorov-Prandtl expression

$$\nu_T = l_d \sqrt{k} \tag{2.78}$$

Eqs. 2.76, 2.78 and the flow equation, Eq. 2.5, constitute a closed system of equations, which together with the boundary conditions must be solved numerically (e.g. by a FDM-scheme). The boundary conditions for the rough wall case are for the velocity

$$u(k_N/30, t) = 0 \tag{2.79}$$

and

$$\frac{\partial u}{\partial z} \to 0 \quad for \quad z \to \infty \tag{2.80}$$

The boundary conditions for the k-equation are

$$k = \frac{1}{\sqrt{c_1}} \nu_T \frac{\partial u}{\partial z} \quad for \quad z = k_N/30 \tag{2.81}$$

and

$$\frac{\partial k}{\partial z} = 0 \quad for \quad z \to \infty \tag{2.82}$$

Eq. 2.81 states that there is local equilibrium close to the wall, while Eq. 2.82 expresses that there is no flux of turbulent kinetic energy at the upper boundary.

Fig. 2.16 demonstrates the usefulness of a one-equation approach, with plots of the time variation in the predicted eddy viscosity at two different levels. The mixing length theory always predicts that the eddy viscosity will fall twice to zero during one wave period, cf. Eq. 2.68. On the other hand, a one-equation model more realisticly predicts that ν_T varies much less during a wave cycle. This is because it takes some time for the eddies to decay after being formed (mainly at the large outer flow velocities). The effect is especially pronounced for small values of a/k_N, a being the amplitude in the near bed motion just outside the boundary layer. Further, the variation in ν_T decreases with increasing distance from the bed. The effect of time variation in ν_T on the velocity profile, as shown in Fig. 2.14, is

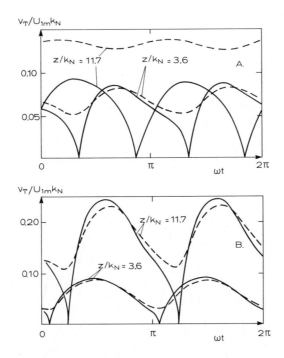

Figure 2.16 Variation in eddy viscosity with time. Dashed line: Predicted by a one-equation model. Full drawn line: Predicted by a mixing-length model. A: $a/k_N = 10^2$. B: $a/k_N = 10^3$. (Justesen and Fredsøe, 1985).

exaggerated, because Eq. 2.67 gives a too large variation in ν_T. Fig. 2.17 shows another outcome of a one-equation model, namely the predicted instantaneous picture of contour plots of turbulent kinetic energy. Here the variation in the x-direction (the direction of wave propagation) is obtained from the time variation by use of the relation

$$\frac{\partial}{\partial x} = -\frac{1}{c}\frac{\partial}{\partial t} \qquad (2.83)$$

It is seen that k is maximum at the bottom, where the main production takes place. The shape of the contour plots is skewed, due to the time lag in the strength of eddies away from the bed.

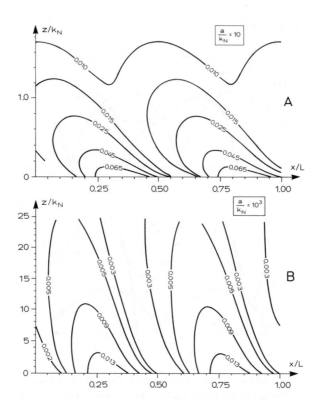

Figure 2.17 Contour plot of turbulent kinetic energy along the bed, predicted by a one-equation model. The wave is moving from left
to right. A: $a/k_N = 10$. B: $a/k_N = 10^3$. (Justesen and
Fredsøe, 1985).

Two-equation models ($k - \varepsilon$ models)

In a two-equation model, the length-scale l is allowed to vary in time and
space instead of being given by a prescribed function like Eq. 2.70. This introduction of a new parameter l requires an additional equation, compared to the
system of equations for the k-equation model. This additional equation is usually
a transport equation for the dissipation ε, and structured as Eq. 2.76 which is a
transport equation for the kinetic energy k.

The transport equation for ε reads

$$\frac{\partial \varepsilon}{\partial t} = \frac{\partial}{\partial z}\left[\frac{\nu_T}{\sigma_D}\frac{\partial \varepsilon}{\partial z}\right] + c_3\varepsilon\frac{\nu_T}{k}\left(\frac{\partial u}{\partial z}\right)^2 - c_4\frac{\varepsilon^2}{k} \tag{2.84}$$

where σ_D, c_3 and c_4 are constants which must be found from experiments, (for a more detailed derivation, see e.g. Launder and Spalding, 1972).

In order to solve Eq. 2.84 together with Eqs. 2.76, 2.77 and 2.5, two additional boundary conditions to the four given by Eqs. 2.79 - 2.82 are required. These are

$$\varepsilon = (c_2)^{3/4}\,\frac{k^{3/2}}{\kappa z}\quad for\quad z \to k_N/30 \tag{2.85}$$

and

$$\frac{\partial \varepsilon}{\partial z} = 0\quad for\quad z \to \infty \tag{2.86}$$

Eqs. 2.85 and 2.86 are equivalent to Eqs. 2.81 and 2.82 but for the dissipation ε replacing the turbulent kinetic energy k. (Rodi, 1980).

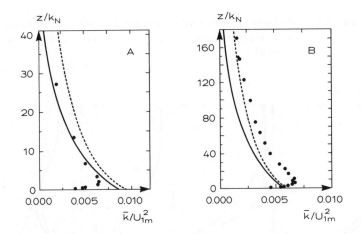

Figure 2.18 Calculated and measured variation of period-averaged turbulent kinetic energy as a function of the distance from the bed A: $a/k_N = 720$. B: $a/k_N = 3700$. —— : Two-equation model. – – – : One-equation model, (Justesen, 1990). ● : Measurements by Jensen et al. (1989).

Fig. 2.18 shows a comparison between the prediction of a one-equation model and a two-equation model with respect to the mean values of k over one wave period, while Fig. 2.19 shows a similar comparison for the instantaneous values. Further, experimental values of k, measured by Jensen et al. (1989), are plotted in the figures. It is seen that the average level of k is lower for the two-equation model.

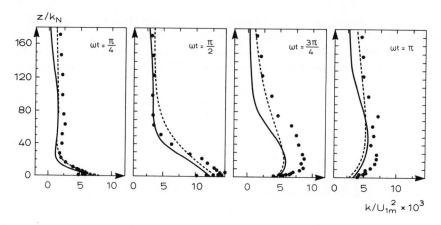

Figure 2.19 Calculated and measured variation of instantaneous turbulent kinetic energy. Symbols as in Fig. 2.18. $a/k_N = 3700$. (Justesen, 1990).

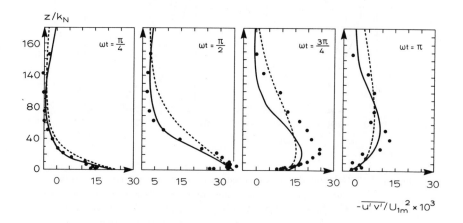

Figure 2.20 Calculated and measured vertical variation of Reynolds stresses in oscillatory flow. $a/k_N = 3700$. Symbols as in Fig. 2.18. (Justesen, 1990).

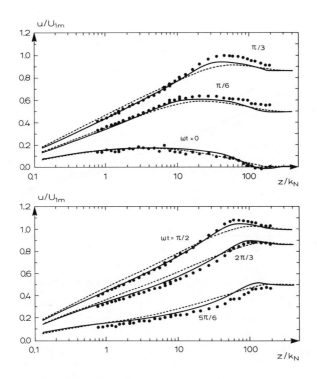

Figure 2.21 Calculated and measured flow velocity. $a/k_N = 3700$. Symbols as in Fig. 2.18.

Fig. 2.20 shows a comparison between the predicted and measured vertical shear stress distribution at different phases. Here, the two-equation model is slightly better than the one-equation model. Fig. 2.21 shows the measured and calculated flow velocity at different phases. Again the two-equation model seems to be slightly better than the one-equation model.

Finally, Fig. 2.22 shows a comparison of the friction factor, calculated by most of the methods described above. It is seen that all methods give nearly the same result. This is promising, because the bed shear stress is one of the most important values for sediment transport calculations. Fig. 2.22 suggests that only a little is gained by using very sophisticated (and computer-expensive) models.

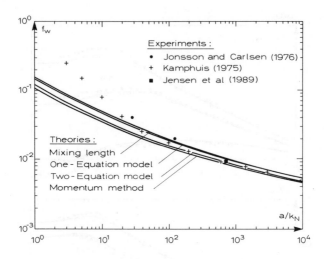

Figure 2.22 Prediction of wave friction factor by different models.

2.4 Streaming

Until now, the wave boundary layer has been treated under the assumption that the flow is uniform in the flow direction. In real waves, however, the flow is non-uniform, as sketched in Fig. 2.23, so the near bed velocity variation in harmonic waves must be changed from Eq. 2.8 to

$$u_0 = U_{1m} \sin(\omega t - kx) \tag{2.87}$$

Here $k = 2\pi/L$ is the wave number.

This means that the boundary layer thickness varies along the bottom in the direction of wave propagation. If the displacement thickness δ^* of the boundary layer

$$u_0 \delta^* = \int_0^\delta (u_0 - u) dz \tag{2.88}$$

is introduced, it is easy to realize from the preceding sections that δ^*, in principle, expands, as shown in Fig. 2.23. From geometrical considerations this expansion is seen to create a vertical flow velocity, (ref. Fig. 2.24) which is small compared to the orbital velocities. This additional vertical velocity attains the value w_∞ outside the boundary layer, while it decreases through the boundary layer to be

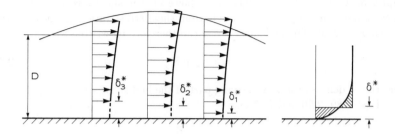

Figure 2.23 Variation in displacement thickness as defined to the right (the two shaded areas must have equal size).

zero at the bed. The existence of this small w_∞ is important because it appears that the time-averaged term $\overline{uw_\infty}$ is different from zero. This means that an additional shear stress will be induced, resulting in a weak additional circulation current. In the following, this additional shear stress term is evaluated.

From the equation of continuity, w_∞ can be expressed by the displacement integrated over the entire boundary layer as

$$w_\infty = -\frac{\partial}{\partial x} \int_0^\delta (u - u_0)dz = \frac{1}{c} \int_0^\delta \frac{\partial}{\partial t}(u - u_0)dz \qquad (2.89)$$

In Eq. 2.89 the relation between spatial and temporal derivatives has been introduced

$$\frac{\partial}{\partial x} = -\frac{1}{c}\frac{\partial}{\partial t} \qquad (2.90)$$

Eq. 2.90 is at present an approximation, because the wave height, due to dissipation (see Section 2.5), decreases slightly in the x-direction. From the linearized flow equation, Eq. 2.5, the last integral in Eq. 2.89 is easily obtained by integrating over the boundary layer thickness

$$\int_0^\delta \rho \frac{\partial}{\partial t}(u_0 - u)dz = -(\tau_b - \tau_\delta) = -\Delta\tau \qquad (2.91)$$

in which $\Delta\tau$ is the increase in instantaneous shear stress through the boundary layer. Now, w_∞ is given by

$$w_\infty = -\frac{\Delta\tau}{\rho c} \qquad (2.92)$$

In the wave boundary layer models described above, τ_δ has been assumed to be zero. However, in general, τ does not vanish at the top of the boundary, even if no resulting current co-exists together with the waves, because a small circulation current over the vertical will always be present in dissipative water waves, cf.

Chapter 5. Outside the boundary layer, however, the shear stress can be taken as a stationary quantity τ_δ, while the bottom shear stress varies significantly with time, cf. Fig. 2.20. The bed shear stress can therefore be written as

$$\tau_b = \tau_\delta + \tilde{\tau}_b \tag{2.93}$$

where the $\sim$ denotes the time-varying part of the bed shear stress.

Eqs. 2.89, 2.90 and 2.91 give

$$w_\infty = -\frac{\tilde{\tau}_b}{\rho c} \tag{2.94}$$

If w_∞ is not 90° out of phase with u_0, additional shear stresses will be induced in the boundary layer.

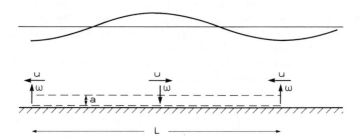

Figure 2.24 Induced vertical velocities due to spatial changes in displacement thickness, as shown in Fig. 2.23. The dashed line indicates the control box used for the momentum equation.

This can be realized from the momentum equation, applied to the control box shown in Fig. 2.24, which is one wave length long. The control box extends from the bed to a level a above the bed. At this level the shear stress is τ_a, the horizontal flow velocity u_a, and the vertical velocity w_a. The time-averaged momentum equation for this box reads

$$\overline{\tau_a} - \overline{\tau_b} = \rho \vec{e} \, \overline{\int_A \vec{v}(\vec{v} \cdot \vec{dA})} = \rho \overline{u_a w_a} \tag{2.95}$$

where $\vec{e}$ is a unit vector in the x-direction.

Above the boundary layer, u_a becomes u_0, w_a becomes w_∞, and $\overline{\tau_a}$ becomes τ_δ, so the bed shear stress is determined by

$$\overline{\tau_b} = -\rho \overline{u_0 w_\infty} + \tau_\delta \tag{2.96}$$

or

$$\Delta \tau = \frac{\rho}{c} \, \overline{u_0 \tilde{\tau}_b} \tag{2.97}$$

Because the term $-\overline{u_0 w_\infty}$ is not zero, it is seen that $\overline{\tau_b}$ will be different from τ_δ, so the shear stress changes through the boundary layer due to the organized transfer of momentum, originating from the displacement effect, as sketched in Fig. 2.23.

Example 2.9:   Laminar and turbulent streaming

Eq. 2.94 was first developed in a modified version by Longuet-Higgins (1957), who applied it to determine the flow induced by the spatial and temporal gradients in the velocity deficit in the laminar case.

The laminar boundary layer flow is given in Example 2.1, Eqs. 2.8, 2.9 and 2.10. Inserting these into Eq. 2.89, gives

$$w_\infty = -\frac{1}{2} U_{1m} k \delta_1 \sin(\omega t - kx) \tag{2.98}$$

where δ_1 is the Stokes length, defined by Eq. 2.10. Longuet-Higgins (1957) assumed the bed shear stress outside the boundary layer to be zero, whereby Eq. 2.98 gives

$$\overline{\tau_b} = -\rho \overline{u_0 w_\infty} = \rho \frac{\pi}{2} \frac{\delta_1}{L} U_{1m}^2 \tag{2.99}$$

The mean induced streaming velocity u_s in a distance a above the bed is found from

$$\tau_a = \nu \frac{\partial u_s}{\partial z}\bigg|_{z=a} = \rho \overline{u_a w_a} + \overline{\tau_b} \tag{2.100}$$

cf. Eq. 2.95. w_a is found from the equation of continuity to be

$$w_a = \frac{1}{c} \int_0^a \frac{\partial}{\partial t}(u - u_0) dz \tag{2.101}$$

in which u is given by

$$u = U_{1m} \sin(\omega t - kx) - U_{1m} \exp\left(-\frac{z}{\delta_1}\right) \sin\left(\omega t - kx - \frac{z}{\delta_1}\right) \tag{2.102}$$

cf. Eq. 2.10. It is now possible to integrate Eq. 2.100 to obtain u_s, as done by Longuet-Higgins (1957). He found that at the top of the boundary layer the mean induced streaming velocity is

$$u_{s,\delta} = \frac{3}{4} \frac{U_{1m}^2}{c} \tag{2.103}$$

i.e. independent of the kinematic viscosity ν.

It is interesting to note that the streaming outside the boundary layer is equal to the Lagrangian drift velocity, except for a constant factor cf. Example 1.2. The drift velocity is given by

$$u_{d,\delta} = \frac{U_{1m}^2}{2c} \tag{2.104}$$

cf. Eq. 1.28. Finally, it can be mentioned that the same principles apply to the turbulent and the laminar case. However, a turbulent model is needed for estimating the eddy viscosity ν_T, which must replace the constant kinematic viscosity ν in Eq. 2.100. Brøker (1985) did such an analysis for a rough wall using the momentum integral method, described in Section 2.2.

The result of that analysis was that the induced streaming at the top of the boundary layer was reduced, compared to the laminar result in Eq. 2.103, and the strength of the streaming became a weak function of the parameter a/k_N. To compare with the laminar result, the turbulent streaming velocity $U_{s,\delta}$ can be written as

$$U_{s,\delta} = f\left(\frac{a}{k}\right) \frac{U_{1m}^2}{c} \tag{2.105}$$

where $f\left(\frac{a}{k}\right)$ is plotted in Fig. 2.25.

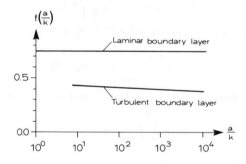

Figure 2.25 Streaming velocity at the top of the boundary layer in the turbulent and the laminar case (after Brøker, 1985).

In the turbulent case, the reduction in the strength of the streaming is related to the fact that the velocity deficit is much smaller in the turbulent boundary layer than in the laminar boundary layer.

2.5   Energy dissipation in the wave boundary layer

Outside the surf zone, together with occasional events of wave-breaking, energy dissipation in the wave boundary layer is responsible for extraction of energy from the wave motion.

Time-averaging over one wave period reveals that the production of energy is equal to the energy dissipation into heat, so the dissipation can be written as

$$\mathcal{D} = \frac{1}{T} \int_0^T \int_0^\infty \tau \frac{\partial u}{\partial z} dz \, dt \qquad (2.106)$$

The instantaneous production term is given by

$$\text{Prod} = \int_0^\infty \tau \frac{\partial u}{\partial z} dz \qquad (2.107)$$

The easiest way to evaluate this term is to realize that the same turbulent flow is created in the boundary layer, either if the outer flow is driven by pressure, as in a u-tunnel (Fig. 2.1), or if the boundary layer is created by oscillating the bed back and forth, the outer water being originally at rest. Considering this last case, Eq. 2.107 is easily integrated to give

$$\text{Prod} = (\tau u)_\infty - (\tau u)_0 - \int_0^\infty \frac{\partial \tau}{\partial z} u \, dz$$

$$= -\tau_b u_0 - \int_0^\infty \frac{\partial u}{\partial t} u \, dz \qquad (2.108)$$

Here the time-averaged value of the last term is equal to zero, the function being periodical, so

$$\mathcal{D} = -\overline{\tau_b \, u_0} \qquad (2.109)$$

Considering the case where the flow above the bed is moving back and forth (instead of the bed), the sign on τ becomes the opposite, so the dissipation is now

$$\mathcal{D} = \overline{\tau_b \, u_0} = \overline{U_f |U_f| u_0} \qquad (2.110)$$

Usually, the energy loss is calculated by use of the so-called energy loss factor f_e. This energy loss factor is defined as follows: it is assumed that the instantaneous bed shear stress is given by

$$\tau_b = \frac{1}{2} \rho f_e u_0^2 \qquad (2.111)$$

where u_0 is the instantaneous outer flow velocity. Eq. 2.110 now becomes

$$\mathcal{D} = \frac{2}{3\pi}\rho f_e U_{1m}^3 \tag{2.112}$$

f_e will differ slightly from f_w, defined by Eq. 2.42, because the temporal variation in τ_b is more complex than given by Eq. 2.111 with f_e being constant. For this reason, the accurate definition of f_e is given by Eq. 2.112. Table 2.1 gives the relation between f_e and f_w for the rough turbulent case. The calculations are carried out by use of a one-equation model (Justesen, 1988a). The table indicates that, for practical purposes, f_e can be taken to be equal to f_w.

Table 2.1 The factors f_e and f_w calculated by a one-equation model.

$\dfrac{a}{k_N}$	f_e	f_w	$\dfrac{f_e}{f_w}$
10^0	0.104	0.124	0.842
10^1	0.0358	0.0376	0.952
10^2	0.0160	0.0157	1.019
10^3	0.00836	0.00807	1.036
10^4	0.00477	0.00446	1.070

Example 2.10: Laminar dissipation. Decay of waves due to dissipation in a laminar wave boundary layer

In the laminar case, the bed shear stress is given by Eq. 2.11

$$\frac{\tau_b}{\rho} = \frac{\nu U_{1m}}{\delta_1}\left[\cos(\omega t) + \sin(\omega t)\right] \tag{2.113}$$

The Stokes length δ_1 is given by Eq. 2.10. Now the dissipation is

$$\mathcal{D} = \overline{\tau_b U_o} = \rho\frac{\sqrt{\nu\omega}}{2\sqrt{2}}\,U_{1m}^2 \tag{2.114}$$

The decay in wave height due to dissipation is found from the continuity equation for the wave energy flux

$$\frac{dE_f}{dx} = -\mathcal{D} \tag{2.115}$$

in which E_f is the wave energy flux, given by

$$E_f = Ec_g \tag{2.116}$$

where c_g is the group velocity, see Eq. 1.23, and E the wave energy. For reasons of simplicity the shallow water wave case is considered,

$$c_g = c = \sqrt{gD} \tag{2.117}$$

where D is the water depth, and c the phase velocity. If the bottom is assumed horizontal, D is constant, and Eqs. 2.116, 2.117, 2.118 and 2.119 give

$$\frac{d}{dx}\left(\frac{1}{8}\rho g H^2 \sqrt{gD}\right) = -\frac{\sqrt{\nu\omega}}{2\sqrt{2}}\, U_{1m}^2 \rho \tag{2.118}$$

In shallow water the velocity amplitude is given by

$$U_{1m} = \frac{H}{2D}c \tag{2.119}$$

which, when inserted into Eq. 2.118, gives

$$\frac{d}{dx}\left(H^2\right) = -\frac{\sqrt{\nu\omega}}{\sqrt{2gD}\,D}H^2 \tag{2.120}$$

or

$$H^2 = H_0^2 \exp\left(-\sqrt{\frac{\nu\omega}{2gD}}\frac{x}{D}\right) \tag{2.121}$$

where H_0 is the wave height at $x = 0$. It is seen that for the distance x_h that the wave has to migrate in order to reduce its height by half, the height is given by

$$\exp\left(-\sqrt{\frac{\nu\omega}{2gD}}\frac{x_h}{D}\right) = \frac{1}{4} \tag{2.122}$$

or

$$x_h = D\sqrt{\frac{2gD}{\nu\omega}}\ln 4 \tag{2.123}$$

For a water depth equal to 10 m, a wave period of 8 sec. and a kinematic viscosity equal to 10^{-6} m^2/s, x_h becomes 219 km.

Example 2.11: Turbulent dissipation

The wave boundary layer, described in the example above, remains laminar up to about

$$RE = \frac{aU_{1m}}{\nu} = \frac{U_{1m}^2 T}{2\pi\nu} \leq 5 \times 10^4 \qquad (2.124)$$

cf. Fig. 2.10. For the waves considered above (D = 10 m, T = 8 sec.) this corresponds to a wave height found by

$$RE = \left(\frac{H}{2D}\right)^2 gD\frac{T}{2\pi\nu} < 5 \times 10^4$$

or

$$H < 0.400 \text{ m}$$

For larger RE-numbers, the boundary layer becomes more and more turbulent. For $RE > 2 \times 10^5$, the boundary layer is nearly fully turbulent, if the bed is assumed to be smooth, cf. Example 2.5. This RE-number is exceeded for $H > 0.80$ m in the preceding numerical example.

For $H = 0.80$ m, the wave decay is calculated from Eqs. 2.111 and 2.112

$$\frac{d}{dx}\left(\frac{1}{8}\rho g H^2 \sqrt{gD}\right) = -\frac{2}{3}\rho f_e \left(\frac{H}{2D}\sqrt{gD}\right)^3 \qquad (2.125)$$

Here $f_e \simeq f_w = 4.96 \times 10^{-3}$, cf. Eq. 2.55 for $RE = 2 \times 10^5$, so

$$\frac{dH}{dx} = -\frac{1}{3}\frac{H^2}{D^2}f_e = 1.06 \times 10^{-5}$$

If the laminar solution, Eq. 2.118, was used at this high RE-number, the gradient in H would be found to be 5.05×10^{-6}, or half the predicted value from the turbulent case.

REFERENCES

Bakker, W.T. (1974): Sand concentration in an oscillatory flow. Proc. Coastal Eng. Conf., pp. 1129-1148.

Brevik, I. (1981): Oscillatory rough turbulent boundary layers. J. Waterway, Port, Coastal and Ocean Eng. Div., ASCE, 107(WW3):175-188.

Brøker, Ida H. (1985): Wave generated ripples and resulting sediment transport in waves. Series paper No. 36, Inst. of Hydrodynamics and Hydraulic Engineering, ISVA, Techn. Univ. of Denmark.

Christoffersen, J.B. and Jonsson, I.G. (1985): Bed friction and dissipation in a combined current and wave motion. Ocean Eng., 12(5):387-423.

Davies, A.G. (1986): A model of oscillatory rough turbulent boundary flow. Est. Coast. Shelf Sci., 23(3):353-374.

Fredsøe, J. (1984): Turbulent boundary layers in wave-current motion. J. Hydraulic Eng., ASCE, 110(HY8):1103-1120.

Grant, W.D. and Madsen, O.S. (1979): Combined wave and current interaction with a rough bottom. J. Geophys. Res., 84(C4):1797-1808.

Hino, M., Kashiwayanagi, M., Nakayama, A. and Hara, T. (1983): Experiments on the turbulence statistics and the structure of a reciprocating oscillatory flow. J. Fluid Mech., 131:363-400.

Jensen, B.L., Sumer, B.M. and Fredsøe, J. (1989): Turbulent oscillatory boundary layers at high Reynolds numbers. J. Fluid Mech., 206:265-297.

Jonsson, I.G. and Carlsen, N.A. (1976): Experimental and theoretical investigations in an oscillatory turbulent boundary layer. J. Hydr. Res., 14(1):45-60.

Justesen, P. and Fredsøe, J. (1985): Distribution of turbulence and suspended sediment in the wave boundary layer. Progress Report No. 62, Inst. of Hydrodynamics and Hydraulic Engineering, ISVA, Techn. Univ. Denmark, pp. 61-67.

Justesen, P. (1988a): Turbulent wave boundary layers. Series Paper 43. Thesis, Techn. Univ. Denmark, Inst. of Hydrodynamics and Hydraulic Engineering, ISVA, Denmark.

Justesen, P. (1988b): Prediction of turbulent oscillatory flow over rough beds. Coastal Eng., 12:257-284.

Justesen, P. (1990): A note on turbulence calculations in the wave boundary layer. Progress Report No. 71, Inst. of Hydrodynamics and Hydraulic Engineering, ISVA, Techn. Univ. Denmark, pp. 37-50, (also in print, J. Hydr. Res.).

Kajiura, K. (1968): A model for the bottom boundary layer in water waves. Bull. Earthquake Res. Inst., 45:75-123.

Kamphuis, J.W. (1975): Friction factor under oscillatory waves. J. Waterways, Port, Coastal and Ocean Eng. Div., ASCE, 101(WW2):135-144.

Laufer, J. (1954): The structure of turbulence in fully developed pipe flow. Rep. 1174, U.S. National Advisory Committee for Aeronautics, National Bureau of Standards, 18 p.

Launder, B.E. and Spalding, D.B. (1972): Mathematical Models of Turbulence. Academic Press, London, 170 pp.

Longuet-Higgins, M.S. (1957): The mechanics of the boundary-layer near the bottom in a progressive wave. Appendix to Russell, R.C.H. and Osorio, J.D.C.: An experimental investigation of drift profiles in a closed channel. Proc. 6th Int. Conf. Coast. Eng., Miami, Fl., 184-193.

Myrhaug, D. (1982): On a theoretical model of rough turbulent wave boundary layers. Ocean Eng., 9(6):547-565.

Nikuradse, J. (1932): Gesetzmässigkeiten der turbulenten Strömung in glatten Rohren. VDI-Forschungsheft 356, Berlin.

Nikuradse, J. (1933): Strömungsgesetze in rauhen Rohren. VDI-Forschungsheft 361, Berlin 1933.

Prandtl, L. (1926): Über die ausgebildete Turbulenz. Verhandlungen des II. Internationalen Kongresses für Technische Mechanik, Zürich 1926, Füssli 1927, pp. 62-75.

Rodi, W. (1980): Turbulence Models and Their Application in Hydraulics. IAHR, Delft, The Netherlands, 104 pp.

Schäffer, H.A. and Svendsen, I.A. (1986): Boundary layer flow under skew waves. Progress Report No. 64, Inst. of Hydrodynamics and Hydraulic Engineering, ISVA, Techn. Univ. Denmark, pp. 13-24.

Sleath, J.F.A. (1987): Turbulent oscillatory flow over rough beds. J. Fluid Mech. 182:369-409.

Spalart, P.R. (1988): Direct simulation of a turbulent boundary layer. J. Fluid Mech. 187:61-98.

Trowbridge, J. and Madsen, O.S. (1984): Turbulent wave boundary layers. 2. Second-order theory and mass transport. J. Geophys. Res., 89(C5): 7999-8007.

Zyserman, J., Justesen, P., Fredsøe, J. and Deigaard, R. (1987): Resulting bed shear stress due to asymmetric waves. Progress Report No. 65, Inst. of Hydrodynamics and Hydraulic Engineering, ISVA, Techn. Univ. Denmark, pp. 31-35.

Chapter 3. Bed friction and turbulence in wave-current motion

In this chapter the findings of Chapter 2 are extended to the more general case, where a current co-exists with the waves. In this case, the turbulence is not restricted to the thin wave boundary layer, but extended to the whole water depth.

The presence of a current together with waves significantly complicates the turbulence structure. In the following the case is considered where the current is driven by a small slope S of the mean water surface. In the case of uniform conditions, the time-averaged vertical shear stress distribution can be found from static considerations to be given by

$$\tau = \rho g D S (1 - z/D) \tag{3.1}$$

In the general case, this shear stress can be divided into two parts, τ_f and τ_s. τ_f is the part that is carried by the Reynolds stresses due to turbulent fluctuations and by the viscous stresses. τ_s is the part that is carried by the organized motion of the flow. In the wave-current case this last contribution occurs in the wave boundary layer due to streaming (see Section 2.4). Usually this streaming contributes only to a small fraction of the total shear stresses near the bed as sketched in Fig. 3.1, and can therefore be neglected.

The flow depth in the combined wave-current motion can be divided into roughly three zones (see Fig. 3.1) (Lundgren, 1972).

In the upper zone (zone I), the turbulence is totally dominated by the current. Here the dominant frequency of the turbulent stresses is usually much

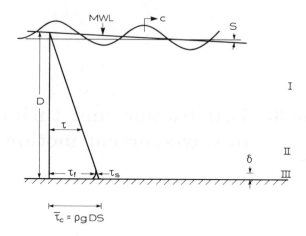

Figure 3.1 Definition sketch and location of different zones of turbulence.

lower than the wave frequency, and the turbulent characteristics are independent of the waves. Zone II is a transition zone, where the turbulent viscosity produced by the wave boundary layer is of the same magnitude as the viscosity pertaining to the current turbulence. In this zone, the dominant frequencies of the turbulence gradually increase with depth and are larger than the wave frequency towards the lower boundary of this zone. Zone III is totally dominated by the wave-produced turbulence. The vertical extent of zone III depends on the relative influences of waves and current. If the current becomes very strong compared to the waves, zone III may totally vanish.

The change in the turbulent structure due to the presence of waves implies that the vertical distribution of the mean current profile will be different compared with the profile without waves. Also, the flow resistance for the current will change due to the presence of waves.

Both the mentioned items are important for the calculation of sediment transport, and are the main topics of this chapter. As in Chapter 2, the start is with a very simplified description in order to obtain a better view of the physics involved.

3.1 Simple considerations on changes in the shape of the velocity profile and increase in bed shear stress

The effect of the presence of the wave boundary layer is easily illustrated:

because of the wave boundary layer the eddy viscosity becomes slightly larger close to the bed than in the case with only a current as illustrated in Fig. 3.2. The eddy viscosity can, roughly speaking, be considered to consist of a contribution ν_c from the current and another contribution ν_w from the waves. From the constitutive equation

$$\frac{\tau}{\rho} = \left(\nu_c + \nu_w\right)\frac{\partial U}{\partial z} \tag{3.2}$$

it is easily realized that for the same value of the bed shear stress, the near bed slope of the current velocity profile U becomes smaller in the combined wave-current case than it does in the current case. Farther away from the bed, the slope is not affected by the waves. As the mean bed shear stress is given by

$$\tau_b = \rho g D S \tag{3.3}$$

it is realized that the presence of waves reduces the water discharge if D and S are kept the same for the two cases with and without waves.

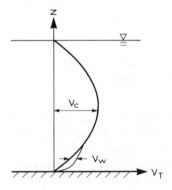

Figure 3.2 Schematic illustration of the current and wave contribution to the time-averaged eddy viscosity.

The purpose of this introductory section is in a simple way to calculate the above described changes in the current velocity profile due to the presence of waves. For simplicity , the case where waves exist together with a weak current is considered first. The present analyses were originally suggested by Fredsøe (1981). The directions of the wave propagation and the current are taken to be the same, unless otherwise stated.

In this first approach, the boundary layer thickness is taken to be a steady value δ and equal to the one determined by the wave motion alone in Chapter 2.

According to Fig. 3.1, the flow is now divided into two zones: an outer zone outside δ, where the turbulence is due to current only, and an inner zone ($z < \delta$), where the turbulence is mainly determined by the waves. No transition zone is included in this simple description. This implies that the strength of the current is assumed to be so weak that the wave boundary layer characteristics (friction factor and boundary layer thickness), are determined by the wave motion alone as an approximation.

Due to the wave motion, a boundary layer is now formed with the wave friction factor f_w and the boundary layer thickness δ, cf. Chapter 2.

At the top of the boundary layer ($z = \delta$), the current velocity U_δ is still unknown at this point.

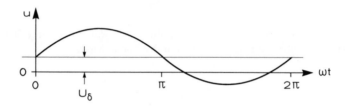

Figure 3.3 Schematic variation in flow velocity with time at the top of the wave boundary layer.

The variation in velocity with time at the level $z = \delta$ is shown in Fig. 3.3. This velocity consists of the steady current contribution U_δ, and an unsteady component u_w from the wave motion (capital letter U denotes in this chapter the steady component while small letter u is the time-varying component). For simplicity, u_w is taken to vary harmonically with time (Airy wave).

With the present assumptions it is possible to determine the average bed shear stress as follows: In pure wave motions, the maximum bed shear stress is determined by

$$\frac{\tau_w}{\rho} = \frac{1}{2} f_w U_{1m}^2 \tag{3.4}$$

The average bed shear stress over one wave period is zero. Henceforth, Eq. 3.4 is assumed to be generally valid for the instantaneous bed shear stress during the wave period, i.e.

$$\frac{\tau_{wc}}{\rho} = \frac{1}{2} f_w u |u| \tag{3.5}$$

in which u is the instantaneous flow velocity given by

$$u = u_w + U_\delta$$
$$= U_{1m} \sin(\omega t) + U_\delta \tag{3.6}$$

From Eqs. 3.5 and 3.6, the resulting mean bed shear stress in the direction of the current is found to be

$$\frac{\overline{\tau_{wc}}}{\rho} = \frac{1}{2} f_w \frac{1}{T} \int_0^T u|u| dt$$
$$= \frac{1}{2} f_w \frac{1}{T} U_{1m}^2 \int_0^T \left(\sin(\omega t) + \frac{U_\delta}{U_{1m}}\right) \left|\sin(\omega t) + \frac{U_\delta}{U_{1m}}\right| dt \tag{3.7}$$

Since $U_\delta/U_{1m} \ll 1$, Eq. 3.7 can be solved analytically as

$$\frac{\overline{\tau_{wc}}}{\rho} \simeq \frac{1}{2} \frac{f_w}{T} U_{1m}^2 \left\{ \int_0^{\frac{T}{2}} \left(\sin(\omega t) + \frac{U_\delta}{U_{1m}}\right)^2 dt \right.$$

$$\left. - \int_{\frac{T}{2}}^{T} \left(\sin(\omega t) + \frac{U_\delta}{U_{1m}}\right)^2 dt \right\}$$

$$\simeq \frac{1}{2} \frac{f_w}{2\pi} U_{1m}^2 \left\{ \int_0^{\pi} \left[\sin^2(\omega t) + 2\frac{U_\delta}{U_{1m}} \sin(\omega t)\right] d(\omega t) \right.$$

$$\left. - \int_{\pi}^{2\pi} \left[\sin^2(\omega t) + 2\frac{U_\delta}{U_{1m}} \sin(\omega t)\right] d(\omega t) \right\}$$

$$= \frac{2}{\pi} f_w U_{1m} U_\delta \tag{3.8}$$

By definition, the current friction velocity U_{fc} is related to $\overline{\tau_{wc}}$ by

$$\overline{\tau_{wc}} = \rho U_{fc}^2 \tag{3.9}$$

so the current friction velocity is found from Eqs. 3.8 and 3.9 giving

$$U_{fc} \simeq \sqrt{\frac{2}{\pi} f_w U_{1m} U_\delta} \tag{3.10}$$

From Eq. 3.8 it is easy to see that the average bed shear stress is significantly increased due to the presence of waves, because f_w is usually much larger than the current friction factor f_c, and U_{1m} is usually much larger than U_δ.

Velocity profiles outside the wave boundary layer

Using Eq. 3.8 makes it possible to estimate the shape of the current velocity profile in the outer zone. Here the turbulence is dominated by the current, so the eddy viscosity can be estimated as

$$\nu_c = \kappa U_{fc} z \tag{3.11}$$

The velocity gradient above the wave boundary layer is given by

$$\frac{dU}{dz} = \frac{\tau}{\rho \nu_c} \simeq \frac{\overline{\tau_{wc}}}{\rho \kappa U_{fc} z}$$

or

$$\frac{dU}{dz} \simeq \frac{U_{fc}}{\kappa z} \tag{3.12}$$

which when integrated gives

$$\frac{U}{U_{fc}} = \frac{1}{\kappa} \ln\left(\frac{z}{k_w/30}\right) \tag{3.13}$$

In Eq. 3.13 the constant k_w is introduced, so the equation follows the usual law of the wall for the rough wall case. Hence, k_w can be interpreted as an apparent 'wave-roughness' felt by the current in the combined wave-current flow, (originally suggested by Grant and Madsen, 1979). By introducing the current velocity at the top of the wave boundary layer U_δ, Eqs. 3.8 and 3.13 give

$$k_w = 30\delta/\exp\left[\kappa U_\delta / \sqrt{\frac{2}{\pi} f_w U_{1m} U_\delta}\right] \tag{3.14}$$

Here δ, f_w and U_{1m} can be calculated from wave data, while U_δ is unknown. However, U_δ can be correlated to the mean flow velocity V by using Eq. 3.14 and averaging over the depth of U from Eqs. 3.10 and 3.13. This gives

$$V = \sqrt{\frac{2}{\pi} f_w U_{1m} U_\delta}\left[6.2 + \frac{1}{\kappa} \ln\left(\frac{D}{k_w}\right)\right] \tag{3.15}$$

where D is the water depth. In the derivation of Eq. 3.15, the usually very small contribution to the total water discharge inside the wave boundary layer is not correctly considered.

The velocity profile inside the wave boundary layer

Inside the wave boundary layer, there are two contributions to the eddy viscosity, one from the wave turbulence ν_w and the other from the current, ν_c. In the present special case, where the directions of wave propagation and current are the same, the combined eddy viscosity can be given by

$$\nu_T = \nu_c + \nu_w = \kappa(U_{fw} + U_{fc})z \tag{3.16}$$

in which U_{fw} is the wave friction velocity. Inside the wave boundary layer, the time-averaged velocity gradient is now given as

$$\frac{dU}{dz} = \frac{\overline{\tau_{wc}}}{\rho\kappa(U_{fw} + U_{fc})z} \tag{3.17}$$

which when integrated gives

$$\frac{U}{U_{fc}} = \frac{1}{\kappa}\frac{U_{fc}}{U_{fw} + U_{fc}}\ln z + c \tag{3.18}$$

The integration constant c is found from the boundary condition

$$U = 0 \ \ for \ \ z = \frac{k_N}{30} \tag{3.19}$$

So, Eq. 3.18 finally gives

$$\frac{U}{U_{fc}} = \frac{1}{\kappa}\frac{U_{fc}}{U_{fw} + U_{fc}}\ln\left(\frac{z}{k_N/30}\right) \tag{3.20}$$

Example 3.1:

In this example, the above findings are illustrated by a numerical example. Let us consider waves with a height of 3 m, a wave period of 8 sec., and a water depth of 10 m. According to the wave tables in Appendix I, sinusoidal wave theory gives a maximum wave-induced bottom velocity (just outside the boundary layer) of $U_{1m} = 1.17$ m/s. With a bed roughness of $k_N = 1$ mm, $fw = 0.0064$ is obtained from Eq. 2.43 and $\delta = 0.036$ m from Eq. 2.45. By choosing the strength of the current at the top level of the wave boundary layer to be (arbitrarily) $U_\delta = 0.15$ m/s, the apparent wave roughness k_w is found from Eq. 3.14 to be 0.114 m or 114 times the physical roughness k_N.

The current friction velocity is found from Eq. 3.10 to be 0.0267 m/s, and the wave friction velocity is found from $U_{fw} = U_{1m}\sqrt{fw/2} = 0.0662$ m/s (cf. Eq. 2.42). The vertical distribution of the flow velocity is now given by Eqs. 3.13 and 3.20 as

$$z > \delta : U = 0.0669 \ \ln(262z) \ \text{(metric units)}$$

$$z < \delta : U = 0.0192 \ \ln(30000z) \ \text{(metric units)}$$

The calculated current distribution is shown in Fig. 3.4, together with the velocity profile for the same water discharge in the absence of waves. The velocity profile is found by first calculating the mean flow velocity from Eq. 3.15 to give $V = 0.46$ m/s, with the friction velocity for the situation with current only being found from the usual flow resistance formula

$$\frac{V}{U_f} = 6.2 + 2.5 \ln\left(\frac{D}{k_N}\right) \tag{3.21}$$

which gives $U_f = 0.0157$ m/s. The velocity profile is then given by

$$\frac{U}{U_f} = \frac{1}{\kappa} \ln\left(\frac{z}{k_N/30}\right) \tag{3.22}$$

The presence of waves decreases the average current velocity near the bed significantly. This is of importance in relation to the transportation of suspended sediment, which is mainly located quite close to the seabed.

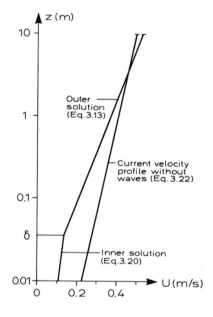

Figure 3.4   Mean velocity profile with and without waves for the same water discharge.

Example 3.2: Increase of the wave length due to the current

The calculations given above can be refined by taking into account the wave lengthening by the current. If the entire wave is transported with the mean flow velocity, then, without the presence of the current, the wave in the preceding example (with a wave length of 70.9 m) will propagate with a celerity c = 8.86 m/s (from the wave table). Hence, c + V = 8.86 + 0.46 = 9.32 m/s.

The wave period T (= 8 sec) may have been obtained from local measurements in the presence of currents. But even if the wave period is known from a location with different current conditions, the wave period is constant for a steady wave field due to the conservation of wave crests.

The higher celerity velocity of the waves on the current results in (with the constant period) an increased wave length given by

$$L_1 = (c + V)\,T = (8.86\,\text{m/s} + 0.46\,\text{m/s}) \times 8\,\text{s} = 74.6\,\text{m}$$

Because the migration velocity of the waves depends on the wave length (cf. the wave table, Appendix I) the new wave length L_1 corresponds to a (higher) migration velocity, which from the wave table is found to be $c = 8.94$ m/s. This new value for c must be inserted in the expression for L_1 giving a new wave length

$$L_2 = (c + V)\,T = (8.94\,\text{m/s} + 0.46\,\text{m/s}) \times 8\,\text{s} = 75.2\,\text{m}$$

One more iteration gives $L_3 = 75.3$ m.

The wave kinematics must be calculated in a frame of reference following the current. In this frame of reference, the wave table gives a wave period

$$T_r = T \times \frac{c + V}{c} = 8\ \text{sec} \times \frac{8.94 + 0.46}{8.94} = 8.41\ \text{sec}$$

which gives a maximum near-bed orbital velocity of

$$U_{1m} = \frac{\pi H}{T_r}\,\frac{1}{\sinh(kD)} = 1.196\,\text{m/s}$$

compared to 1.17 m/s if the current is not taken into account. Usually, this effect can be neglected.

If the wave direction and the direction of the current are not parallel, the above calculations must be based on that component of the current which is in the direction of wave propagation.

Example 3.3:   Three-dimensional wave-current flow

It is not difficult to extend the above findings to the general three-dimensional case, where the direction of wave propagation forms an angle γ to the current direction, see Fig. 3.5.

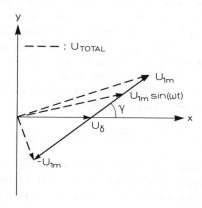

Figure 3.5 Instantaneous velocity at the top of the wave boundary layer.

In a horizontal $x - y$ coordinate system with the x coordinate in the current direction, the instantaneous flow velocity just outside the wave boundary layer can be written as

$$\vec{U} = \begin{pmatrix} U_x \\ U_y \end{pmatrix} = \begin{pmatrix} U_\delta + U_{1m} \sin(\omega t) \cos \gamma \\ U_{1m} \sin(\omega t) \sin \gamma \end{pmatrix} \quad (3.23)$$

For this purpose, the angle, γ, between the current and the direction of the wave propagation can be assumed to lie in the range from 0 to 90 degrees.

The instantaneous bed shear stress is still given by Eq. 3.5 and is assumed to have the same direction as the flow velocity vector just outside the wave boundary layer

$$\vec{\tau}/\rho = \frac{1}{2} f_w \vec{U} |\vec{U}| \quad (3.24)$$

The instantaneous bed shear stress in the x-direction is

$$\tau_x/\rho = \frac{1}{2} f_w \sqrt{U_\delta^2 + U_{1m}^2 \sin^2(\omega t) + 2 U_\delta U_{1m} \sin(\omega t) \cos \gamma}$$

$$[U_\delta + U_{1m} \sin(\omega t) \cos \gamma] \quad (3.25)$$

The mean shear stress in the current direction is then

$$\overline{\tau_{wc}}/\rho = \frac{1}{2}f_w U_{1m}^2 \frac{1}{T}\int_0^T \sqrt{\kappa_1^2 + \sin^2(\omega t) + 2\kappa_1 \sin(\omega t)\cos\gamma}$$

$$\left[\kappa_1 + \sin(\omega t)\cos\gamma\right]dt \tag{3.26}$$

where $\kappa_1 = U_\delta/U_{1m}$.

The integral in Eq. 3.26, as in Eq. 3.7, is an elliptic integral which cannot be solved analytically. In general, it must be solved numerically. In the present case where the wave motion close to the bed is assumed to be dominant, the asymptotic expression can be applied for small values of κ_1:

$$\frac{1}{T}\int_0^T \sqrt{\sin^2(\omega t) + \kappa_1^2 + 2\kappa_1 \cos\gamma \sin(\omega t)} \left[\cos\gamma \sin(\omega t) + \kappa_1\right]dt$$

$$\sim \kappa_1 \frac{4}{\pi}\frac{1+\cos^2\gamma}{2} \tag{3.27}$$

$\overline{\tau_{wc}}/\rho$ can thus be written as

$$\overline{\tau_{wc}}/\rho = \frac{2}{\pi}f_w U_{1m}U_\delta \frac{1+\cos^2\gamma}{2} \tag{3.28}$$

The current shear velocity, U_{fc}, is hence given by (cf. Eq. 3.9)

$$U_{fc} = \sqrt{\overline{\tau_{wc}}/\rho} = \left(\frac{2}{\pi}f_w U_{1m}U_\delta \frac{1+\cos^2\gamma}{2}\right)^{\frac{1}{2}} \tag{3.29}$$

Note that the average shear stress vector will be in the same direction as the current only for $\gamma = 0°$ and $\gamma = 90°$. This side-effect is discussed in detail by Grant and Madsen (1979). The maximum angle between the two directions is approximately 20°.

From Eq. 3.29 it is easily seen that U_{fc} is a maximum for $\gamma = 0°$. From Eq. 3.14, which in its general form (from Eq. 3.13) reads

$$k_w = 30\delta/\exp\left[\kappa U_\delta/U_{fc}\right] \tag{3.30}$$

it is then seen that the apparent wave roughness is maximum when the wave direction is the same as the direction of the current.

3.2   The integrated momentum method applied to combined wave-current motion

In order to simplify the problem it was assumed in the previous section that the near-bed current represented by U_δ was sufficiently weak when compared to the wave-induced near-bed flow velocity U_{1m}, so the wave boundary layer characteristics did not change due to the presence of the current. In order to extend the findings of the previous section to a more general case where no restrictions are introduced on the ratio U_δ/U_{1m}, a reasonably simple model based on the same momentum equation concept as applied in Section 2.1 will be outlined here. For simplicity, only the case of a hydraulically rough bed is considered, which is the most important as far as sediment transport investigations are concerned.

Kinematic description

As in Section 2.1, the idea is to make a reasonable guess on the shape of the velocity profile, and insert this guess into the momentum equation. From Section 3.1, it is natural to guess the shapes of the velocity profiles to be logarithmic, but with one slope near the bed where the turbulence from the wave motion is of importance, and another one far from the bed where the turbulence is due to the current alone. The velocity profiles are described in a similar manner to Section 3.1:

The instantaneous velocity profile $\vec{u}$ consists of two parts: a steady component $\vec{U}_c$ due to the mean current, and an unsteady component $\vec{u}_w$ due to the wave motion.

The unsteady component is described in the same way as in the previous section: outside a small boundary layer at the bottom which is still called the wave boundary layer, the velocity from the unsteady flow is calculated by potential theory. So, in the case of a sinusoidal wave, the velocity just outside the boundary layer is given by

$$u_o = U_{1m} \sin(\omega t) \tag{3.31}$$

The thickness of the wave boundary layer, δ, is no longer prescribed as in Section 3.1. Also, δ is allowed to vary with time.

Inside the boundary layer, the instantaneous velocity profile (the sum of the steady and unsteady component) is assumed to be logarithmic and is given by

$$\frac{u}{u_f} = \frac{1}{\kappa} \ln\left(z/\left(\frac{k_N}{30}\right) \right) \tag{3.32}$$

The steady component $\vec{U}_c$ is described as follows: outside the wave boundary layer the turbulence is due to the mean current. Here, the usual logarithmic velocity distribution

$$\frac{U_c}{U_{fc}} = \frac{1}{\kappa} \ln\left(z \middle/ \left(\frac{k_w}{30}\right)\right) \tag{3.33}$$

is adopted. As in Section 3.1, k_w is the apparent bed roughness which is different from the grain roughness, as the wave boundary layer acts as a larger roughness element. U_{fc} in Eq. 3.33 is the current friction velocity which will be determined later.

Inside the wave boundary layer, the mean current velocity profile is assumed to be given by

$$\frac{U_c}{U_{fo}} = \frac{1}{\kappa} \ln\left(z \middle/ \left(\frac{k_N}{30}\right)\right) \tag{3.34}$$

where U_{fo} is the current friction velocity for the inner profile. The boundary condition at the top of the wave boundary layer ($z = \delta + k/30$) requires that the vectorial sum of the potential flow velocity Eq. 3.31, and the mean current profile Eq. 3.34, be equal to the instantaneous velocity at $z = \delta + k/30$ given by Eq. 3.32. If the angle between the mean current direction and direction of wave propagation is called γ, this condition becomes

$$\left[\frac{u_f}{\kappa} \ln\left(\frac{\delta + k_N/30}{k_N/30}\right)\right]^2 = \left[\frac{U_{fo}}{\kappa} \ln\left(\frac{\delta + k_N/30}{k_N/30}\right) + U_o \cos\gamma\right]^2 + \left[u_o \sin\gamma\right]^2 \tag{3.35}$$

cf. Fig. 3.6, where U_δ is the mean current velocity at the distance δ above the bed (given by Eq. 3.34).

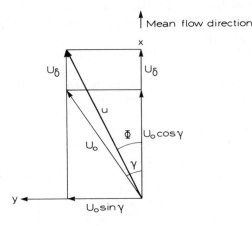

Figure 3.6 Definition sketch of γ and Φ.

If, as in Eq. 2.32, a quantity Z defined by

$$Z = \frac{\kappa U_o}{u_f^*} \tag{3.36}$$

is introduced, where u_f^* in the present case is given by

$$\frac{1}{u_f^*} = \frac{U_{fo}\cos\gamma}{u_f^2 - U_{fo}^2} + \sqrt{\frac{U_{fo}^2\cos^2\gamma}{(u_f^2 - U_{fo}^2)^2} + \frac{1}{u_f^2 - U_{fo}^2}} \tag{3.37}$$

Eq. 3.35 yields the same relationship as Eq. 2.33, i.e.

$$\delta = \frac{k_N}{30}\left(e^Z - 1\right) \tag{3.38}$$

Also, the angle Φ between the instantaneous flow direction in the boundary layer and the mean current direction (see Fig. 3.6) is needed for later use: from Fig. 3.6 it is seen that

$$\cos\Phi = \frac{u_o\kappa\cos\gamma + U_{fo}Z}{U_f Z} \tag{3.39}$$

Finally, using Eqs. 3.37 and 3.38, we obtain

$$\sin\Phi = \frac{u_f^*}{u_f}\sin\gamma \tag{3.40}$$

from Eq. 3.39.

Dynamic description

The momentum equation for the boundary layer in the direction perpendicular to the mean current direction (the y-direction, see Fig. 3.6) is given by

$$\int_{k_N/30}^{\delta + k_N/30} \rho\frac{d}{dt}(U\sin\Phi - u_o\sin\gamma)dz = -\tau_b\sin\Phi \tag{3.41}$$

where the first term under the integral on the left-hand-side represents the acceleration in the y-direction. The second term is the pressure gradient from the wave outside the wave boundary layer. The right-hand-side of Eq. 3.41 represents the bed shear stress component in the y-direction.

The principle for the study of the wave boundary layer development is the same as that previously presented for the case of pure wave motion in combined wave/current motion. The development of the boundary layer for $0 \le \omega t \le \pi$ is first studied. At $t = 0$, the flow pattern is identical to the resulting mean current velocity profile, u_0 being zero, and the small phases between the outer flow and the boundary layer flow due to inertia effects being disregarded.

At $\omega t = \pi$, the system returns to the original flow situation ($u_0 = 0$), and a new wave boundary layer, similar to the former one, develops during the next half wave period.

Example 3.4: Numerical solution of the integrated momentum equation

The development of the boundary layer is calculated from Eq. 3.41. After insertion of Eq. 3.40 followed by integrating Eq. 3.41 gives

$$-u_f u_f^* + \frac{k_N}{30}\left(e^Z - 1\right)\frac{dU_o}{dt} = \frac{k_N}{30\kappa}\left[e^Z(Z-1)+1\right]\frac{du_f^*}{dt} \tag{3.42}$$

u_f^* is defined in Eq. 3.36, from which (as in Eq. 2.17)

$$\frac{du_f^*}{dt} = \frac{u_f^*}{U_o}\frac{du_o}{dt} - \frac{u_f^*}{Z}\frac{dZ}{dt} \tag{3.43}$$

is obtained. Inserting Eq. 3.43 into Eq. 3.42 and rearranging yields

$$\frac{dZ}{dt} = \frac{Z(1+Z-e^Z)}{e^Z(Z-1)+1}\frac{1}{u_o}\frac{du_o}{dt} + \frac{30\kappa}{k_N}\sqrt{\frac{\kappa^2 u_o^2 + Z^2 U_{fo}^2 + 2Z\kappa U_{fo}u_o\cos\gamma}{e^Z(Z-1)+1}} \tag{3.44}$$

Like Eq. 2.20, Eq. 3.44 is singular at $t = 0$, which by Taylor expansion can be written as

$$\frac{dZ}{d(\omega t)} = \beta_1\left(\left(\frac{1}{Z}+\frac{\kappa^2}{2}\left(\frac{U_{1m}}{U_{fo}}\right)^2\frac{(\omega t)^2}{Z^3}\right) + \kappa\frac{U_{1m}}{U_{fo}}\frac{\omega t}{Z^2}\cos\gamma\right) - \frac{Z}{\omega t} \tag{3.45}$$

where

$$\beta_1 = \frac{60\kappa}{\omega}\frac{U_{fo}}{k_N} \tag{3.46}$$

The solution to Eq. 3.45 for small values of ωt is

$$Z = \sqrt{\frac{2}{3}\beta_1}\sqrt{\omega t} \tag{3.47}$$

Eq. 3.45 must be solved numerically for larger values of ωt.

The variation in Z depends on the two parameters: β_1 and U_{1m}/U_{fo}. However, β_1 can be written as (cf. Eq. 3.46)

$$\beta_1 = \frac{60\kappa}{\omega}\frac{U_{1m}}{k_N}\frac{U_{fo}}{U_{1m}} = 60\kappa\frac{a}{k_N}\bigg/\frac{U_{1m}}{U_{fo}} \tag{3.48}$$

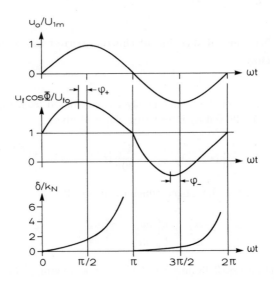

Figure 3.7 Variation of u_o, u_f, and δ during a wave period for $a/k_N = 10$ and $U_{1m}/U_{fo} = 10$. φ_+ is the phase between maximum shear stress and maximum velocity, φ_- between minimum shear stress and minimum velocity.

where a is the free stream particle amplitude. Hence, Z alternatively depends on the two quantities a/k_N and U_{1m}/U_{fo}.

Fig. 3.7 presents an example of the variation in u_f and boundary layer thickness δ for specific values of a/k_N and U_{1m}/U_{fo} for the special case where the direction of wave propagation is the same as the mean current direction.

When the variation of u_f with time is obtained, applying the momentum equation in the mean flow direction (see Fig. 3.6) gives the mean bed shear stress $\overline{\tau}$ in this direction. The momentum equation in this direction is

$$\int_{k_N/30}^{\delta+k_N/30} \rho \frac{d}{dt}(U \cos \Phi - u_o \cos \gamma)dz = -\tau_b \cos \Phi + \tau_{wc} \qquad (3.49)$$

where τ_{wc} is the current shear stress outside the wave boundary layer, which is directed in the x-direction. Integration of Eq. 3.49 with respect to time over one wave period yields

$$\overline{\tau_{wc}} = \rho U_{fc}^2 = \frac{1}{T}\int_o^T \tau_b \cos \Phi \, dt = \frac{1}{T}\int_o^T \rho u_f^2 \cos \Phi \, dt \qquad (3.50)$$

because the left-hand-side of Eq. 3.40 becomes zero when it is integrated over a wave period (no acceleration of the mean flow velocity).

The apparent roughness k_w is determined by matching the inner and the outer mean current profile in the mean thickness of the boundary layer δ_m. This is done by combining Eqs. 3.33 and 3.34 to give

$$\frac{k_w}{k_N} = \left(\frac{30\delta_m}{k_N}\right)^{1 - U_{fo}/U_{fc}} \tag{3.51}$$

δ_m here is the mean value of $\delta(\omega t = \pi/2)$ and $\delta(\omega t = 3\pi/2)$ (see Fig. 3.7).

Model results

In Fig. 3.8 the variations in k_w/k_N as the function of U_{1m}/U_{fc} are shown for different values of a/k_N.

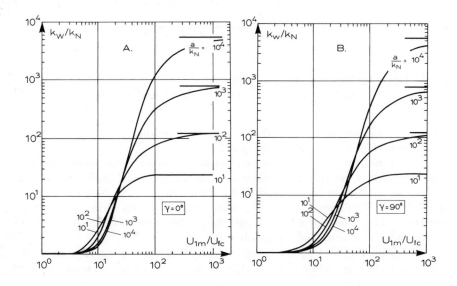

Figure 3.8 Variation of k_w/k_N with U_{1m}/U_{fc} for different values of a/k_N.
Fig. 3.8A: $\gamma = 0°$, Fig. 3.8B: $\gamma = 90°$.

Fig. 3.8A shows the variations for $\gamma = 0°$, while Fig. 3.8B shows the corresponding variation for $\gamma = 90°$. It is interesting to note that the curves cross each other, which means that the apparent roughness for a strong current (i.e. small values of U_{1m}/U_{fc}) is the largest for small values of a/k_N (i.e. fast oscillations). The converse applies for the case for a weak current. Fig. 3.9

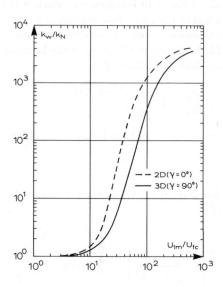

Figure 3.9 Variation of k_w/k_N with U_{1m}/U_{fc} for two-dimensional (broken line) and three-dimensional (full-drawn line) flow. $a/k_N = 10^4$.

presents a comparison between the apparent bed roughness for two- and three-dimensional flow for the same value of a/k_N.

It turns out that the apparent roughness always is the largest for two-dimensional flow, which is also indicated by the simple model in Example 3.2.

Fig. 3.10 shows the variation of δ_m/k with U_{1m}/U_{fc} for both 2D ($\gamma = 0°$) and 3D ($\gamma = 90°$) flow: The variations turn out to be nearly identical, the three-dimensional boundary layer being slightly larger than the two-dimensional, except at the two limits (weak and strong current).

Finally, Fig. 3.11 shows the variation in U_{fc}/U_{fo} with U_{1m}/U_{fc} for $\gamma = 0°$ and $90°$. From Figs. 3.8, 3.10 and 3.11, it is easy to describe the flow as explained in the example below.

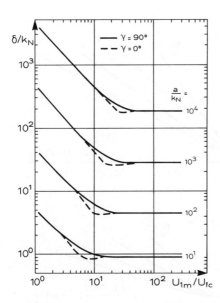

Figure 3.10 Variation of wave boundary layer thickness δ/k_N with U_{1m}/U_{fc}.

Example 3.5: Application of diagrams to find current velocity profiles

If the boundary layer thickness δ_m is small compared with the water depth, Eq. 3.33 yields

$$\frac{V}{U_{fc}} = \frac{1}{\kappa}\left[\ln\left(\frac{D}{k_w/30}\right) - 1\right] \tag{3.52}$$

where V is the mean current velocity.

Figs. 3.8 and 3.11 are now applied in the following way: assuming that V, D, U_{1m}, a, and k are known quantities, an initial value of k_w can be found by a suitable choice of U_{fc} in Fig. 3.8. If the chosen value of U_{fc} is not equal to that obtained from Eq. 3.52, a new value of U_{fc} must be chosen until convergence is achieved.

If the inner solution is of interest, U_{fo} can be easily found from Fig. 3.11.

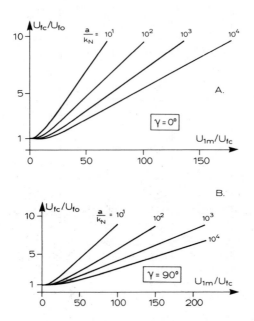

Figure 3.11 Variation of U_{fc}/U_{fo} with U_{1m}/U_{fc}. Fig. 3.11A: $\gamma = 0°$, Fig. 3.11B: $\gamma = 90°$.

Example 3.6: Variation in the current friction factor and maximum bed shear stress.

From the graphs in Figs. 3.8, 3.10 and 3.11 it is possible to determine the friction factor for the flow. The friction factor for the mean current motion is defined by

$$f_c = 2\left(\frac{U_{fc}}{V}\right)^2 \tag{3.53}$$

where V/U_{fc} is determined by Eq. 3.52. Besides the data already given in Figs. 3.8, 3.10 and 3.11, it is also necessary to know the dimensionless water depth D/k_N. The variation of f_c is shown in Fig. 3.12 for $\gamma = 0°$ and $\gamma = 90°$ for a specific value of D/k_N and a/k_N: in a strong current, f_c approaches f_{co}, which is the friction factor in a pure current, obtained from Colebrook-White's formula (1937). This is obtained by integration over the depth of Eq. 2.20. For a weak current, f_c increases to a higher value, which is given by the limit,

$$\sqrt{2/f_c} \rightarrow \frac{1}{\kappa}\left[\ln(D/\delta_w) - 1\right] \tag{3.54}$$

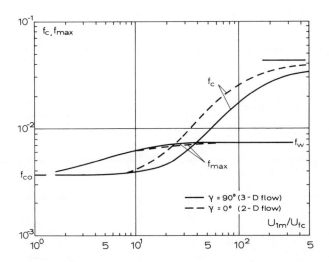

Figure 3.12 Variation of mean current friction factor f_c and maximum friction factor f_{max} with U_{1m}/U_{fc}. $a/k_N = 10^3$, $D/k_N = 10^3$.

The parameter U_{1m}/U_{fc} shown in Fig. 3.12 can be replaced by the more easily applicable parameter U_{1m}/V by use of Eq. 3.52. Fig. 3.13 shows examples of such diagrams, and clearly indicates what the strengths of waves and current must be in order to have wave- or current-dominated flow resistance.

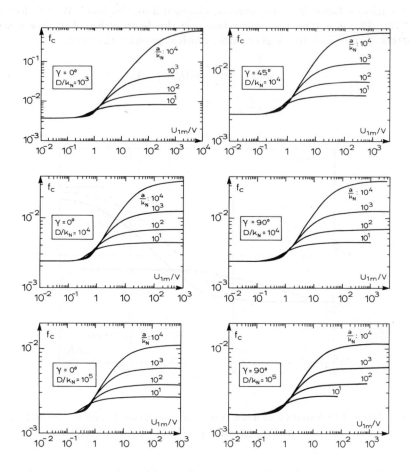

Figure 3.13 Variation of the mean current friction factor with U_{1m}/V for different combinations of a/k_N and D/k_N.

3.3   Refined modelling of the wave-current motion

In the two previous sections, the focus has been on bed friction and flow resistance. For this purpose, an integrated approach like the application of the momentum equation will be sufficient.

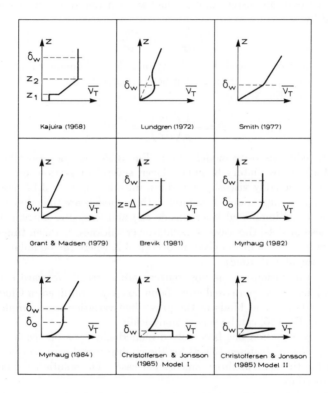

Figure 3.14 Schematical mean eddy viscosity distributions for time-invariant
eddy viscosity models. (After Justesen, 1988).

However, as outlined in Chapter 2 for the pure wave boundary layer, several more refined models exist in which the exact shape of the velocity profile and much more detailed information on the vertical distribution of the turbulence can be obtained. However, most of these models involve numerical calculations to a large extent. In the model described in Section 3.1, no computation was needed at all. And in the model described in Section 3.2, the only numerical work was the solution of a first-order ordinary differential equation.

The following briefly describes how the more refined models are constructed. In co-directional wave-current motion, the flow equation can be written as

$$\frac{\partial u}{\partial t} = -\frac{1}{\rho}\left(\frac{\partial p}{\partial x}\right)_w - \frac{1}{\rho}\left(\frac{\partial p}{\partial x}\right)_c + \frac{\partial}{\partial x}\left(\nu_T \frac{\partial u}{\partial z}\right) \tag{3.55}$$

In Eq. 3.55 the pressure gradient is divided into two, namely one originating from the mean current motion, and another due to the wave motion and determined by

$$\frac{1}{\rho}\left(\frac{\partial p}{\partial x}\right)_w = -\frac{\partial u_o}{\partial t} \tag{3.56}$$

where u_o is the wave-induced flow velocity outside the boundary layer.

The mean pressure gradient can be correlated to the slope S of the mean water surface (cf. Eq. 3.1) by

$$P = -\frac{1}{\rho}\left(\frac{\partial p}{\partial x}\right)_c = -gS \tag{3.57}$$

In the eddy viscosity models, ν_T in Eq. 3.55 must be prescribed. As for the case of a pure oscillatory boundary layer, several suggestions on the vertical distribution of the eddy viscosity have been introduced. Fig. 3.14 shows several examples of these suggestions. Common to these models are that they distinguish between the wave-dominated layer (inside δ_w) and the outer current-dominated layer. In these models, the wave boundary layer thickness is taken time-invariant (in contradiction to the model given in Section 3.2, where the boundary layer was allowed to expand with time).

The computations of the flow pattern are a straight-forward extension of the calculations for the pure oscillatory boundary layer, outlined in Chapter 2. In the model by Grant and Madsen, the prescribed variation is so simple that the flow equation, Eq.3.55, can be solved analytically.

The flow equation has been solved numerically by a one-equation closure (cf. Section 2.3) by Justesen (1988) for the co-directional case and by Davies et al. (1988) for the general three-dimensional case. The solution is found by an iterative procedure:

An initial guess at the mean current profile is calculated, and one period of oscillation is computed. The shear stresses τ_i are averaged over this period, and the deviation from the desired distribution $\Delta\bar{\tau}$, cf. Eq. 3.1, is determined by

$$\Delta\bar{\tau} = \tau - \frac{1}{2\pi}\int_0^{2\pi}\tau_i(z,\omega t)d(\omega t) \tag{3.58}$$

The velocity profile is now corrected by the quantity $\Delta\bar{U}$ given by

$$\Delta\bar{U}(z) = \int_{k_N/30}^z \frac{\Delta\bar{\tau}(z)}{\rho\bar{\nu}_T(z)}dz \tag{3.59}$$

where $\bar{\nu}_T(z)$ is the mean eddy viscosity during one period. This procedure is repeated until acceptable convergence is obtained, usually after 5-10 periods of calculations. Otherwise the solution procedure is similar to that used for the one-equation model for pure wave motion.

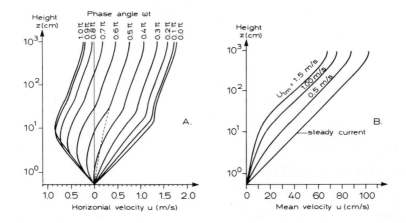

Figure 3.15 Vertical profiles of flow velocity in co-linear case. A: instantaneous profiles at different phase angles. $U_{1m} = 1$ m/s.
B: Time-averaged current profiles. $D = 10$ m, $k_N = 0.5$ cm and $P = 0.035$ cm/s². After Davies et al. (1988).

Figs. 3.15A and 3.15B show the instantaneous and time-averaged flow velocities calculated by Davies et al. (1988). In Fig. 3.15B, the mean pressure gradient P has been kept constant, while the near-bed wave-induced orbital motion has been given different strengths, thereby giving rise to different flow resistances for the current. Hence, the mean flow velocity will decrease with increasing values of U_{1m}.

It can be seen that the profile of the mean flow velocity shown in Fig. 3.15B confirms the existence of an inner and an outer logarithmic layer as shown in Fig. 3.3.

A comparison of different wave-current theories has been made by Dyer and Soulsby (1988) by testing the different methods through numerical runs. The predictions are listed in the table below.

Table 3.1 Comparison of wave-current interaction theories (from Dyer and Soulsby (1988), slightly modified). $k_N = 150$ mm.

Input variables.

Depth-averaged mean velocity V (cm/s)	79.5	67.2	61.2	73.4	78.4
Wave orbital velocity amplitude U_{1m} (cm/s)	50	100	150	100	100
Angle between waves and current γ (°)	0	0	0	45	90

Mean bed shear-stress $\bar{\tau}$ (dyn/cm^2)

Eddy-viscosity model (Grant and Madsen)	46.7	47.3	46.0	54.8	57.3
Momentum-deficit integral (Fredsøe)	31.5	31.9	32.9	34.0	34.3
k-equation model (Davies-Soulsby)	33.8	33.8	33.8	33.8	33.8

Max bed shear-stress τ_{max} (dyn/cm^2)

Eddy-viscosity model	176.9	377.5	632.6	362.9	317.6
Momentum-deficit integral	120.0	243.5	414.4	240.3	202.9
k-equation model	130.3	279.6	488.5	269.6	241.6

It can be seen from the table that the relatively simplified method, developed in Section 3.2, gives quite similar results to the more sophisticated turbulence models like that by Davies et al. (1988).

However, one of the advantages of the k-equation model is that it can predict the level of turbulent kinetic energy as a function of time and space. This is illustrated in Fig. 3.16A. Fig. 3.16B shows the calculated variation in eddy viscosity from the same model.

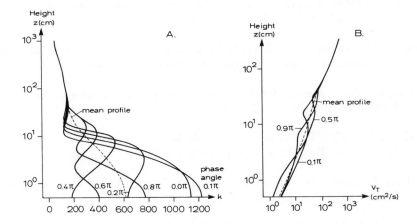

Figure 3.16 Vertical profiles at different phase angles of (A) turbulent kinetic
energy and (B) eddy viscosity for the co-linear case. Parameter
settings are the same as in Fig. 3.15. The dashed lines indicate
the mean distributions. (After Davies et al., 1988).

REFERENCES

Brevik, I. (1981): Oscillatory rough turbulent boundary layers. J. Waterway, Port,
Coastal and Ocean Eng. Div., ASCE, 107(WW3):175-188.

Christoffersen, J.B. and Jonsson, I.G. (1985): Bed friction and dissipation in a
combined current and wave motion. Ocean Eng., 12(5):387-423.

Colebrook, C.F. and White, C.M. (1937): Experiments with Fluid Friction in
Roughened Pipes. Proc. Roy. Soc., London, A 161, pp 367-381.

Davies, A.G., Soulsby, R.L. and King, H.L. (1988): A numerical model of the
combined wave and current bottom boundary layer. J. Geophys. Res.,
93(C1):491-508.

Dyer, K.R. and Soulsby, R.L. (1988): Sand transport on the continental shelf.
Ann. Rev. Fluid Mech., 20:295-324.

Fredsøe, J. (1981): Mean current velocity distribution in combined waves and
current. Progress Report No. 53, Inst. of Hydrodynamics and Hydraulic
Engineering, ISVA, Techn. Univ. Denmark, pp. 21-26.

Fredsøe, J. (1984): Turbulent boundary layers in wave-current motion. J. Hydraulic. Eng., ASCE, 110(HY8):1103-1120.

Grant, W.D. and Madsen, O.S. (1979): Combined wave and current interaction with a rough bottom. J. Geophys. Res., 84(C4):1797-1808.

Kajuira, K. (1968): A model for the bottom boundary layer in water waves. Bull. Earthquake Res. Inst., 45:75-123.

Lundgren, H. (1972): Turbulent currents in the presence of waves. Proc. 13th Int. Conf. Coast. Eng., Vancouver, Canada, 623-634.

Myrhaug, D. (1982): On a theoretical model of rough turbulent wave boundary layers. Ocean Eng., 9(6):547-565.

Myrhaug, D. (1984): A theoretical model of combined wave and current boundary layers near a rough sea bottom. Proc. 3rd., Offshore Mechanics and Arctic Engineering, OMAE Symp., New Orleans, LA., 1:559-568.

Smith, J.D. (1977): Modelling of sediment transport on continental shelves. In: The Sea, Vol. 6, Ed. by E.D. Goldberg et al. Wiley - Interscience, New York, pp. 539-577.

Chapter 4. Waves in the surf zone

The surf zone is the name of shallow water areas where waves break, for example on a beach. The wave-breaking is associated with a conversion of the energy from ordered wave energy to turbulence and to heat. The surf zone is the area with the most intense sediment transport because of the high intensity of the turbulence and the shallow water which makes agitation of sediment from the bottom easy. Furthermore, the wave-breaking generates strong currents, as described in Chapter 5 and 6, which transport the sediment along the coast.

4.1 Wave-breaking

As waves propagate into shallower water, the process of shoaling leads to increasing wave heights. This process cannot continue, and at a certain location the wave breaks. The wave-breaking will typically take place when the wave height is about 0.8 times the local water depth. The waves break because their steepness becomes very large as the depth becomes shallower. The forward wave orbital velocity at the crest becomes large, and the crest topples because it is unstable. While the shoaling process is characterized by a very small energy loss, the wave-breaking is associated with a very large loss of wave energy. The surf zone along the beach is where the wave energy flux from offshore is dissipated to turbulence and heat. Due to the strong energy dissipation, the wave height decreases towards the shore in the surf zone.

The breaking waves can be divided into several different types, the three most important are, Fig. 4.1:

- Spilling breakers
- Plunging breakers
- Surging breakers

Spilling breakers are characterized by the forward slope of the wave top becoming unstable. A plume of water and air bubbles slides down the slope from the crest. The volume of the plume increases, and it travels with the wave as a surface roller. Spilling breakers are also found among waves in deep water, where their energy dissipation is an important part of the energy budget for wind-generated waves.

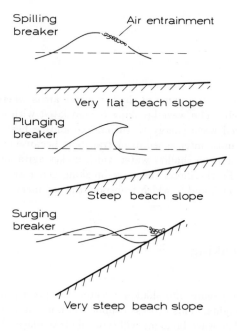

Figure 4.1 Breaker types.

For a plunging breaker the crest of the wave moves forward and falls down at the trough in front of it as a single structured mass of water or a jet. The impact of the jet generates a splash-up of water which continues the breaking process and creates large coherent vortices, which can reach the bottom and stir up considerable amounts of sediment. The flow caused by entrained air further spreads the sediment over the vertical, and clouds of suspended sediment are often observed at a location of plunging breakers.

Possible modes of the generation of the splash-up have been described by Peregrine (1983). Later experiments have shown that the water of the jet continues through the water surface to generate the vortices, and that the splash-up consists of water originally from the point of plunging (Bonmarin, 1989). The flow field of the plunging and splash-up, as described by Peregrine (1983), is shown in Fig. 4.2 together with the vortices generated by the plunging.

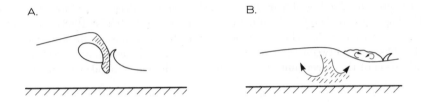

Figure 4.2 A: The geometry of the plunging breaker, after Peregrine (1983). B: The continuation of the wave-breaking and the generated vortices.

In surging breakers it is not the crest of the wave that becomes unstable. It is the foot of the steep front that rushes forward, causing the wave crest to decrease and disappear.

The occurrence of the different types of breakers depends on the character of the incoming waves and of the beach. The most important factor is the slope of the beach and the steepness of the incoming waves. Spilling breakers occur at very gentle beach slopes and relatively steep incoming waves, while plunging breakers are found for steeper bed slopes and less steep waves. Surging breakers are found on very steep beaches.

Galvin (1968) found a relationship between the wave geometry and the breaker type. The waves can be characterized by the surf similarity parameter ξ (Battjes, 1974), which is the ratio between the beach slope and the square root of the wave steepness. The wave steepness can be calculated from the deep water wave height H_0 or the wave height at breaking H_b. In both cases the deep water wave length L_0 is used in the expressions for ξ

$$\xi_0 = \tan \beta / \sqrt{H_0/L_0} \quad , \quad \xi_b = \tan \beta / \sqrt{H_b/L_0} \qquad (4.1)$$

where $\tan \beta$ is the beach slope. The following domains were found from Galvin's experimental data:

Spilling breakers : $\xi_o < 0.5$ or $\xi_b <$ 0.4

Plunging breakers : $0.5 < \xi_o < 3.3$ or $0.4 < \xi_b < 2.0$

Surging breakers : $3.3 < \xi_o$ or $2.0 < \xi_b$ (4.2)

Another example of criteria for the different breaker types is given in Fig. 4.3, which is made by Kjeldsen (1968). This diagram is based on data from Iversen (1952), analyzed by Patrick and Wiegel (1955), and on Kjeldsen's own data. The parameters used in Fig. 4.3 are the beach slope and the deep water wave steepness, the diagram can therefore be directly compared with Galvin's (1968) deep water relations based on ξ_0. Galvin's relations are shown as dashed lines in Fig. 4.3. The difference between the two criteria is considerable, and may be taken as an indication of the uncertainty involved in the prediction of breaker characteristics.

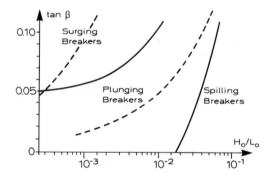

Figure 4.3 Breaker type criteria according to Kjeldsen (1968): fully drawn lines, and Galvin (1968): dashed lines.

After a wave has broken as a spilling or plunging breaker, a transition occurs. For the spilling breaker the surface roller grows, and the wave height decreases rapidly. For the plunging breaker the jet of water that plunges down pushes up a very turbulent mass of water which continues the wave-breaking process. In both cases the wave is transformed into a bore-like broken wave, and this inner part of the surf zone (Fig. 4.4) can be described as a series of periodic bores (Svendsen et al., 1978). The ratio between the local wave height and mean water depth decreases from the value of about 0.8 at the point of wave-breaking to become almost constant at about 0.5 in the inner zone. Fig. 4.5 shows the experimental results of Horikawa and Kuo (1966) at different beach slopes (the slope was constant for each beach tested) together with the empirical relation by Andersen and Fredsøe (1983)

$$\frac{H}{D} = 0.5 + 0.3 \exp\left(-0.11\frac{\Delta x}{D_b}\right) \tag{4.3}$$

where Δx is the distance inshore of the breaking point, and D_b is the water depth at the breaking point.

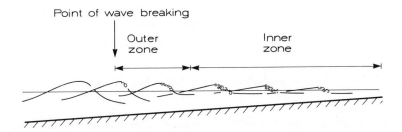

Figure 4.4 The outer and the inner parts of the surf zone.

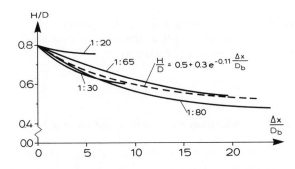

Figure 4.5 Variation in wave height after breaking on a sloping bottom. Measurements from Horikawa and Kuo (1966).

The conditions of the outer zone, with rapid transition of the waves, have been very difficult to describe, especially for plunging breakers, and practically no theories are available which can be used as basis for a hydrodynamic description or sediment transport calculations. The inner zone with the borelike waves has been a subject of intense research activity during the last decade. This has led to the development of theories which are very useful as the first steps for understanding and calculating the hydrodynamics and sediment transport in the surf zone. As no model is available for the outer surf zone, models for the inner surf zone are as a first approach often extended to cover conditions all the way out to the point of wave-breaking. In many cases, this is not entirely satisfactory, and may require field or laboratory data for calibration for a reasonable prediction to be obtained; still at present it may be the only feasible way of bridging the gap between the inner surf zone and the incoming waves. A further complication is that the incoming waves often are irregular, which means that the point of wave-breaking and consequently the location of the transition to inner zone conditions are varying and are particular for each individual wave.

Example 4.1:   The wave height at the breaking point

If it is assumed that the waves at the point of breaking can be described by linear shallow water wave theory, a rough estimate of the wave height at breaking can be obtained.

The wave height and length at deep water are H_0 and L_0. L_0 is found as (cf. Appendix I):

$$L_0 = \frac{g}{2\pi} T^2 \tag{4.4}$$

where g is the acceleration of gravity and T is the wave period.

The wave height and water depth at the point of wave-breaking are H_b and D_b. The criterion for wave-breaking is taken as

$$\frac{H_b}{D_b} = 0.8 \tag{4.5}$$

The energy dissipation from deep water to the point of breaking is assumed to be small.

The shore-normal energy flux at deep water E_{f0} is therefore equal to the energy flux at wave-breaking E_{fb}. E_{f0} and E_{fb} are found by deep water and shallow water wave theory, respectively,

$$E_{f0} = \frac{1}{16} \rho g H_0^2 c_0 = \frac{1}{16} \rho g H_0^2 \frac{g}{2\pi} T \tag{4.6}$$

and

$$E_{fb} = \frac{1}{8} \rho g H_b^2 c_b = \frac{1}{8} \rho g H_b^2 \sqrt{g D_b} \tag{4.7}$$

As $E_{f0} = E_{fb}$ and $D_b = H_b/0.8$ then

$$\frac{H_0}{H_b} = \left(2\sqrt{H_0}\frac{2\pi}{gT}\sqrt{1.25g} \right)^{2/5} = \left(2\sqrt{2.5\pi}\sqrt{\frac{H_0}{L_0}} \right)^{2/5} \tag{4.8}$$

or

$$\frac{H_b}{H_0} = 0.50\left(\frac{L_0}{H_0} \right)^{1/5} \tag{4.9}$$

Komar and Gaughan (1972) compared this expression with field and laboratory data and found better agreement by changing the constant of Eq. 4.9 to an empirical value of 0.56

$$\frac{H_b}{H_0} = 0.56\left(\frac{L_0}{H_0} \right)^{1/5} \tag{4.10}$$

The reason for this adjustment is that linear shallow water theory is inaccurate near the breaking point. The waves are more peaked at the crest and more flat at the trough than a sine-curve, cf. cnoidal wave theory or higher order wave theory. The wave energy flux of such a less 'full' wave profile is somewhat less than predicted by linear wave theory for a wave of a given height.

4.2 Modelling of variation of wave height and water level in the surf zone

In order to describe the sediment transport it is necessary to know the wave height, the water motion in the waves, the bed shear stress, the turbulence level, and the currents which are induced in the surf zone. An important factor in this context is therefore the energy dissipation which on one hand is used to calculate the wave height decay, and on the other hand is a measure of the production of turbulence in the surf zone.

First of all, the water motion of the waves in surf zone will be considered. So far, it has not been possible to develop any detailed models of this rather complex flow phenomenon, so the description of the dynamics of the surf zone must therefore be based on highly simplified models for the broken waves. A successful conceptual model, which was first applied by Svendsen (1984) and (1984a), splits the water into two parts. The main part of the water body is part of the wave motion, i.e. the water particles move back and forth with a wave orbital motion, whilst the breaking is modelled by a surface roller on the front face of each wave. The surface roller is modelled as a mass of water which follows the wave fronts and consequently, the water particles of the surface roller have a horizontal velocity equal to the wave celerity c. The mass of the water in the surface roller is given by the cross-sectional area A of the roller. Fig. 4.6A shows the structure of the wave according to the model. The elevation of the upper boundary of the water in the wave motion is called η. Where there is no surface roller, η is the water surface elevation, whilst in the presence of a surface roller, η is the elevation of the lower boundary of the surface roller.

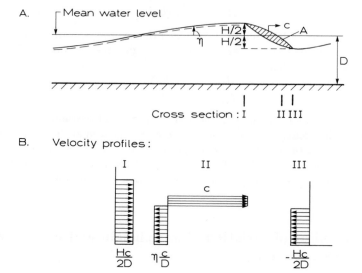

Figure 4.6 A: The structure of a broken wave with a surface roller.
B: Velocity profiles of the three cross sections indicated.

The orbital motion of the broken wave is determined by the simplest possible theory, linear shallow water wave theory. All quantities are given only at their lowest order.

The wave celerity is therefore estimated as

$$c = \sqrt{gD} \tag{4.11}$$

and the horizontal orbital velocity is given by

$$u = \eta \frac{c}{D} \tag{4.12}$$

Fig. 4.6B gives three examples of the velocity profiles in a broken wave, at the wave top (I), at the wave trough (III), and at an intermediate location where the strong forward velocity c of the surface roller can be seen (II). In accordance with the linear shallow water wave theory the pressure distribution is assumed to be hydrostatic, and the vertical orbital velocity is given by

$$w = \frac{z}{D}\frac{\partial \eta}{\partial t} \qquad (4.13)$$

The profile of a broken wave will normally deviate from a sinusoidal curve. A better approximation is often obtained by assuming that η varies linearly between the wave top and wave trough, a so-called "saw-tooth profile".

Energy flux and radiation stress in broken waves

Both the surface rollers and the shape of the wave profile are of significance for the dynamics of the surf zone, as can be illustrated by the energy flux and the radiation stress associated with broken waves, Svendsen (1984).

When the energy flux is calculated, the contributions from the water in the wave motion and from the surface rollers can be found separately and summed to give the total. The energy flux of the wave motion E_{fw} can be found from the assumptions of linear shallow water wave theory

$$E_{fw} = \frac{1}{T}\int_0^T \int_0^D p^+ u\,dz\,dt = \frac{1}{T}\int_0^T \int_0^D \rho g\eta u\,dz\,dt =$$

$$D\rho g\eta\left(c\frac{\eta}{D}\right) = \rho g c\overline{\eta^2} = B\rho g c H^2 \qquad (4.14)$$

where H is the wave height, and p^+ is the deviation of the pressure from the average (hydrostatic) pressure. B is a coefficient which depends on the wave profile. If a sinusoidal wave profile is assumed, B is found as

$$B = \frac{\overline{\eta^2}}{H^2} = \frac{1}{H^2}\overline{\left(\frac{H}{2}\sin(kx - \omega t)\right)^2} = \frac{1}{8} \qquad (4.15)$$

where k is the wave number and ω is the cyclic frequency. For a saw-tooth profile B is $1/12$.

The energy flux associated with the passage of a single surface roller is given as

$$\rho A \frac{c^2}{2} \qquad (4.16)$$

and the mean energy flux is found as

$$E_{fr} = \rho \frac{Ac^2}{2T} \qquad (4.17)$$

The total energy flux is therefore given as

$$E_f = E_{fw} + E_{fr} = B\rho g c H^2 + \rho \frac{Ac^2}{2T} \tag{4.18}$$

The thrust or radiation stress of the waves is an important factor for the dynamics of the surf zone. The radiation stress can also be found as one part due to the wave motion and another associated with the surface rollers.

As described in Section 1.3, the radiation stress associated with the wave motion is found from the pressure force and from the momentum flux.

Using the shallow water approximation, the contribution from the pressure can be found from the expression for the hydrostatic pressure

$$F_p = \frac{1}{T} \int_0^T \frac{1}{2}\rho g(D + \eta)^2 - \frac{1}{2}\rho g D^2 dt = \frac{1}{2}\rho g\overline{\eta^2} = \frac{1}{2}B\rho g H^2 \tag{4.19}$$

and the momentum flux can be written

$$F_m = \frac{D}{T} \int_0^T \rho u^2 dt = D\rho\left(\frac{c\eta}{D}\right)^2 = \rho g\overline{\eta^2} = B\rho g H^2 \tag{4.20}$$

The volume of the rollers is so small that they do not contribute significantly to the pressure force. The momentum flux associated with the passage of a single surface roller is given as

$$\rho Ac \tag{4.21}$$

and the mean momentum flux due to the rollers is therefore

$$F_r = \rho\frac{A}{T}c \tag{4.22}$$

The total radiation stress is therefore

$$S_{xx} = \frac{3}{2}B\rho g H^2 + \rho\frac{A}{T}c \tag{4.23}$$

The subscript xx signifies that Eq. 4.23 represents a normal stress in the x-direction (cf. Fig. 1.7).

Energy dissipation

The similarities between the front of a broken wave and a bore or a hydraulic jump, Fig. 4.7, have led to the idea of expressing the energy loss in the surf zone through the dissipation in a bore (Le Méhauté, 1962). It has been found that this

Figure 4.7 The similarity between a broken wave and a bore.

procedure gives reasonable estimates of the energy loss, and several models use the bore analogy to calculate the dissipation.

The use of the bore analogy assumes a constant velocity distribution over the vertical, and no curvature of the water surface (hydrostatic pressure distribution). Deviations from this assumption may require some correction terms to be introduced in the expressions for the energy loss when calibrating a model to field or laboratory data, as discussed by Svendsen and Madsen (1981).

The energy loss in a stationary hydraulic jump is given by

$$\rho g q \Delta H \tag{4.24}$$

where q is the specific discharge, and ΔH is the head loss of the jump. In the following, all expressions are made with the assumption of a uniform velocity distribution over the vertical.

The head loss ΔH is given by

$$\Delta H = \frac{H^3}{4D^2 - H^2} \tag{4.25}$$

where H is the difference between the water depths at each side of the jump, and D is the average water depth through the jump, Fig. 4.8.

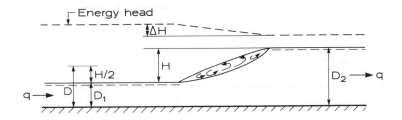

Figure 4.8 The hydraulic jump. Definition sketch.

By changing the frame of reference to a broken wave in the surf zone, the quantity q can be expressed as

$$q = cD \qquad (4.26)$$

The rate of energy loss at each wave front is therefore found as

$$\rho g q \Delta H = \rho g c \frac{DH^3}{4D^2 - H^2} \qquad (4.27)$$

The spacing of the wave fronts is one wave length L, and the mean energy dissipation per unit bed area is given by

$$\overline{\mathcal{D}} = \frac{\rho g c D H^3}{L(4D^2 - H^2)} = \frac{\rho g D H^3}{T(4D^2 - H^2)} \qquad (4.28)$$

The propagation velocity of a bore is very close to the linear shallow water celerity $\sqrt{gD}$, the difference being of second order in terms of (H/D).

Example 4.2:   The volume of the surface rollers

A problem which has not yet been completely solved is the determination of the volume of the surface rollers. An accurate knowledge of the amount of water in the rollers is important because the velocity of the rollers is so large compared to the orbital velocities in the wave motion. The analogy between the broken wave and a hydraulic jump can be used to make an estimate of the cross-sectional area A of the rollers.

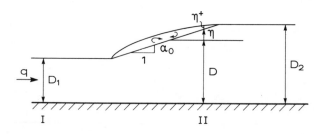

Figure 4.9 Engelund's (1981) model for the hydraulic jump.

Engelund (1981) made a simple dynamic model of the hydraulic jump, which is based on the momentum equation. In a hydraulic jump the roller is stationary, and the roller is included in the momentum equation as a layer of "dead" water which contributes to the hydrostatic pressure, but does not contribute to the momentum flux. The idea is to calculate the local height η^+ of this "dead water", so that the momentum equation is fulfilled exactly at each cross section. The horizontal projection between cross section I and II in Fig. 4.9 then gives

$$\frac{1}{2}\rho g D_1^2 + \rho \frac{q^2}{D_1} = \frac{1}{2}\rho g (D + \eta + \eta^+)^2 + \rho \frac{q^2}{(D+\eta)} \tag{4.29}$$

where $(D+\eta)$ is the local depth under the surface roller, and η^+ is the local roller thickness. D_1 and D_2 can be written

$$D_1 = D - H/2 \ , \ D_2 = D + H/2 \tag{4.30}$$

Inserting Eq. 4.30 into Eq. 4.29, using that $\eta^+ = 0$ for $D + \eta = D_2$, gives

$$q = \sqrt{gD(D^2 - H^2/4)} \tag{4.31}$$

which inserted into Eq. 4.29 gives

$$\frac{1}{2}\rho g (D - H/2)^2 + \rho \frac{gD(D^2 - H^2/4)}{D - H/2}$$

$$= \frac{1}{2}\rho g (D + \eta + \eta^+)^2 + \rho \frac{gD(D^2 - H^2/4)}{(D + \eta)} \tag{4.32}$$

and

$$\frac{\eta^+}{D} = \frac{D+\eta}{D} + \sqrt{\left(3 + \left(\frac{H}{2D}\right)^2\right) - 2\left(1 - \left(\frac{H}{2D}\right)^2\right) / \left(\frac{D+\eta}{D}\right)} \tag{4.33}$$

In the following it is assumed that the boundary between the stagnant water of the roller and the flowing water below is a straight line with a slope of α_0. η is in this way given as

$$\eta = xH/l_r = x\alpha_0 \tag{4.34}$$

where l_r is the length of the surface roller.

The flow under the surface roller is divergent, and resembles in some respects the flow in a straight walled diffusor. By analyzing the velocity distribution in diffusor flows Engelund found that the streamline between the main flow and the

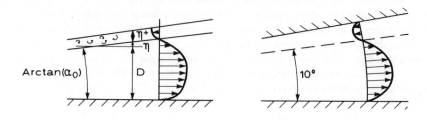

Figure 4.10 The analogy between the flow in a hydraulic jump and a separated diffusor flow.

separating flow formed an angle of 10° with the wall bounding the main flow.
From this, analogy α_0 was estimated to be $\tan 10°$, Fig. 4.10.

The cross-sectional area of the surface roller is therefore found as

$$A = \int_{-l_r/2}^{l_r/2} \eta^+ dx = \frac{D^2}{\alpha_0} \int_{\eta=-H/2}^{\eta=H/2} \frac{\eta^+}{D} d\left(\frac{\eta}{D}\right) =$$

$$\frac{D^2}{\alpha_0}\left[-\frac{1}{2}\left(\frac{D+\eta}{D}\right)^2 + \sqrt{\frac{D+\eta}{D}}\sqrt{\left(3+\left(\frac{H}{2D}\right)^2\right)\frac{D+\eta}{D} - \left(2-\frac{1}{2}\left(\frac{H}{D}\right)^2\right)}\right.$$

$$\left.-\frac{2-\frac{1}{2}\left(\frac{H}{D}\right)^2}{\sqrt{3+\left(\frac{H}{2D}\right)^2}}\ln\left(\sqrt{\frac{D+\eta}{D}}+\sqrt{\frac{D+\eta}{D}-\left(2-\frac{1}{2}\left(\frac{H}{D}\right)^2\right)\Big/\left(3+\left(\frac{H}{2d}\right)^2\right)}\right)\right]_{\eta=-H/2}^{\eta=H/2}$$

(4.35)

A is shown in Fig. 4.11 as a function of H/D. With an accuracy within a
few per cent the surface roller volume obtained by this model can be calculated as

$$A = \frac{H^2}{\alpha_0}\frac{H/D}{4}$$

(4.36)

No measurements of the surface roller volumes in the surf zone have been
published. Duncan (1981) has made measurements of rollers in waves that have
been generated by a towed hydrofoil. Svendsen (1984) approximated these results
with the relation

$$A = 0.9H^2$$

(4.37)

For the range of wave heights in the surf zone: $H/D = 0.5 - 0.8$, and for
$\alpha_0 = \tan 10°$ the difference between the two expressions is less than 30%, and both
expressions may be used to estimate the amount of water in the surface rollers.

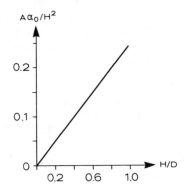

Figure 4.11 The cross-sectional area of the surface roller according to the model of Engelund (1981).

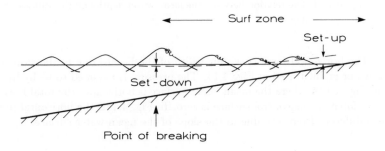

Figure 4.12 Variation of wave height and the mean water surface in the surf zone.

4.2.1 The wave set-up and set-down

The strong energy dissipation in the surf zone and the associated decrease of the wave height towards the shore give a gradient in the radiation stress. The decrease in the radiation stress is balanced by a slope of the mean water surface (averaged over a wave period), the wave set-up, as illustrated in Fig. 4.12. The magnitude of the wave set-up can be determined by a horizontal projection of the momentum equation.

The wave set-up or set-down is measured relative to the still water level, i.e. the water surface there would be without any waves. The depth from the still water level to the bed is called D_0, and the wave set-up or set-down is called ΔD, giving the relation $D = D_0 + \Delta D$ between the mean water depth and the still water depth; as illustrated in Fig. 4.13.

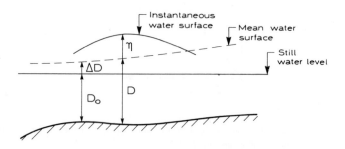

Figure 4.13 The relation between the mean water depth and the still water depth.

The horizontal force (per unit width of the beach) due to the slope of the mean water is illustrated in Fig. 4.14. The pressure is assumed to be hydrostatic, and in Fig. 4.14A where the mean water surface is horizontal, the total horizontal pressure force on the control surface is zero. In Fig. 4.14B the horizontal pressure force is different from zero due to the slope of the mean water surface.

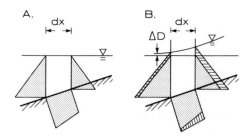

Figure 4.14 The pressure on a control volume. A: horizontal water surface. B: with a wave set-up.

The net pressure force can be found by applying the Gauss theorem to the control volume: the integral of the pressure acting on the surface is equal to the pressure gradient integrated over the volume, giving a net horizontal force of

$$dP = -\int_V \frac{\partial p}{\partial x} dV = -\int_V \rho g \frac{\partial(\Delta D)}{\partial x} dV = -\rho g D \frac{\partial(\Delta D)}{\partial x} dx \qquad (4.38)$$

The force due to the sloping mean water surface balances the force due to the gradient in the wave height: the gradient in the radiation stress. The momentum equation gives

$$\frac{dS_{xx}}{dx} + \frac{dP}{dx} = 0 \qquad (4.39)$$

or

$$\frac{dS_{xx}}{dx} = -\rho g D \frac{d(\Delta D)}{dx} \qquad (4.40)$$

When the wave height and the bed level are known across the coastal profile, Eq. 4.40 can be used to describe the variation of the mean water surface.

It should be noted that the momentum equation Eq. 4.40 does not include terms due to bed friction, so that the radiation stress and the mean pressure gradient balance each other completely. This is allowed for the purpose of calculating the mean water surface. It will be shown in Chapter 6 that the mean bed shear stress is small as long as there is no significant net current in the cross-shore direction.

Example 4.3:   A simple model for the wave set-up

The magnitude of the wave set-up in the surf zone can be calculated analytically by introducing some simplifying assumptions:

1. The wave height in the surf zone is found as a constant times the local water depth

$$H = KD \qquad (4.41)$$

2. The waves are described by the linear shallow water wave theory.

The first assumption is in fact an implicit method for modelling the energy dissipation in the surf zone, and this example is therefore anticipating Section 4.2.2 by introducing one of the simplest possible models for the wave height and mean water level in the surf zone.

Under these assumptions Eq. 4.40 reads

$$\rho g D \frac{d(\Delta D)}{dx} = -\frac{dS_{xx}}{dx} = -\frac{d}{dx}\left(\frac{3}{16}\rho g H^2\right) = -\frac{3}{16}\rho g K^2 2D \frac{dD}{dx} \qquad (4.42)$$

or

$$\Delta D = -\frac{3}{8}K^2 D + \text{constant} \tag{4.43}$$

The constant is determined from the boundary condition. ΔD is chosen to be zero at the breaker line $(D = D_b)$

$$\Delta D = \frac{3}{8}K^2(D_b - D) \tag{4.44}$$

The maximum wave set-up is found at the coastline

$$\Delta D_{\max} = \frac{3}{8}K^2 D_b = \frac{3K}{8}H_b \approx 0.2H_b \sim 0.3H_b \tag{4.45}$$

The set-up can thus be up to about one quarter of the incoming wave height.

Example 4.4:   The wave set-down

Outside the surf zone the wave height variation is due to the shoaling process, which as a first approximation can be calculated by assuming that the energy dissipation is zero, i.e. that the energy flux is constant. The shoaling causes a variation in S_{xx}, and the mean water surface elevation outside the surf zone can be found from equation 4.40.

As energy dissipation has been neglected, Bernoulli's equation is applicable outside the surf zone and can also be used to calculate the mean water elevation (Dean and Dalrymple, 1984). At the free water surface the Bernoulli equation reads

$$\frac{u^2 + w^2}{2g} + \eta - \frac{1}{g}\frac{\partial \phi}{\partial t} = C(t) \tag{4.46}$$

where ϕ is the velocity potential. The base level for the Bernoulli equation is taken to be the still water level, which at every location is given as $z = D_0$. In this example, η is measured from the still water level. Equation 4.46 can be expressed by values taken at $z = D_0$ by making a Taylor expansion, only second order terms are maintained.

$$\left(\frac{u^2 + w^2}{2g}\right)_{z=D_0} + \eta - \frac{1}{g}\left(\frac{\partial \phi}{\partial t}\right)_{z=D_0} - \frac{1}{g}\eta\left(\frac{\partial^2 \phi}{\partial z \partial t}\right)_{z=D_0} = C(t) \tag{4.47}$$

Time averaging gives

$$\overline{\eta} = -\left(\overline{\frac{u^2 + w^2}{2g}}\right)_{z=D_0} + \frac{1}{g}\overline{\eta\left(\frac{\partial^2 \phi}{\partial z \partial t}\right)}_{z=D_0} + \overline{C(t)} \tag{4.48}$$

Insertion of u, w, η and ϕ from linear wave theory gives

$$\bar{\eta} = -\left(\frac{1}{2}\left(\frac{kHg}{2\omega}\right)^2 + \frac{1}{2}\left(\frac{kHg}{2\omega}\tanh(kD_0)\right)^2\right)\frac{1}{2g}$$

$$+\frac{H}{2}\left(\frac{Hgk}{2\omega}\tan(kD_0)\right)\frac{\omega}{2g} + \overline{C(t)} \tag{4.49}$$

The dispersion relation

$$\frac{L}{T} = \sqrt{gD_0\frac{\tanh(kD_0)}{kD_0}} \tag{4.50}$$

gives that

$$\frac{k^2g}{\omega^2} = \frac{T^2}{L^2}g = \frac{k}{\tanh(kD_0)} \tag{4.51}$$

which inserted in Eq. 4.49 gives

$$\bar{\eta} = \frac{H^2}{4}\left(-\frac{1}{4}\frac{k}{\tanh(kD_0)} + \frac{1}{4}k\tanh(kD_0)\right) + \overline{C(t)} =$$

$$-\frac{H^2}{4}\frac{k}{2\sinh(2kD_0)} + \overline{C(t)} \tag{4.52}$$

Choosing $\bar{\eta} = 0$ at deep water, $\overline{C(t)}$ shall be taken as zero and Eq. 4.52 implies that the mean water surface will become slightly depressed as the waves propagate into shallow water, therefore the term 'set-down'. Applying the shallow water wave theory, Eq. 4.52 becomes

$$\bar{\eta} = -\frac{H^2}{16D_0} \tag{4.53}$$

and at the breaker line $(H_b = 0.8D_b)$ $\bar{\eta}$ is found to be

$$\bar{\eta}_b = -0.05H_b \tag{4.54}$$

This set-down should be included as a boundary condition for the calculation of set-up, if the wave set-up is taken relative to the mean water level at deep water.

4.2.2 Surf zone models

The approximations introduced in Example 4.3 can be taken as an example of a very simple surf zone model which gives the variation of the wave height and the mean water level across the surf zone. Such models are of considerable interest because the local value of the wave height and the water depth will be the starting point for a long range of different calculations, such as sediment transport and wave-driven currents. For more detailed hydrodynamic modelling the simple approximation given by Eq. 4.41 is not sufficient. As can be seen from Fig. 4.5, the decay of the wave height, even on a beach with constant slope, is more complex than indicated by the constant ratio between wave height and water depth used in Eq. 4.41. When more complicated bed topographies are considered like for instance a barred coast, it is necessary to develop a model which includes the most important physical processes involved.

If the analysis is restricted to a two-dimensional uniform coast with normally incident waves, a situation which can be represented in a wave flume, then the water surface elevation and the wave height can be determined by simultaneous solution of the energy equation and momentum equation.

The *momentum equation* was described in section 4.2.1, and expresses the balance between the pressure force from the slope of the mean water surface and the gradient of the radiation stress (Eq. 4.40).

The *energy equation* expresses the variation in the energy flux, cf. Fig. 4.15. The energy which is dissipated at the section between x and $x + dx$ must be supplied by a larger energy flux into the section than out of it, for stationary conditions the energy equation reads

$$\frac{dE_f}{dx} = -\overline{\mathcal{D}} \tag{4.55}$$

where $\overline{\mathcal{D}}$ is the mean energy dissipation rate per unit bed area. An example of the determination of $\overline{\mathcal{D}}$ was given in section 4.2, with the assumption that the energy dissipation at each wave front corresponds to that in a bore or a hydraulic jump, so

$$\overline{\mathcal{D}} = \frac{\rho g D H^3}{T(4D^2 - H^2)} \tag{4.56}$$

Roughly speaking, it can be said that the energy equation is used to describe the decay of the wave height due to the loss of energy, and that the momentum equation then calculates the slope of the mean water surface. Actually, the two equations will often be coupled, and the models must be solved together numerically. Each of the two equations are accurate for describing the wave height and mean water profile. This has been tested by insertion of measured quantities from

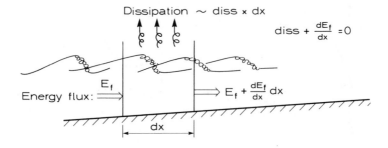

Figure 4.15 The relation between the energy flux and the energy
dissipation.

laboratory experiments, Stive and Wind (1982). It has, however, proven to be
a difficult task to provide a sufficiently accurate description of the processes and
quantities, such as the water surface profile and the pressure and velocity field in
the waves, in order to make the models give reliable predictions.

Several models have been developed based on the ideas described above.
In the following, a few models which are based on different principles and with
different purposes are outlined and discussed.

Example 4.5:   The model by Battjes and Janssen

A model was developed by Battjes and Janssen (1978) with the purpose of
describing natural irregular waves over complex beach topographies with longshore
bars. The basis for the statistical description of the wave heights is the Rayleigh
distribution (cf. Section 1.4). The local maximum wave height $H_{\max}$ is determined
as a constant times the local water depth, and the fraction Q_b of the waves that
are actually broken is given by the number of waves which, according to the
Rayleigh distribution, would have been larger than $H_{\max}$. The energy dissipation
is described by the bore analogy using $H_{\max}$ as the height of the bore and taking
only the fraction Q_b of the waves that are broken into account. Because of the
many uncertainties associated with the conditions in the surf zone, Battjes and
Janssen (1978) chose to use simple linear wave theory and to introduce coefficients
which can be calibrated in order to improve the accuracy of the model. The energy
flux is calculated by the H_{rms}-value of the wave heights and the energy equation
is used to calculate the shoreward gradient of H_{rms}. H_{rms} is calculated from the
Rayleigh distribution, which is truncated so that no wave heights exceed $H_{\max}$.
The momentum and energy equation is then used to predict the variation of $\bar{\eta}$ and

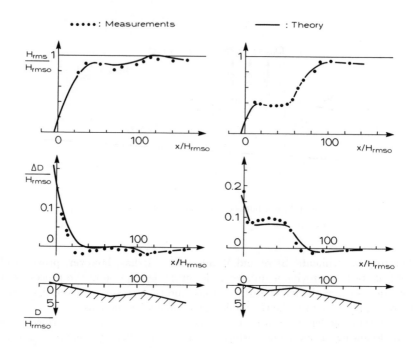

Figure 4.16 Examples of predictions of ΔD and H_{rms}, after Battjes and
Janssen (1978). $\cdots$: measured data. — : numerical results.

H_{rms} across the coastal profile by solving the two equations step by step, starting
at an offshore location with known boundary conditions.

The model by Battjes and Janssen (1978) has proven to be a good predictor
for conditions with irregular waves and rather complex topographies. A detailed
calibration of the model has been carried out by Battjes and Stive (1985). The
model is able to reproduce the conditions at a longshore bar, where wave-breaking
ceases in the trough inshore of the bar. Fig. 4.16 shows an example of measured
and calculated wave heights and mean water surface on a barred profile. H_{rmso} is
the deep water value of H_{rms}.

Example 4.6:   The model by Svendsen

Svendsen (1984) based his model on the same main principles as Battjes
and Janssen (1978), but the approach was different. This model is formulated for
regular waves, and the emphasis is put on the formulation of the wave hydrodynamics, for instance by introducing the effect of the surface rollers in the energy

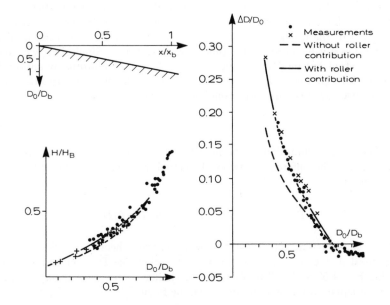

Figure 4.17 Measured and calculated wave height and set-up in the inner surf zone, after Svendsen (1984). Beach slope: 1:34. Deep water wave steepness: $H_0/L_0 = 0.024$.

and momentum equations. This gives a more refined model which does not require as much calibration. It can be used in the inner surf zone where hydrodynamic wave description is valid, but it cannot model the transition through the outer zone from the point of wave-breaking to the inner surf zone, and it is not able to model complex profiles with longshore bars because the pause in wave-breaking inshore of the bar crest is not included in the model. Fig. 4.17 shows an example of measured and calculated wave heights and mean water surface in the inner surf zone. It can be seen how the roller contribution gives a significant improvement in the description of the wave set-up. The boundary conditions for the model are taken at $x = 0.8x_b$ well inshore the point of wave breaking, where the inner surf zone is expected to begin.

Example 4.7: The model by Dally, Dean and Dalrymple

Dally et al. (1985) have developed a model which does not rely on the bore analogy for describing the energy dissipation. They base their description on an analysis of the experimental results from Horikawa and Kuo (1966). These experiments showed that when a breaking or broken wave enters an area with horizontal bed, the breaking continues, and the wave height decreases, until the height reaches a value of about 0.4 times the water depth. Similarly it was found that on constant but moderate slopes the wave height asymptotically approaches the ratio 0.5 D, cf. Fig. 4.5 and Eq. 4.3. The model therefore requires that there is a stable wave energy flux corresponding to a stable wave height of

$$H = KD \tag{4.57}$$

The main assumption of the model is then that the energy dissipation is proportional to the difference between the actual wave energy flux and the stable wave energy flux

$$\overline{\mathcal{D}} = \frac{K_1}{D}\left(\frac{\rho g}{8}H^2\sqrt{gD} - \frac{\rho g}{8}K^2D^2\sqrt{gD}\right) \tag{4.58}$$

where K_1 is a constant, and linear shallow water wave theory is applied. Eq. 4.58 is only used if it gives a positive value of $\overline{\mathcal{D}}$, otherwise $\overline{\mathcal{D}}$ is set equal to zero. By using Eq. 4.58 and still applying linear shallow water wave theory, a simple differential equation for determining H can be found

$$\frac{d}{dx}\left(H^2\sqrt{D}\right) = -\frac{K_1}{D}\left(H^2\sqrt{D} - K^2D^2\sqrt{D}\right) \tag{4.59}$$

This equation can then be combined with the momentum equation to form a surf zone model. A calibration on the tests of Horikawa and Kuo (1966) gave the values $K = 0.40$ and $K_1 = 0.15$. This model is able to reproduce the pause in wave-breaking at a finite wave height on a horizontal bed where the models, using the bore analogy for calculating the energy dissipation, give a continuing gradual decrease in the wave height due to wave-breaking.

Example 4.8:   Direct simulation of the waves in the surf zone

A completely different approach can be made by solving the flow equations in the surf zone directly. This type of model takes the consequence of the drastic variation of waves over the short distance of the surf zone width, and abandons the idea of describing wave properties like wave height, wave length and period.

It is not yet feasible to solve the complete three-dimensional flow equations, and the depth-integrated versions must be applied. Numerical models solving the depth-integrated flow equations in two horizontal dimensions have for many years been used to simulate wave conditions outside the surf zone; for example to describe the disturbance inside a harbour due to incoming waves. Typically such a model will solve the equations numerically in a rectangular grid, using a finite difference scheme. The model then calculates the water surface elevation and the velocity in each grid point for each time step. The distance between the grid points and the length of the time steps must be small compared to the wave length and wave period. In order to obtain stable solutions for finite amplitude waves, the deviation from the hydrostatic pressure distribution must be taken into account by including terms describing the effect of vertical accelerations and curvature of the streamlines. The resulting equations are called the Boussinesq equations, and models which solve these equations are called Boussinesq models. The details of Boussinesq models are described by Abbott (1979), Abbott et al. (1978), and by Madsen and Warren (1984).

The strength of the Boussinesq models is that they are flow models which describe the water motion, and not models which describe wave parameters. This means that, once a model is constructed, it is 'only' a matter of prescribing the right boundary conditions to make it simulate complex phenomena like waves on a current, irregular waves, or waves with finite length of the crests. To extend the Boussinesq models to cover the surf zone would therefore be an important contribution to coastal hydrodynamics. Until now, models have only been developed for one horizontal dimension, simulating normally incident waves on a long uniform coast. The main issue of surf zone modelling is to describe the strong energy dissipation due to the wave-breaking. Deigaard (1989) has made an approach based on ideas similar to the model for the hydraulic jump of Engelund (1981), cf. Example 4.2. In the following, this model, which has some promising features, is described in detail.

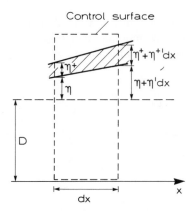

Figure 4.18 The control surface for determining the pressure force from the roller.

Energy is extracted from the wave motion due to the work done by the forces between the surface roller and the water in the wave beneath it. The interaction and energy exchange between the surface roller and the wave have been analyzed by Cointe and Tulin (1986) for deep water waves. Deigaard and Fredsøe (1989) made a detailed description of the transfer of energy in the surf zone from the wave energy flux, which is evenly distributed over the vertical to the zone near the surface where the energy dissipation occurs, mainly in the shear layer beneath the surface roller. In the numerical flow model, the effect of surface rollers with known geometry is included as an additional force in the depth-integrated flow equations. As a first approximation, the effect of a surface roller with local thickness η^+ can be calculated by assuming a hydrostatic pressure distribution. The net horizontal pressure force on the control surface of Fig. 4.18 is found as

$$dP = \rho g \left(\frac{1}{2}(D + \eta + \eta^+)^2 - \right.$$

$$\frac{1}{2} \left(D + \eta + \frac{\partial \eta}{\partial x} dx + \eta^+ + \frac{\partial \eta^+}{\partial x} dx \right)^2 \right) =$$

$$-\rho g \left((D + \eta) \frac{\partial \eta}{\partial x} dx + \eta^+ \frac{\partial \eta}{\partial x} dx + (D + \eta) \frac{\partial \eta^+}{\partial x} dx \right) \qquad (4.60)$$

where η is the instantaneous surface elevation of the water taking part in the wave motion. The additional pressure force from the roller is thus

$$dP^+ = -\rho g \left(\eta^+ \frac{d\eta}{dx} + (D + \eta) \frac{d\eta^+}{dx} \right) dx \qquad (4.61)$$

This pressure term is then included in the momentum equation, and will modify the dynamics of the wave motion, simulating the interaction between waves and rollers and extract energy from the waves.

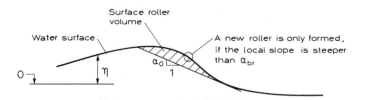

Figure 4.19 Principles for determining the geometry of the surface roller.

The geometry of the surface roller at each time step in the simulation is determined by a simple procedure, cf. Fig. 4.19. The breaker criterion is expressed by the local slope of the water surface. If a surface roller is not already present, it will only be formed if the local water surface is steeper than α_{br}. The lower boundary of the roller is assumed to be a straight line with the slope α_0 where $\alpha_0 \leq \alpha_{br}$. The surface roller starts where the water surface has the same slope as the lower boundary of the roller, Fig. 4.19 and the breaking will cease when the maximum slope at the wave front becomes smaller than α_0. The momentum contribution given by Eq. 4.61 is multiplied by a calibration factor, K_r, which for example shall compensate for inaccuracies in the determination of the surface roller geometry.

The procedure for calculating the effect of surface rollers during the time step from $t = n\Delta t$ to $t = (n+1)\Delta t$ is as follows

- By extrapolation from the previous time step, the water surface profile $\eta(x)$ is estimated at time $t = (n+1/2)\Delta t$.
- From the water surface profile the location and thickness of the surface rollers are determined as described above.
- The momentum contribution from the surface roller is calculated by Eq. 4.61 for each grid point, and the flow equations are solved including this term.

Fig. 4.20 shows some simulation results for a beach with a constant bed slope of 1:30. The beach rises from a still water depth of 1.5 m to 0.3 m. At the right boundary the waves are absorbed, whilst at the left boundary, regular waves are generated with the height 0.5 m and period 4 s. Figs. 4.20A and B show the water surface profile and the surface rollers at different times from 20 s to 24 s after the start of the simulation. The calculations have been made with $\alpha_{br} = 0.33$, $\alpha_0 = 0.176 \, (= \tan 10°)$, and with $K_r = 1.5$. The grid size is $\Delta x = 0.25$ m, and $\Delta t = 0.05$ s.

Fig. 4.21 shows results from a barred coastal profile. The bed slopes are 1:40, the bar crest level is -0.75 m and the trough is at the level -1.0 m. The incoming waves and all model parameters are unchanged from the constant slope simulation. In Fig. 4.21 it can be seen how the wave-breaking ceases inshore of the bar (no surface rollers) and that smaller waves are formed between the main wave crests in the trough region.

The two examples illustrate some properties of the model which makes it attractive for further development. Firstly, it appears physically realistic to relate the energy loss to the properties of the surface rollers, which, at least in the inner surf zone, play a dominant role in the wave transformation. Secondly, this model determines whether or not the waves break from the local conditions in time and space. This can be of importance for describing why the wave-breaking ceases inshore of a bar, and in the case of irregular waves where a single point of breaking cannot be defined a priori. This model is also able to reproduce the complete cessation of wave-breaking at a wave height of about 0.4 D, as the breaking wave propagates into an area with a horizontal bed.

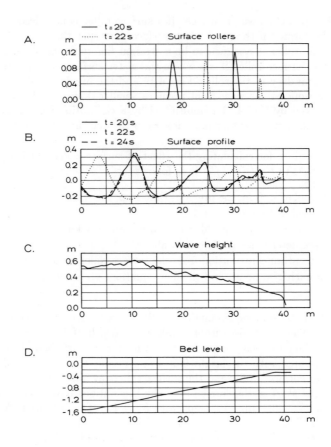

Figure 4.20 Results from a constant slope profile. A: Location of surface
rollers. B: Water surface profiles. C: Wave height distribution.
D: Model topography.

Many new developments are, however, still required before a practical design
tool is available in the form of Boussinesq models for the surf zone. The main
drawbacks of the present model are that at the beginning of wave-breaking a large
surface roller is immediately introduced, which causes some reflection of wave
energy. It is more realistic to have a gradual development of the surface roller.
The momentum of the surface roller is also of importance for the dynamics of the
surf zone (Svendsen, 1984), and to include transfer of momentum to the surface
rollers from the waves would be an improvement of the model.

The outer surf zone with plunging breakers and rapid transition of the
waves cannot be expected to be described in a fully satisfactory manner by any

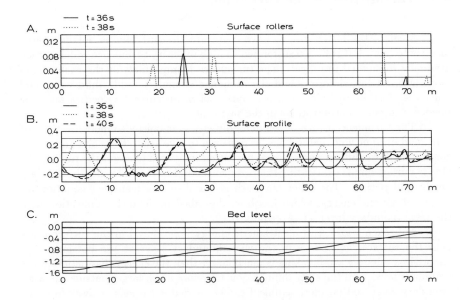

Figure 4.21 Results from a barred coastal profile. A: Location of surface
rollers. B: Water surface profile. C: Model topography.

Boussinesq model and depth-integrated model certainly cannot represent a plunging breaker. The representation of the outer zone may, however, be calibrated in order to provide the best possible connection between the incoming waves and the inner surf zone.

The areas where Boussinesq models of the surf zone can be applied are manifold. A simple one-dimensional model as presented above can be developed to study low period oscillations induced by wave groups (surf beat) and their significance for sediment transport and coastal morphology. By incorporating surface rollers in a two-dimensional model, a combined model for waves and depth-integrated wave-driven currents can be formulated, where the wave motion and the mean currents in the surf zone are modelled simultaneously. As described in Chapter 5, the state of the art with respect to modelling wave-driven currents is to operate with two models (more or less closely coupled), one for the waves and one for the current.

4.3   Turbulence in the surf zone

Most of the heavy loss of wave energy in the surf zone is not dissipated directly to heat, but is first transformed to high intensity turbulence which is then gradually dissipated into heat. The turbulence is important for the exchange of momentum (eddy viscosity) and the exchange of matter (eddy diffusivity) and must therefore be included in a description of mean current velocities and suspended sediment concentrations in the surf zone.

The turbulence is one of the more difficult phenomena to model in the surf zone, and until now no method has been proven to give accurate results under field conditions. Two different approaches have been suggested to assess the turbulence. One is purely empirical, relating the turbulence level to the water depth, the wave height and period. The other approach is based on the transport equation for turbulent kinetic energy, which has been described in detail in connection with the wave boundary layer, cf. Appendix II.

In broken waves, turbulent kinetic energy is produced in the surface roller near the surface, as well as in the wave boundary layer at the bottom. The energy loss due to wave-breaking is much larger than the energy dissipation in the boundary layer, and the eddy diffusivity is therefore much larger outside the wave boundary layer, so that more sediment can be carried in suspension away from the bed.

4.3.1   Time- and depth-averaged model

The first application of the turbulent energy equation to surf zone conditions was made by Battjes (1975). Its purpose is to quantify the cross-shore turbulent exchange of momentum, and it is therefore sufficient with a depth-integrated and time-averaged description.

It is assumed that the turbulent kinetic energy k and the energy dissipation rate ϵ are uniformly distributed over the depth D. A situation at equilibrium is considered, which means that the rate of conversion of energy from one form to another is constant: the loss of wave energy is equal to the production of turbulence which in turn is equal to the dissipation of turbulent energy. The rate of energy dissipation per unit bed area is written as

$$\overline{\mathcal{D}} = \rho D \bar{\epsilon} \tag{4.62}$$

where the turbulent energy dissipation per unit volume is modelled by the expression

$$\epsilon = c_2 \frac{k^{3/2}}{\ell_d} \tag{4.63}$$

ℓ_d is the length scale of the turbulence and is taken to be proportional to the water depth. Eqs. 4.62 and 4.63 then give the following relation

$$\overline{\mathcal{D}} \propto \rho k^{3/2} \tag{4.64}$$

or

$$k \propto \left(\frac{\overline{\mathcal{D}}}{\rho}\right)^{2/3} \tag{4.65}$$

The energy dissipation is estimated from the wave energy loss, cf. Eq. 4.55, using linear shallow water theory and assuming that the wave height is proportional to the water depth

$$H = KD \tag{4.66}$$

Eq. 4.65 can then be written as

$$\overline{\mathcal{D}} = -\frac{dE_f}{dx} = -\frac{d}{dx}\left(\frac{1}{8}\rho g H^2 \sqrt{gD}\right) =$$

$$-\rho g^{3/2}\frac{1}{8}K^2\left(\frac{5}{2}D^{3/2}\frac{dD}{dx}\right) \tag{4.67}$$

By combining Eqs. 4.65 and 4.67 a measure of the turbulent energy is obtained

$$k \propto gD\left(\frac{dD}{dx}\right)^{2/3} \tag{4.68}$$

Example 4.9:   Turbulent energy based on the bore analogy

As an alternative, the wave energy loss could be assessed by the bore analogy, cf. Section 4.2. In this case the dissipation can be written

$$\overline{\mathcal{D}} = \frac{\rho g D}{T}\frac{H^3}{(4D^2 - H^2)} \approx \frac{\rho g D H^3}{4D^2 T} \tag{4.69}$$

giving

$$k \propto H^2\left(\frac{g}{DT}\right)^{2/3} \tag{4.70}$$

4.3.2   Variation of turbulence in time and space

When the purpose of the turbulence modelling is to give the basis for suspended sediment transport modelling or calculation of velocity profiles, then the detailed distribution of the turbulence in time and space is of importance. Laboratory measurements of the turbulence in broken waves indicate that the turbulence intensity decreases towards the bed, particularly for the vertical component which is of importance for sediment suspension and velocity profiles. The detailed distribution of the turbulence under broken waves has been modelled by Deigaard et al. (1986) using the k-equation. Justesen et al. (1986) combined the description of the turbulence from the wave-breaking with the turbulence produced in the wave boundary layer.

The energy dissipation in the wave boundary layer is very small compared to the energy loss due to wave-breaking. Nevertheless, it is of vital importance to include the boundary layer when dealing with sediment transport. The suspended sediment concentration is determined from the balance between settling and turbulent diffusion. The vertical turbulent exchange factor due to the broken waves decreases to zero at the bed, so the turbulence in the wave boundary layer is therefore the major factor in determining the concentration profile near the bed where the main part of the suspended sediment is found, even in the surf zone. The wave boundary layer provides the contact between the bed and the zone of high turbulence intensity away from the bed. According to this description, no grains are carried in suspension without being agitated and lifted through the wave boundary layer by the turbulence there. The wave boundary layer therefore acts as a 'bottleneck' which must be passed by the sediment going into suspension.

The transport equation for the turbulent kinetic energy is written, cf. Appendix II

$$\frac{dk}{dt} = \frac{\partial}{\partial z}\left(\frac{\nu_T}{\sigma_k}\frac{\partial k}{\partial z}\right) + \frac{\text{PROD}}{\rho} - c_2\frac{k^{3/2}}{\ell_d} \tag{4.71}$$

Eq. 4.71 includes the time variation in k and the vertical diffusion, whilst the horizontal diffusion is neglected under the assumption that the horizontal gradients are small compared to the vertical.

The term 'PROD' stands for the production of the turbulent kinetic energy. It has two contributions. The first is the well known production which is obtained from the flow equation

$$\text{PROD} = \tau\frac{\partial u}{\partial z} \tag{4.72}$$

This contribution is dominant in the wave boundary layer. The second contribution is due to the wave-breaking; the estimation of this term is discussed in the following. The description of wave boundary layers by turbulence models is

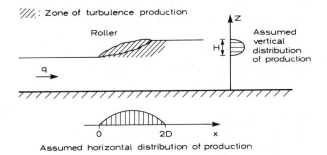

Figure 4.22 Location of production of assumed horizontal distribution of turbulence in a hydraulic jump.

described in Section 2.3, and in the following only the turbulence originating from the wave-breaking is treated in detail.

It is not yet possible to model the details of the flow pattern in the vicinity of the surface rollers, and Eq. 4.72 can therefore not be used to calculate the production of turbulence in the surface rollers. Instead the total production rate is estimated from the analogy with a bore or a hydraulic jump, and the location of the production is prescribed on basis of measurements in a hydraulic jump made by Rouse et al. (1958). It is found that the production of turbulence in a hydraulic jump is concentrated in a zone below the surface with a thickness of about the height of the jump. In the horizontal direction the production is concentrated downstream from the toe of the jump over a distance of approximately two times the water depth. The actual distribution of the production is taken to be parabolic in the vertical as well as the horizontal direction, as indicated in Fig. 4.22.

When moving from the hydraulic jump, which is stationary but with a strong through-flow of water, to the broken waves, which are propagating but have a negligible net flow, a conversion must be made from the horizontal coordinate in the jump to the time in the rollers. Under the broken waves, consider a column of water extending from the bed to the surface. With each passing wave this water column is moved back and forth with the horizontal wave orbital motion and it is stretched in the vertical due to the vertical orbital motion. Similarly, each time a wave front passes, the water column receives a strong production of turbulence near the water surface.

With these assumptions, and the parabolic variation in time and space indicated in Fig. 4.22, the distribution of the production in time and space is given by

$$\text{PROD} = E_{\text{loss}} \frac{36}{(H\, \delta T)^2} z^* \left(1 - \frac{z^*}{H}\right) t \left(1 - \frac{t}{\delta T}\right) \qquad (4.73)$$

where z^* is a vertical coordinate, zero at the surface and directed downwards, δT is the part of the wave period during which turbulent production takes place, and E_{loss} is the energy loss per unit bed area during one wave period: $\Delta H \rho g D$, cf. Eq. 4.28. The time in Eq. 4.73 is reset to start at zero each time a wave front passes the water column under consideration. The different elements in the description of the turbulence in the broken waves are shown in Fig. 4.23.

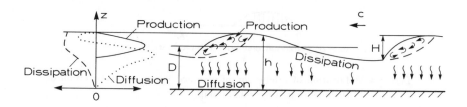

Figure 4.23 Definition sketch describing elements involved in modelling of turbulence (to the left, time-averaged values of production, vertical diffusion, and dissipation of turbulent energy are shown).

With a one-equation turbulence model the length scale of the turbulence ℓ_d is not calculated by the model but must be prescribed. In the upper part of the water column the presence of the bed has minor influence on the turbulence, and here the conditions are similar to those of free turbulence, and the length scale is taken to be constant ℓ_{max}. ℓ_{max} is of the order 0.1 D.

Close to the bed the length scale of the turbulence is reduced due to the proximity of the bed. In this region the model uses a linear variation with the distance from the bed, similar to the near-bed variation of ℓ_d in the case of boundary layer flow. The length scale of the turbulence is written as

$$\ell_d = \begin{cases} c_2^{1/4} \, \kappa z & \text{for } 0 < z \le \ell_{max}/\kappa c_2^{1/4} \\[2mm] \ell_{max} & \text{for } \ell_{max}/\kappa c_2^{1/4} < z \le D \end{cases} \tag{4.74}$$

where κ is the von Kármán constant.

The conditions close to the bed are only represented by a rather crude approximation in the model. It is the vertical turbulent velocity fluctuations that are damped by the proximity of the bed. The horizontal fluctuations are not reduced, except in the oscillatory wave boundary layer. This means that the turbulence becomes very anisotropic close to the bed. When dealing with suspended sediment or velocity profiles over the vertical, it is the vertical velocity fluctuations that are of importance, and the present turbulence model can, with the chosen length scale distribution, be expected to give a more realistic representation of the vertical turbulent fluctuations close to the bed than of the horizontal.

In the model the solution of the transport equation is simplified by neglecting the horizontal oscillation and stretching the water column. This means that the convective terms at the right-hand-side of Eq. 4.71 are neglected, and that the equation can be solved in a rectangular domain in time and space. The equation is solved in the region $0 \leq z \leq D$ from $t = 0$ until a periodic solution is obtained. The boundary conditions are zero vertical flux at the surface

$$\frac{\partial k}{\partial z} = 0 \qquad \text{at } z = D \qquad (4.75)$$

and vanishing turbulence at the bed

$$k = 0 \qquad \text{at } z = 0 \qquad (4.76)$$

The latter boundary condition is equivalent to setting the dissipation of turbulence at the bed equal to the production, which is zero for surface-generated turbulence. This is in contrast to the situation with turbulence generated by bed shear stress, where k approaches a finite value at the bed because the production becomes infinite.

Fig. 4.24 shows examples of calculated distributions in time and space of the turbulence under broken waves. The calculations are compared to measurements made by Stive (1980). The tests of Stive (1980) have been made on a plane beach with a slope of 1:40. Two test series have been made, test 1 with a wave period $T = 1.79$ s and a deep-water wave height of $H_0 = 0.159$ m, and test 2 with $T = 3.00$ s and $H_0 = 0.142$ m. The turbulence is represented by the root mean square of the horizontal component of the turbulent fluctuations. Fig. 4.25 shows the profiles of the measured and calculated time-averaged turbulent kinetic energy. Fig. 4.26 gives an example of the distribution of the turbulent kinetic energy showing the structure near the bed, where the turbulence from wave-breaking interacts with the turbulent wave boundary layer, as calculated by the model of Justesen et al. (1986).

Example 4.10: The distribution of turbulence close to the bed

This example considers the conditions close to the bed in the surf zone, but neglects the turbulence from the wave boundary layer.

Close to the bed the one-equation turbulence model gives only a weak variation of the turbulence with time, and the near-bed variation in k can be found by steady state calculations. The length scale close to the bed is given by

$$\ell_d = c_2^{1/4} \kappa z \qquad (4.77)$$

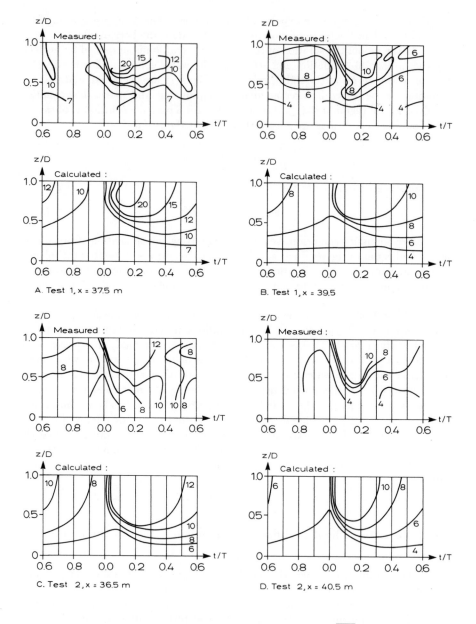

Figure 4.24 Measured and calculated distributions of $\sqrt{\overline{u'^2}}$. Measurements by Stive (1980). In test 1 wave-breaking occurs at $x = 35.5$ m, and in test 2 at $x = 33.5$ m.

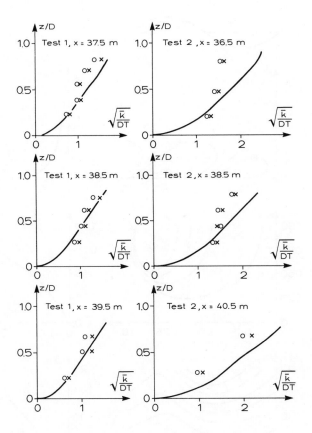

Figure 4.25 Measured and calculated time-averaged turbulent kinetic energy. Measurements by Stive (1980). In test 1 wave-breaking occurs at $x = 35.5$ m and in test 2 at $x = 33.5$ m.

The boundary condition away from the bed is given by

$$k = k_0 \quad \text{at } z = z_0$$

There is no production of turbulence near the bed, and the transport equation for k reads

$$0 = \frac{\partial}{\partial z}\left(\frac{\nu_T}{\sigma_k}\frac{\partial k}{\partial z}\right) - c_2\frac{k^{3/2}}{\ell_d} \tag{4.78}$$

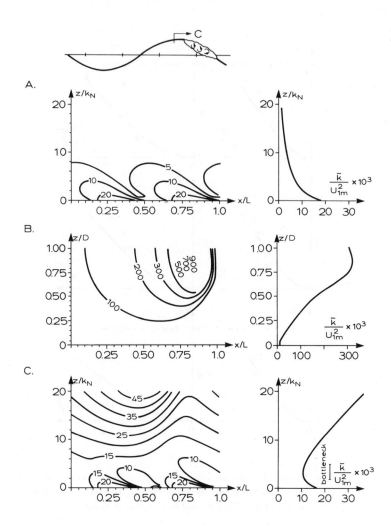

Figure 4.26 Distribution of turbulent kinetic energy in broken and unbroken waves. Contour plots show the temporal variation of k,
and the corresponding time-averaged levels are depicted on the
right. A: Unbroken waves with only bottom-generated turbulence; B: Total depth in broken wave; C: Near-bed region in
broken wave: $a/k = 10^2, gT^2/D = 300, H/D = 0.5$. Turbulence levels indicated are $\sqrt{k}/U_{1m} \times 10^3$.

It is assumed that k varies with a power α of z

$$\frac{k}{k_0} = \left(\frac{z}{z_0}\right)^\alpha \tag{4.79}$$

Using that $\nu_T = \ell_d\sqrt{k}$, Eq. 4.78 can be written in non-dimensional form

$$\frac{\partial}{\partial(z/z_0)}\left(\frac{c_2^{1/4}\,\kappa}{\sigma_k}\frac{z}{z_0}\sqrt{k/k_0}\,\frac{\partial\,k/k_0}{\partial z/z_0}\right) - c_2\frac{(k/k_0)^{3/2}}{\ell_d/z_0} = 0 \tag{4.80}$$

inserting Eq. 4.79 gives

$$\frac{\partial}{\partial(z/z_0)}\left(\frac{c_2^{1/4}\,\kappa}{\sigma_k}\frac{z}{z_0}\left(\frac{z}{z_0}\right)^{\alpha/2}\alpha\left(\frac{z}{z_0}\right)^{\alpha-1}\right) - \frac{c_2(z/z_0)^{3\alpha/2}}{c_2^{1/4}\,\kappa\,z/z_0} = 0 \tag{4.81}$$

or

$$\frac{c_2^{1/4}\,\kappa}{\sigma_k}\alpha\frac{3}{2}\alpha\left(\frac{z}{z_0}\right)^{3\alpha/2-1} - \frac{c_2}{c_2^{1/4}\,\kappa}\left(\frac{z}{z_0}\right)^{3\alpha/2-1} = 0 \tag{4.82}$$

giving

$$\alpha^2 = \frac{c_2}{c_2^{1/4}\,\kappa}\frac{\sigma_k}{c_2^{1/4}\,\kappa}\frac{2}{3} = \frac{2}{3}\frac{c_2^{1/2}\sigma_k}{\kappa^2} \tag{4.83}$$

with $c_2 = 0.08$, $\sigma_k = 1$ and $\kappa = 0.4$ this gives

$$\alpha = 1.09$$

It is seen that according to this model the turbulent kinetic energy varies almost linearly with the distance from the bed. Similarly the velocity fluctuations vary almost as

$$\sqrt{\overline{w'^2}} \propto \sqrt{z} \tag{4.84}$$

and the turbulent diffusion coefficient varies as

$$\nu_T \propto z^{3/2} \tag{4.85}$$

Example 4.11:   Discussion of the principles behind modelling turbulence in spilling breakers

The turbulence model based on the k-equation is an example of an engineering approach to modelling the turbulence and flow in spilling breakers the model being validated by comparison with experimental results. There are, however, some basic characteristics of the turbulence which should be taken into account when evaluating the present results or considering the development of more sophisticated models.

When considering the turbulence in the surf zone (spilling breakers and broken waves) one can distinguish between four different zones, each with distinct turbulent characteristics. The four zones are, cf. Fig. 4.27:

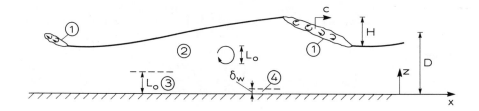

Figure 4.27 The four characteristic zones of turbulence in waves in the surf zone.

Zone 1. The area near the surface roller with an intense production of turbulent kinetic energy due to the shear between the water in the surface roller and in the wave motion.

Zone 2. The area where the turbulence, remaining from the production zone (Zone 1), spreads downwards and dissipates gradually until a new wave front passes with intensive production in the roller.

Zone 3. The zone where the structure of the turbulence is influenced by the proximity of the bed.

Zone 4. The turbulent wave boundary layer where production of turbulence is associated with the periodic bed shear stress. The thickness δ_w of the wave boundary layer is small compared to the length scale of the outer turbulence, and Zone 4 is embedded in Zone 3.

In the following, the characteristics of the four zones are discussed with the k-equation model as the starting point.

Zone 1 is characterized by the very intense production of turbulence, which takes place in the shear layer at the lower boundary of the surface roller. At this location the ordered wave energy is transformed to turbulence. The length scale of the locally produced turbulence is small, and the dissipation of turbulent energy is considerable. In a shear layer about two third of the turbulence generated appears to be dissipated locally (Bradshaw and Ferris, 1965). Zone 1 is located near the water surface, and a detailed description of the turbulence is not required for the prediction of the circulation current or sediment transport which is most sensitive to the near-bed conditions. The turbulence in Zone 1 is therefore not modelled in any detail. Zone 1 is included only as a source of turbulence, corresponding to the turbulence surviving the strong local dissipation.

The turbulence in Zone 2, 3 and 4 is modelled by the transport equation for turbulent kinetic energy, Eq. 4.71.

In *Zone 2* the vertical spreading and decay of the turbulence are modelled by applying the constant length scale $\ell_{\max}$ of order O(0.1D). In reality, the length scale of the turbulence cannot be expected to be constant, but rather to increase with distance (in time and space) from the location of production, similar to grid-generated turbulence. $\ell_{\max}$ is therefore an average value which is calibrated to give satisfactory results. The formulation of the transport equation is based on the assumption of homogeneous, isotropic turbulence and the measurements by Stive (1980) indicate that the deviation of the turbulence from isotropic and homogeneous conditions is no more severe than other flow situations which have been successfully described by the k-equation. All in all there is reason to expect that the distribution of turbulence in Zone 2 can realistically be described by the k-equation.

In *Zone 3* the turbulent eddies are influenced by the proximity of the bed. Here the flow belongs to a class of situations described as 'turbulence in a box', which are related by having externally generated turbulence close to a surface inhibiting velocity fluctuations normal to it. This problem of shear-free boundary layers has received considerable attention during the last twenty years, both with experimental (Uzkan and Reynolds, 1967; Thomas and Hancock, 1977; McDougall, 1979; and Brumley and Jirka, 1987) and theoretical investigations (Hunt and Graham, 1978; Birigen and Reynolds, 1981; Hunt, 1984 and 1984A).

The difference between a free and a fixed surface lies primarily in the boundary condition for the surface-parallel velocity fluctuations. At a clean free surface the fluctuating shear stress is zero, whilst at a fixed surface the velocities are zero, giving a near surface boundary layer with fluctuating shear stresses. In the present case the difference between the two boundaries may be neglected, as any fluctuating shear stresses are expected to be small compared to the oscillatory shear stresses in the wave boundary layer.

The thickness of the layer influenced by the bed is scaled by the integral length scale of the turbulence away from the bed. Inside this layer the vertical velocity fluctuations and the vertical length scale decrease reaching zero at the bed. The horizontal velocity fluctuations increase slightly towards the bed. In

Zone 3 the turbulence thus becomes very anisotropic close to the bed, and can only be described by the k-equation as a rather crude approximation. In many cases, for the description of sediment transport or mean current velocity profiles, it is the vertical turbulent exchange factor which is of greatest importance, and which should be used for comparison with the results of the k-equation.

In the model Zone 3 is represented by making the length scale vary linearly with the distance from the bed. The variation in ℓ_d is the same as for boundary layers with a wall shear stress. This model gives a decreasing turbulent kinetic energy and eddy viscosity towards the bed, and in the absence of the wave boundary layer (Zone 4) k and ν_T would become zero at the bed, as described in Example 4.10.

The model by Hunt and Graham (1978) predicts the mean square of the vertical velocity fluctuations $\overline{w'^2}$ to vary as $z^{2/3}$, a result supported by several experiments, e.g. Thomas and Hancock (1977) and Brumley and Jirka (1987). With respect to the length scale the results are less consistent, but it is reasonable to assume that the length scale varies linearly with z, cf. Hunt (1984) and (1984A). This gives a vertical turbulent exchange factor varying as (Hunt, 1984A)

$$\nu_T \propto z^{1/3} z = z^{4/3} \tag{4.86}$$

It is thus seen that the variation of the vertical turbulent exchange factor, obtained by considering the anisotropic turbulence close to a wall, does not deviate significantly from the ν_T predicted by the present application of the k-equation (cf. Example 4.10). This variation of ν_T in the surf zone is also supported by measurements by Okayasu et al. (1988), who determined ν_T from measurements of the velocity profile of the mean circulation current and of the turbulent Reynolds stresses. These measurements clearly show that ν_T decreases towards zero at the bed, but it is not possible to determine with any certainty which power of z gives the best fit.

Zone 4 is dominated by the turbulence generated by shear stresses in the oscillatory wave boundary layer. The conditions in this shear-dominated boundary layer is in better agreement with the assumptions behind the k-equation, and the turbulence model can be expected to give reliable predictions, as has been found in the case of an oscillatory boundary layer without an outer source of turbulence, Justesen (1988).

It can be argued that the turbulent velocity fluctuations of the externally generated turbulence and of the shear-generated turbulence from the wave boundary layer are statistically independent, Bradshaw (1974), Hunt (1984). Therefore the most correct procedure may be to model the two phenomena independently and add the results, rather than using a coupled model, where a non-linear equation, such as the k-equation is applied to describe the two sources of turbulence simultaneously. In view of the very crude assumptions which have been made in connection with the use of the k-equation, this aspect seems to be unimportant. It may, however, be taken as an indication that not much is achieved by introducing more advanced models, as for example a two-equation turbulence model. Any improvements to the present rather crude approach may require the application of much more advanced models, which for instance can take the anisotropy of the turbulence into account.

REFERENCES

Abbott, M.B. (1979): Computational Hydraulics, Elements of the Theory of Free Surface Flows. Pitman Advanced Publishing Program. Pitman Publishing Ltd., London. 326 pp.

Abbott, M.B., Petersen, H.M. and Skovgaard, O. (1978): On the numerical modelling of short waves in shallow water. J. Hydr. Res., 16(3):173-203.

Andersen, O.H. and Fredsøe, J. (1983): Transport of suspended sediment along the coast. Progress Report No. 59, Inst. of Hydrodynamics and Hydraulic Engineering, ISVA, Techn. Univ. of Denmark, pp. 33-46.

Battjes, J.A. (1974): Computation of set-up, longshore currents, run-up and overtopping due to wind-generated waves. Dept. of Civil Engrg., Delft Univ. Tech., Report No. 74-2, 241 pp.

Battjes, J.A. (1975): Modelling of turbulence in the surf zone. Proc. Symp. Modelling Techniques, ASCE, San Francisco, pp. 1050-61.

Battjes, J.A. and Janssen, J.P.F.M. (1978): Energy loss and set-up due to breaking in random waves. Proc. 16th Coastal Eng. Conf., Hamburg, pp. 569-87.

Battjes, J.A. and Stive, M.J.F. (1985): Calibration and verification of a dissipation model for random breaking waves. J. Geophys. Res., 90:9159-67.

Biringen, S. and Reynolds, W.C. (1981): Large eddy simulation of the shear-free turbulent boundary layer. J. Fluid Mech., 103:53-63.

Bonmarin, P. (1989): Geometric properties of deep-water breaking waves. J. Fluid Mech., 209:405-433.

Bradshaw, P. (1974): Effect of free-stream turbulence on turbulent shear layers. I.C. Aero Report 74-10, Dept. of Aeronautics, Imperial College of Science and Technology, London, pp. 1-13.

Bradshaw, P. and Ferris, D.H. (1965): The spectral energy balance in a turbulent mixing layer. National Physical Laboratory, Aerodynamics Division, Report 1144.

Brumley, B.H. and Jirka, G.H. (1987): Near-surface turbulence in a grid-stirred tank. J. Fluid Mech., 183:235-263.

Cointe, R. and Tulin, M.P. (1986): An analytical model for steady spilling breakers. Proc. NSF Symposium on Air-Sea Interaction, Taiwan.

Dally, W. R., Dean, R.G. and Dalrymple, R.A. (1985): Wave height variation across beaches of arbitrary profile. J. Geophys. Res., 90(C6):11917-11927.

Dean, R.G. and Dalrymple, R.A. (1984): Water Wave Mechanics for Engineers and Scientists. Prentice-Hall Inc., Englewood Cliffs, New Jersey, 353 pp.

Deigaard, R. (1989): Mathematical modelling of waves in the surf zone. Progress Report No. 69, Inst. of Hydrodynamics and Hydraulic Engineering, ISVA, Techn. Univ. Denmark, pp. 47-60.

Deigaard, R., Fredsøe, J. and Hedegaard, I.B. (1986): Suspended sediment in the surf zone. J. Waterway, Port, Coastal and Ocean Engineering, ASCE, 112(1):115-128.

Deigaard, R. and Fredsøe, J. (1989): Shear stress distribution in dissipative water waves. Coastal Eng., 13:357-378.

Duncan, J.H. (1981): An experimental investigation of breaking waves produced by a towed hydrofoil. Proc. R. Soc., London, Ser. A, 377, pp. 855-866.

Engelund, F. (1981): A simple theory of weak hydraulic jumps. Progress Report No. 54, Inst. of Hydrodynamics and Hydraulic Engineering, ISVA, Techn. Univ. Denmark, pp. 29-32.

Galvin, C.J. (1968): Breaker type classification on three laboratory beaches. J. Geophys. Res., 73(12):3651-3659.

Horikawa, K. and Kuo, C.T. (1966): A study on wave transformation inside surf zone. Proc. 10th Coastal Eng. Conf., Tokyo, pp. 217-233.

Hunt, J.C.R. (1984): Turbulence structure in thermal convection and shear-free boundary layers. J. Fluid Mech., 138:161-184.

Hunt, J.C.R. (1984A): Turbulence structure and turbulent diffusion near gas-liquid interfaces. W. Brusaert and G.H. Jirka (eds.). Gas Transfer at Water Surfaces, D. Reidel Publ. Co., pp. 67-82.

Hunt, J.C.R. and Graham, J.M.R. (1978): Free-stream turbulence near plane boundaries. J. Fluid Mech., 84:209-235.

Iversen, H.W. (1952): Laboratory study of breakers. U.S. National Bureau of Standards, Circular No. 521, pp. 9-31.

Justesen, P. (1988): Turbulent wave boundary layers. Series Paper 43, Inst. Hydrodynamics and Hydraulic Engineering, ISVA, Techn. Univ. Denmark, pp. 1-226.

Justesen, P., Fredsøe, J. and Deigaard, R. (1986): The bottleneck problem for turbulence in relation to suspended sediment in the surf zone. Proc. 20th Coastal Eng. Conf., Taipei, pp. 1225-1239.

Kjeldsen, S.P. (1968): Bølge brydning, fysisk beskrivelse. (Wave breaking, physical description, in Danish), M.Sc. thesis. Techn. Univ. Denmark, 110 pp.

Komar, P.D. and Gaughan, M.K. (1972): Airy wave theory and breaker height prediction. Proc. 13th Coastal Eng. Conf., Vancouver, pp. 405-418.

Le Méhauté , B. (1962): On non-saturated breakers and the wave run-up. Proc. 8th Coastal Eng. Conf., pp. 77-92.

Madsen, P.A. and Warren, I.R. (1984): Performance of a numerical short-wave model. Coastal Eng., 8(1):73-93.

McDougall, T.J. (1978): Measurements of turbulence in a zero-mean-shear mixed layer. J. Fluid Mech., 94:409-431.

Okayasu, A., Shibayama, T. and Horikawa, K. (1988): Vertical variation of undertow in the surf zone. Proc. 21st Coastal Eng. Conf., Costa del Sol, Malaga, Spain, pp. 478-491.

Patrick, D.A. and Wiegel, R. L. (1955): Amphibian fractors in the surf. Proc. 1st Conf. on Ships and Waves, The Engineering Foundation Council on Wave Research and the American Soc. Naval Architects and Marine Engineers, pp. 397-422.

Peregrine, D.H. (1983): Breaking waves on beaches. Ann. Rev. Fluid Mech., 15:149-178.

Rouse, H., Siao, T.T. and Nagaratnam, S. (1958): Turbulence characteristics of the hydraulic jump. Proc. ASCE, 84(HY1):1-30.

Stive, M.J.F. (1980): Velocity and pressure field of spilling breakers. Proc. 17th Coastal Eng. Conf., Sydney, pp. 547-566.

Stive, M.J.F. and Wind, H.G. (1982): A study of radiation stress and set-up in the nearshore region. Coastal Engineering, 6:1-25.

Svendsen, I.A. (1984): Wave heights and set-up in a surf zone. Coastal Engineering, 8(4):303-329.

Svendsen, I.A. (1984a): Mass flux and undertow in a surf zone. Coastal Engineering, 8(4):347-365.

Svendsen, I.A., Madsen, P. and Hansen, J.B. (1978): Wave characteristics in the surf zone. Proc. 16th Coastal Eng. Conf., Hamburg, pp. 520-539.

Svendsen, I.A. and Madsen, P. (1981): Energy dissipation in hydraulic jumps and breaking waves. Progress Report No. 55, Inst. of Hydrodynamics and Hydraulic Engineering, ISVA, Techn. Univ. Denmark, pp. 39-47.

Thomas, N.H. and Hancock, P.E. (1977): Grid turbulence near a moving wall. J. Fluid Mech., 82:481-496.

Uzkan, T. and Reynolds, W.C. (1967): A shear-free turbulent boundary layer. J. Fluid Mech., 28:803-821.

Chapter 5. Wave-driven currents

In the surf zone the flux of wave energy and the radiation stress decrease in the shoreward direction and vanish at the shoreline. Chapter 4 described how the radiation stress gradient for normally incident waves creates a slope of the mean water surface in the surf zone: the wave set-up. Generally, however, the change in the wave momentum flux cannot be balanced by a pressure gradient from a sloping mean water surface. In addition to the pressure gradient, shear stresses are required, which can only be associated with a mean current. Breaking waves can drive strong currents in the surf zone, and the wave-driven currents are important for the sediment transport and morphological development in the coastal region.

The earliest theories for wave-driven currents concerned the depth-integrated velocity. More recently, the velocity profile has been included in the description, but until now no fully three-dimensional model has been developed which can be used to calculate the complete velocity field in a coastal area.

5.1   The longshore current

The situation with a current along a straight uniform coast with uniform wave conditions was one of the first to be treated by rational methods, and it is still important, for example in connection with modelling of sediment transport along the coast. The forces involved in the longshore current are found by considering the force balance on a prismatic control surface with the sides dx and dy, Fig. 5.1. In this case x is the shore-normal coordinate, and y is the shore-parallel coordinate. Due to the uniform conditions it is not necessary to consider momentum fluxes at the two sides parallel to the x-axis.

The force balance is established in the x-direction as well as the y-direction. In the following the different elements in the force balance are discussed.

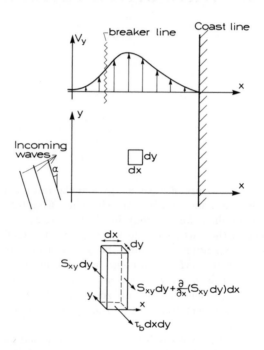

Figure 5.1 Top: The coordinate system and the control volume.
Bottom: The force balance between the radiation stress gradient and the bed shear stress.

Wave-induced forces

As described in Section 1.3 the radiation stress, or momentum flux associated with the wave motion, can be written as a tensor. The flux of shore-normal momentum is written S_{xx}, where

$$S_{xx} = F_p + F_m \cos^2 \alpha \qquad (5.1)$$

F_p is the pressure part and F_m the momentum part of the radiation stress. α is the angle between the shore and wave crests.

The flux of shore-parallel momentum, the shear component of the radiation stress, is written S_{xy}, where

$$S_{xy} = F_m \sin \alpha \cos \alpha \qquad (5.2)$$

The shore-normal force on the control surface is found to be

$$S_{xx}dy - \left(S_{xx} + \frac{dS_{xx}}{dx}dx\right)dy = -\frac{dS_{xx}}{dx}dxdy \qquad (5.3)$$

The shore-parallel force is

$$S_{xy}dy - \left(S_{xy} + \frac{dS_{xy}}{dx}dx\right)dy = -\frac{dS_{xy}}{dx}dxdy \qquad (5.4)$$

Pressure gradients

The mean water surface can only have a slope in the x-direction due to the uniform condition, and there is therefore only a pressure force in the shore-normal direction

$$P_x = -\rho g D \frac{d\Delta D}{dx}dxdy \qquad (5.5)$$

where ΔD is the difference between the mean water depth D and the still water depth D_0 as described in Section 4.2.1.

Bed shear stress

Due to the uniform condition the depth-integrated mean current must be zero in the shore-normal direction. As a good approximation the bed shear stress τ_b can be taken to be parallel to the depth-integrated flow velocity. The force on the control surface from the bed shear stress is therefore

$$T_y = -\tau_b dxdy \qquad (5.6)$$

Momentum exchange

Momentum is exchanged across the coastal profile not only because of the turbulence created by wave-breaking, but also due to some of the organized water motion in the surf zone. The exchange is commonly expressed as a product of the velocity gradient and a momentum exchange coefficient

$$M_{xy} = -\rho E D \frac{dV_y}{dx} \qquad (5.7)$$

where E is the momentum exchange coefficient, and V_y is the depth-integrated longshore current velocity.

The force on the control surface is given as

$$-\frac{dM_{xy}}{dx}dxdy = \frac{d}{dx}\left(\rho ED\frac{dV_y}{dx}\right)dxdy \tag{5.8}$$

The force balance in the x- and y-direction

There are only two terms in the x-direction, the radiation stress gradient and the pressure gradient, giving

$$\frac{dS_{xx}}{dx} = -\rho gD\frac{d\Delta D}{dx} \tag{5.9}$$

This is the equation for the wave set-up, which was encountered previously as Eq. 4.40. The only addition is that the wave direction now has to be taken into account when determining the cross-shore radiation stress.

There are three terms in the momentum balance equation for the y-direction

$$-\frac{dS_{xy}}{dx} - \tau_b + \frac{d}{dx}\left(\rho ED\frac{dV_y}{dx}\right) = 0 \tag{5.10}$$

This is the equation for a steady uniform longshore current, expressing the equilibrium between the driving (radiation stress gradient) retarding (bed shear stress) and redistributing forces. During the last 20 years Eq. 5.10, with only relatively minor modifications, has formed the basis for practically all rational approaches to describing uniform longshore currents.

Example 5.1:   An analytical longshore current model

As a baseline example the solution to Eq. 5.10, presented by Longuet-Higgins (1970, 1970a), is considered. This model has the advantage that it can be solved analytically and is still frequently used, although later improvements have been introduced for several of the elements in the solution.

The wave motion is described by linear shallow water wave theory and the wave height in the surf zone is taken to be a constant times the local water depth

$$H = KD \tag{5.11}$$

where K is taken as 0.8. The wave set-up is neglected.

The shear component of the radiation stress is

$$S_{xy} = F_m \sin\alpha\cos\alpha = \frac{1}{8}\rho g H^2 c \cos\alpha \frac{\sin\alpha}{c} = E_{fx}\frac{\sin\alpha}{c} \qquad (5.12)$$

where E_{fx} is the cross-shore wave energy flux. The wave direction is determined by refraction and according to Snell's law the term

$$\frac{\sin\alpha}{c} \qquad (5.13)$$

is constant. It is thus seen from Eq. 5.12 that the shear radiation stress varies with the shore-normal energy flux, or the energy dissipation $\overline{\mathcal{D}}$

$$-\frac{dS_{xy}}{dx} = -\frac{dE_{fx}}{dx}\frac{\sin\alpha}{c} = \overline{\mathcal{D}}\frac{\sin\alpha}{c} \qquad (5.14)$$

It can also be seen that the driving force is zero when there is no energy dissipation. In fact, this result is much more universal than indicated here. It can be shown that, when no wave energy is dissipated, the radiation stress gradients can be balanced by variations in the mean water level without driving any mean current, even for complex, non-uniform topographies and wave conditions (Dingemans et al., 1987).

Using linear wave theory, the driving force can be calculated

$$-\frac{\partial S_{xy}}{\partial x} = -\frac{\partial}{\partial x}\frac{\rho g H^2}{8}c\cos\alpha\frac{\sin\alpha}{c} =$$

$$-\frac{\rho g}{8}\frac{5}{2}K^2 D\sqrt{gD}\frac{dD}{dx}\cos\alpha\frac{\sin\alpha}{c} \approx$$

$$-\frac{5}{16}\rho\,K^2(gD)^{3/2}\frac{dD}{dx}\frac{\sin(\alpha_0)}{c_0} \qquad (5.15)$$

where it has been assumed that α is small in the surf zone ($\cos\alpha \approx 1$) and the constant in Eq. 5.13 is calculated from the deep water values α_0 and c_0.

The bed friction is estimated using a quadratic resistance law, based on the combined wave orbital velocity and mean current velocity

$$\vec{\tau}_b = \rho C_r \vec{V}\,|\,\vec{V}\,| \qquad (5.16)$$

where the depth-averaged velocity vector is given by

$$\vec{V} = \left\{ \begin{array}{c} U_{1m}\cos\alpha\cos(\omega t) \\ V_y + U_{1m}\sin\alpha\cos(\omega t) \end{array} \right\} \qquad (5.17)$$

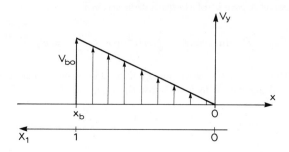

Figure 5.2 The velocity distribution according to Longuet-Higgins (1970), neglecting momentum exchange.

The time-averaged bed shear stress in the y-direction is, for $\alpha << 1$

$$\tau_b = \overline{\rho C_r \, V_y \sqrt{V_y^2 + U_{1m}^2 \cos^2(\omega t)}} \tag{5.18}$$

Assuming that $V_y << U_{1m}$, the time-averaging gives

$$\tau_b = \frac{2}{\pi}\rho C_r U_{1m} V_y =$$

$$\frac{1}{\pi}\rho C_r \sqrt{gD} \, \frac{H}{D} V_y = \frac{1}{\pi}\rho C_r \sqrt{gD} \, KV_y \tag{5.19}$$

If the momentum exchange is neglected, the longshore current can be calculated from the balance between the driving and the retarding force

$$-\frac{5}{16}\rho \, K^2(gD)^{3/2}\frac{dD}{dx}\frac{\sin(\alpha_0)}{c_0} = \frac{1}{\pi}\rho \, C_r\sqrt{gD} \, KV_y \tag{5.20}$$

or

$$V_y = \frac{5\pi}{16}\frac{K}{C_r}gD\tan\beta\frac{\sin(\alpha_0)}{c_0} \tag{5.21}$$

where $\tan\beta$ is the bed slope. For a beach profile with constant bed slope, the current velocity is proportional to the local water depth D, and the velocity distribution is triangular, as shown in Fig. 5.2. The longshore current velocity at the breaker line where the depth is D_b, calculated without momentum exchange, is an important parameter in the following calculations

$$V_{b0} = \frac{5\pi}{16}\frac{K}{C_r}gD_b\tan\beta\frac{\sin(\alpha_0)}{c_0} \tag{5.22}$$

The momentum exchange coefficient is estimated as the product of a length scale and a velocity scale. The length scale is taken to be proportional to the distance from the shore (or the local water depth). The velocity scale is taken to be proportional to the wave celerity, $c = \sqrt{gD}$, giving

$$E = K_e |x| \sqrt{gD} = K_e \frac{D}{\tan\beta} \sqrt{gD} \qquad (5.23)$$

where K_e is a constant. With this momentum exchange coefficient the force balance in Eq. 5.10 reads

$$\frac{5}{16} \rho \ K^2 (gD)^{3/2} \tan\beta \frac{\sin(\alpha_0)}{c_0} = \frac{\rho}{\pi} C_r \sqrt{gD} K V_y$$

$$- \frac{d}{dx}\left(\rho K_e \frac{D^2}{\tan\beta} \sqrt{gD} \frac{dV_y}{dx} \right) \qquad (5.24)$$

When the dimensionless velocity $V_1 = V_y / V_{0b}$ is introduced, Eq. 5.24 can be written

$$D = D_b V_1 + \frac{5}{2}\pi \frac{K_e}{K} \frac{D_b}{C_r} D \frac{d\,V_1}{dx} - \pi \frac{K_e}{K C_r} \frac{D_b}{\tan\beta} \frac{D^2}{dx^2}\frac{d^2 V_1}{dx^2} \qquad (5.25)$$

Introducing the dimensionless coordinate X_1, which is defined as

$$X_1 = D/D_b = -x/x_b \qquad (5.26)$$

the problem can be characterized by a single parameter P

$$P = \pi \frac{K_e \tan\beta}{K C_r} \qquad (5.27)$$

and Eq. 5.25 now reads

$$V_1 - \frac{5}{2}P X_1 \frac{dV_1}{dX_1} - P X_1^2 \frac{d^2 V_1}{dX_1^2} = X_1 \qquad (5.28)$$

Eq. 5.28 is valid in the surf zone. Outside the surf zone the driving force is zero, and the right-hand-side of Eq. 5.28 is therefore taken to be zero for $X_1 > 1$.

Eq. 5.28 has the particular solution

$$V_1 = A X_1 = \frac{1}{1 - 5P/2} X_1 \qquad (5.29)$$

and the homogeneous solution

$$V_1 = C_1 X_1^{P_1} + C_2 X_1^{P_2} \qquad (5.30)$$

where C_1 and C_2 are arbitrary constants, and P_1 and P_2 are given by

$$P_1, P_2 = -\frac{3}{4} \pm \sqrt{\frac{9}{16} + \frac{1}{P}} \tag{5.31}$$

V_1 is thus found to be

$$V_1 = \begin{cases} AX_1 + C_1 X_1^{P_1} + C_2 X_1^{P_2} & 0 \le X_1 < 1 \\ C_3 X_1^{P_1} + C_4 X_1^{P_2} & 1 \le X_1 \end{cases} \tag{5.32}$$

The arbitrary constants are found from the condition that V_1 must be bounded for $0 \le X_1 < \infty$, and the requirement that V_1 and $\frac{dV_1}{dX_1}$ are continuous at $X_1 = 1$, giving

$$C_1 = A(P_2 - 1)/(P_1 - P_2) \ ; \ C_2 = 0;$$

$$C_3 = 0 \ ; \ C_4 = A(P_1 - 1)/(P_1 - P_2) \tag{5.33}$$

The velocity profiles for different values of P are shown in Fig. 5.3. The parameter P describes the significance of the momentum exchange and it can be seen how the velocity profile approaches the triangular distribution given by Eq. 5.21 as P decreases to zero.

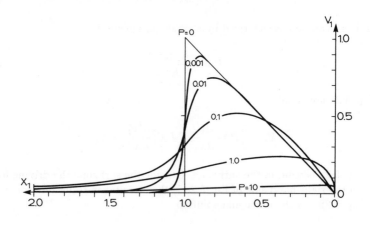

Figure 5.3 Theoretical form of the longshore current $V_1 = V_y/V_{0b}$ as a function of $X_1 = -x/x_b$ and the lateral mixing parameter $P = \pi \frac{K_e \tan \beta}{K \ C_r}$. After Longuet-Higgins (1970a).

5.2   Further developments of longshore current models

Most of the recent models for the longshore current are composed of the same elements as the model by Longuet-Higgins (1970), which was described in the previous section. The new developments have taken the form of improvement in the individual terms in the force balance, described by Eq. 5.10. The later more refined models can therefore be described by considering each term of the force balance.

5.2.1   The driving force

The basic formulation of the forces induced by wave-breaking has not been changed since the first models were introduced, based on the radiation stress concept, Bowen (1969), Thornton (1970) and Longuet-Higgins (1970, 1970a). The developments have mainly concerned the description of the wave breaking.

The description of irregular waves gives an important modification to the distribution of the longshore driving force. When the incoming waves have different heights, they will also break at different water depths. Therefore the energy dissipation and the longshore driving force will be more evenly distributed than for a description based on regular waves, where the distribution of the driving force is discontinuous at the point of wave-breaking. With irregular waves the driving force will increase gradually from zero at a large water depth, where the first waves begin to break. In the inner surf zone the driving force will approach the regular wave force, because practically all waves are breaking/broken here, see Fig. 5.4. The smoothing of the velocity profile due to irregular waves is considerable, and when irregular waves are considered it becomes much less important to model the cross-shore momentum exchange correctly.

A model including irregular waves has been developed by Battjes (1974) using a spectral description. The surf zone model by Battjes and Janssen (1978) has also been used successfully as a basis for longshore current calculations, cf. Section 4.4.

Fig. 5.5 shows a comparison between longshore current velocities calculated from regular and irregular waves. The irregular wave calculations have been made using wave statistics based on the Rayleigh distribution of the wave heights. It can be noted how much the irregularity of the waves smooths down the current velocity profile.

The directional spreading of the incoming waves also has significance for the driving force. Application of realistic distributions of the wave energy direction can give a reduction by up to one half of the total driving force, integrated over the

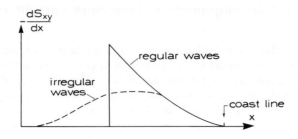

Figure 5.4 Distribution of the longshore driving force, regular waves and irregular waves.

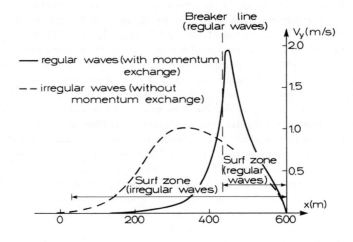

Figure 5.5 Longshore current velocities calculated with regular and irregular waves. Deep water wave height: $H_{rms} = 1.2$m, $T = 7.5$s, $\alpha_0 = 45°$. Constant beach slope: $\tan \beta = 0.01$. After Zyserman (1989).

coastal profile. It should be noted that the total longshore force is unaffected by the application of an irregular wave description assuming unchanged wave energy flux, or the use of more sophisticated wave theories in the surf zone, cf. Eq. 5.12, which describes the relation between the energy dissipation and the longshore driving force. For a given wave energy flux from offshore, the total longshore force is the same.

More advanced wave descriptions have not led to significantly improved longshore current models. This is partly because the finer wave descriptions will only redistribute the driving force, as described above, and partly because the problem of combining the incoming wave field with a wave model for the inner surf zone has not yet been solved satisfactorily.

In addition to depth refraction, refraction due to the longshore current can be included in the wave description (Dalrymple, 1980; Southgate, 1989). The current refraction may change the location of the breaker line and thus the distribution of the driving force over the profile.

Other driving forces can be included in a model, e.g. wind shear stress and a longshore water surface gradient S. The wind will give two components. An onshore component to be included in the set up

$$\tau_w \cos(\alpha_w) = \frac{1}{2}\rho_a f_w U_{10}^2 \cos(\alpha_w) \qquad (5.34)$$

where τ_w is the surface shear stress from the wind, α_w is the angle between the wind and the x-axis, U_{10} is the wind velocity 10 m above the sea level, and f_w is the corresponding friction factor, ρ_a is the air density.

The longshore component reads

$$\tau_w \sin(\alpha_w) = \frac{1}{2}\rho_a f_w U_{10}^2 \sin(\alpha_w) \qquad (5.35)$$

A coastal current will often be driven by a mean longshore water surface slope of S, which gives a longshore driving force of

$$\tau_s = \rho g D S \qquad (5.36)$$

In many cases, the contributions from wind shear stress and coastal currents can be neglected because they are relatively small compared to the wave-driven forces. This is discussed further in Example 5.3.

Example 5.2: The effect of directional spreading.

In order to illustrate the effect of directional spreading of the incoming wave energy, a simplified example is considered. The deep water waves are assumed to be composed of two wave trains with the same characteristics, except for the wave direction, Fig. 5.6. The mean wave direction is α_0. One wave train has the direction $\alpha_0 + \Delta\alpha$, the other $\alpha_0 - \Delta\alpha$. The two wave trains are statistically independent, and the combined wave energy and radiation stress is found as the sum of the two contributions

$$S_{xy} = \frac{1}{2}F_m \sin\big(2(\alpha_0 + \Delta\alpha)\big) + \frac{1}{2}F_m \sin\big(2(\alpha_0 - \Delta\alpha)\big) =$$

$$\frac{1}{2}F_m\Big(\sin\big(2(\alpha_0 + \Delta\alpha)\big) + \sin\big(2(\alpha_0 - \Delta\alpha)\big)\Big) \qquad (5.37)$$

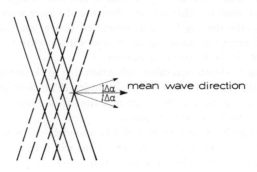

Figure 5.6 A wave field composed of two crossing wave trains.

where $\frac{1}{2}F_m$ is the momentum part of the radiation stress for each of the wave trains. The ratio between the actual S_{xy} and the value for no directional spreading S_{xy0} is therefore

$$\frac{S_{xy}}{S_{xy0}} = \frac{\sin\big(2(\alpha_0 + \Delta\alpha)\big) + \sin\big(2(\alpha_0 - \Delta\alpha)\big)}{2\sin(2\alpha_0)} = \cos(2\Delta\alpha) \qquad (5.38)$$

S_{xy}/S_{xy0} is shown in Fig. 5.7. It is seen that S_{xy} becomes zero for $\Delta\alpha = 45^\circ$, where the two wave trains cross each other at a right angle.

The directional spreading of waves is normally described by use of an energy distribution $K(\Delta\alpha)$ which describes the density of the wave energy as a function of the direction of propagation $\Delta\alpha$ measured relative to the mean wave direction.

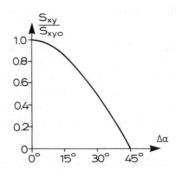

Figure 5.7 Spreading factor versus $\Delta\alpha$ for two crossing wave fields.

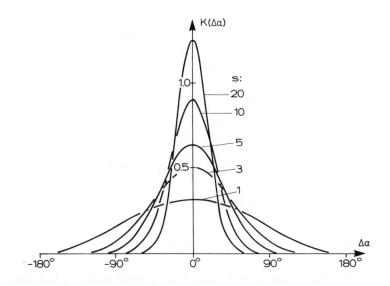

Figure 5.8 Spreading function versus $\Delta\alpha$.

The function $K(\Delta\alpha)$ must be normalized so that the total energy is still described by the non-directional wave spectrum

$$\int_{-\pi}^{\pi} K(\Delta\alpha)d(\Delta\alpha) = 1 \tag{5.39}$$

In the general situation the directional spreading may vary with the frequency in the spectrum, and K is then specified as a function of both the frequency and $\Delta\alpha$. Several different empirical functions have been developed for describing the distribution of the directionally spread energy. A commonly used function is

$$K(\Delta\alpha) = \frac{2^{2s}\left(\Gamma(s+1)\right)^2}{2\pi\,\Gamma(2s+1)}\cos^{2s}(\Delta\alpha/2) \tag{5.40}$$

The directional distribution of the wave energy is illustrated in Fig. 5.8 for different values of the power s. Fig. 5.9 shows the reduction of the shear component of the radiation stress as a function of s. For natural storm waves s is often estimated to be about 8. According to Fig. 5.9 this corresponds to a reduction of the longshore driving force by a factor 0.6.

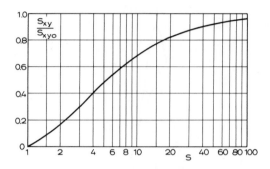

Figure 5.9 Spreading factor versus s.

Example 5.3:   Calculation of driving forces from waves, wind shear stress, and coastal current

Consider a coast with a constant slope 1:50 and the following incident wave properties. The wave height at the point of breaking is $H_b = 1$ m and period $T = 5$ s. The waves are breaking at a water depth $D_b = H_b/0.8 = 1.25$ m and the deep water wave direction is $\alpha_0 = 20°$. Wave set-up is neglected. The waves are described by linear wave theory. The wave angle at the point of breaking is found by a refraction calculation to be $\alpha_b \approx 8°$.

The total longshore driving force is found as

$$S_{xy} = \frac{1}{2} \sin(2\alpha_b) \, F_{mb} = 158 \text{ N/m} \tag{5.41}$$

The wind friction factor is estimated by assuming a logarithmic velocity profile and a surface roughness of 5 cm

$$f_w \approx \frac{2}{\left(2.5 \, \ln\left(\frac{30 \times 10 \text{ m}}{0.05 \text{ m}}\right)\right)^2} = 4.23 \times 10^{-3} \tag{5.42}$$

The width of the surf zone is $1.25 \cdot 50$ m $= 62.5$ m. The wind speed which would be required to give the same longshore driving force in the surf zone is then estimated from Eq. 5.35, assuming that the wind direction corresponds to the deep water wave direction

$$\tau_w \sin 20° = 158/62.5 \text{ N/m}^2 = 2.53 \text{ N/m}^2 \tag{5.43}$$

giving

$$U_{10}^2 = \frac{2.53 \text{ N/m}^2 \times 2}{\rho_a f_w \sin(20°)} = 2740 \text{ m}^2/\text{s}^2 \tag{5.44}$$

or $U_{10} = 52$ m/s, which is an extremely strong wind compared to the modest wave height of 1 m.

The total longshore driving force in the surf zone associated with a coastal current is obtained by integrating Eq. 5.36 across the surf zone

$$\int_0^{62.5\text{m}} \rho g \frac{x}{50} S \, dx' = \frac{\rho g}{2} \frac{(62.5 \text{ m})^2}{50} S \tag{5.45}$$

where x is the distance from the shoreline. The water surface slope which gives a driving force equal to the radiation stress is found as

$$S = 158 \text{ N/m} \times 2 \times 50/(\rho g(62.5 \text{ m})^2) = 4.13 \times 10^{-3} \tag{5.46}$$

For a water depth of 10 m and a bed roughness of 5 cm, a water surface slope of $4.13 \cdot 10^{-3}$ corresponds to a current velocity of

$$V_y = \sqrt{gDS} \times 2.5\left(\ln\left(\frac{10 \text{ m} \times 30}{0.05 \text{ m}}\right) - 1\right) = 3.9 \text{ m/s} \tag{5.47}$$

which is also an extraordinarily strong current. It can be seen from these two examples that in many cases the contribution from the wind and coastal currents to the driving forces in the surf zone are of minor importance.

5.2.2 The flow resistance

While the main principles behind the description of the longshore driving force have remained unchanged, the flow resistance term has been subject to a considerable improvement.

The original approach of using a quadratic flow resistance term for the combination of the mean current and the wave orbital velocity was not a good approximation because of the large difference between the thickness of the wave boundary layer and of the mean current boundary layer. Due to the thin wave boundary layer, the friction factor for the oscillatory motion is larger than for the mean motion.

An improvement has been to introduce the models for combined wave-current boundary layers, as described in Chapter 3. An example of such an application has been presented in the model by Deigaard et al. (1986a).

In the present section, only the depth-integrated current velocity V_y has been considered. It is, however, not possible to analyze the flow resistance in any detail without looking at the vertical distribution of the shear stress, the eddy viscosity, and the current velocity. The models for wave-current boundary layers do not, however, take the strong turbulence from wave-breaking into account. The increase of the eddy viscosity, due to the breaking-induced turbulence, will, for a given shear stress distribution, give a more vertical velocity profile with a smaller depth-integrated current velocity, compared to the logarithmic mean velocity profile of the wave-current boundary layer models developed for use outside the surf zone. The flow resistance is therefore increased by the turbulence from the wave-breaking.

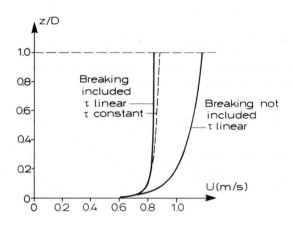

Figure 5.10 Example of velocity profiles for a longshore current calculated with and without eddy viscosity contribution from wave-breaking. D = 2 m, H = 1 m, T = 6 s, $\tau_b = 1.8$ N/m^2.

Deigaard et al. (1986a) calculated the velocity profile with and without the breaking-induced turbulence. The turbulence was calculated as described in Chapter 4. The velocity profiles are shown in Fig. 5.10. The depth-integrated velocity is reduced by about 20% when the effect of breaking-induced turbulence is included. The use of the wave-current boundary layer model alone, without taking the additional turbulence into account, will thus cause an error of about 20% in the calculated current velocity, a difference which is small compared to the general uncertainty associated with longshore current calculations. The difference between the two velocity profiles is largest far from the bed, and if the purpose is to calculate sediment transport rates, the difference is even smaller, because the suspended sediment is mainly concentrated near the bed.

The two solid velocity profiles in Fig. 5.10 are calculated from a triangular shear stress distribution. As will be seen later in Chapter 6, a constant shear stress over the vertical is in fact a better approximation. The use of a constant shear stress gives an increase to the depth-integrated velocity of about 2%, when the breaking-induced turbulence is included, as indicated with the broken curve in Fig. 5.10.

5.2.3 The momentum exchange

Considerable effort has been made to improve the description of the cross-shore exchange of longshore momentum, but none of the solutions has been entirely satisfactory. Still, it must be kept in mind that this term only redistributes the momentum without contributing to the total balance, and that the effect of irregular incoming waves will smooth out the velocity profile so much that the accuracy of the modelling of the momentum exchange becomes much less important.

Two objections may be brought against the formulation of the momentum exchange used by Longuet-Higgins (1970). First, E is related to a global parameter, the distance from the shore, rather than to local parameters. Secondly, E increases with the distance from the shore, also outside the surf zone where a decrease of momentum exchange can be expected. The first objection is less severe, considering the assumptions made. With a constant beach slope and a fixed ratio between the wave height and the water depth in the surf zone both the wave height and the water depth in the surf zone are proportional to the distance from the shore.

Many attempts have been made to improve the description of momentum exchange, but in all cases the exchange has been formulated with an exchange factor and the gradient in the depth-integrated current.

Jonsson et al. (1976) calculated E as a function of the local wave conditions. The depth-integrated turbulence model for the surf zone, described in Chapter 4, was actually developed by Battjes (1975) in order to model E. The problem with a discontinuous E at the breaker line was later overcome by including horizontal diffusion of the turbulent kinetic energy (Battjes, 1983).

A contribution to the momentum exchange, which has been largely neglected, originates from the ordered mean flow in the surf zone, the circulation current. A complete analysis of this mechanism has not yet been carried out. But in Example 5.4 the contribution from the water in the surface rollers has been assessed.

Example 5.4:   Momentum exchange in the surf zone, due to organized water motion

Using Engelund's (1981) model, cf. Example 4.2, the cross-sectional area of the surface rollers can be approximated by

$$A = \frac{H^3}{4D\alpha_0} \tag{5.48}$$

The acceleration of the surface roller in the x-direction towards the coast is small, and the horizontal forces acting on the roller must be close to a balance. Three forces acting on the surface roller are considered: gravity ($\rho g A$), a pressure force, P_r, and a shear force, T_r, on the surface between the roller and the water below, Fig. 5.11.

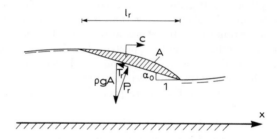

Figure 5.11 The forces acting on the surface roller.

Equilibrium between these forces gives the integrated shear stress along the length l_r of the roller (assuming α_0 to be small)

$$T_r \approx \rho g A \alpha_0 = \frac{\rho g H^3}{4D} \tag{5.49}$$

The momentum exchange for a situation with normally incident waves acting with a shore-parallel current is determined as follows:

The longshore current is locally described by a linear variation (Fig. 5.12)

$$V_y = V_{y0} + V_y' x \tag{5.50}$$

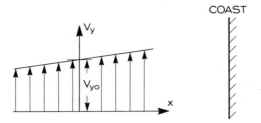

Figure 5.12 The longshore velocity distribution.

    The shore-parallel velocity component of the water in the surface roller is called V_r. The shore-normal velocity of the roller is the wave celerity, c. The shear force acting on the surface roller is assumed to be directed against its velocity relative to the water below. When V_r is not equal to V_y, the shear stress acting on the surface roller will thus form an angle to the shore-normal direction. Assuming that $V_y - V_r$ is small compared to c, this angle can be approximated by $(V_y - V_r)/c$.
    The shore-parallel momentum equation for the surface roller (per unit length of the wave front) then reads

$$\rho A \frac{dV_r}{dt} = T_r \frac{(V_y - V_r)}{c} \tag{5.51}$$

The translation of the roller in the x-direction with the velocity c gives

$$\frac{dV_r}{dt} = c \frac{dV_r}{dx} \tag{5.52}$$

which inserted in Eq. 5.51 gives

$$V_r = C_1 \exp\left(-\frac{T_r}{\rho c^2 A}x\right) + V_y'\left(x - \frac{\rho c^2 A}{T_r}\right) + V_{y0} \tag{5.53}$$

    In the following the vanishing transient part of the solution is neglected, i.e. $C_1 = 0$.
    The flux of shore-parallel momentum through a shore-parallel cross section (the shear force) has two contributions from the organized motion considered: from the surface rollers and from the average return flow. Due to continuity considerations the discharges of these opposing flows are of the same magnitude and equal to: $q_x = A/T$. The time-averaged shear force is calculated as

$$\frac{T_{xy}}{\rho} = -\frac{1}{T}\int_0^T q_x V_y dt = -\frac{1}{T}\left(AV_r - AV_y\right) =$$

$$\frac{A}{T}\left(V_y - V_y'(x - \frac{\rho c^2 A}{T_r}) - V_{y0}\right) = V_y'\frac{A}{T}\frac{\rho c^2 A}{T_r} = EDV_y' \tag{5.54}$$

Where the momentum exchange coefficient E has been introduced. E is thus found
to be

$$E = \frac{A^2 c^2 \rho}{TDT_r} = \frac{A}{T\alpha_0} = D\sqrt{gD}\frac{(H/D)^3}{\left(\sqrt{\frac{gT^2}{D}}4\alpha_0^2\right)} \tag{5.55}$$

The dimensionless exchange factor, $E/D\sqrt{gD}$ for α_0 equal to $\tan 10°$ is
shown in Fig. 5.13 as a function of the relative wave height and dimensionless
wave period. The result may be compared to the magnitude of the turbulent eddy
viscosity, which is estimated to be a few per cent of $D\sqrt{gD}$ (Stive and Wind,
1986).

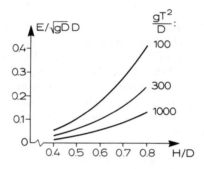

Figure 5.13 The dimensionless momentum exchange coefficient as a function of H/D and gT^2/D.

5.2.4 Longshore current on barred coastal profiles

The assumption of a constant beach slope is often not a good approximation;
sandy coasts exposed to a wave climate will often build up barred profiles. Fig.
5.14 shows an example from the Danish North Sea coast with three bars. A
longshore current model should therefore be able to describe the absence of wave-
breaking in the trough inshore of the bar, which for example can be done by using
the model by Battjes and Janssen (1978), Chapter 4. The calculated velocity
profiles in Fig. 5.14 is made by the model by Deigaard et al. (1986a). Here the
height of breaking and broken waves are determined by Eq. 4.3 (letting the depth
at the crest of the bar govern the height of the broken waves in the trough inshore

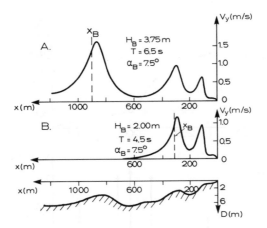

Figure 5.14 Calculated longshore current profiles on coast with three bars. Wave conditions are given at the outer breaking point. A: breaking on all bars; B: breaking at two inner bars.

of the bar). According to this simple formulation the reformed waves immediately inshore of a bar will have half the height the water depth over the bar.

The driving force dS_{xy}/dx is only significant on the offshore face of the bars and at their crests. The calculated velocity profiles have therefore a local maximum at each bar. This characteristic velocity distribution is often difficult to confirm by field measurements. An example of longshore current measurements on a barred beach, from the Super Duck experiments, is shown in Fig. 5.15; these measurements presented by Whitford and Thornton (1988) show a clear tendency for the strongest longshore current velocities to be located near the bar crest. One of the main reasons for deviations from the theoretical velocity profile may be that the beach topography deviates from the assumed uniform coast with straight shore-parallel depth contours. If the topography is irregular, complex circulation currents can be generated, which have a very significant effect on the momentum exchange in the surf zone.

5.3 Wave-driven currents on a non-uniform beach profile

In the preceding section a longshore wave-driven current was considered. The influence of different effects was discussed: irregular waves, wind, momentum exchange, barred profiles, etc.. A much more drastic change in the current pattern

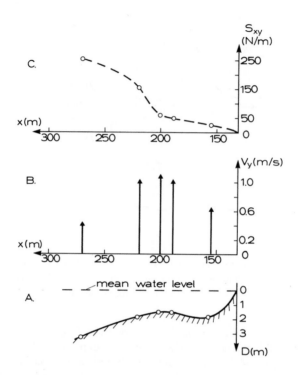

Figure 5.15 Measurements on a barred profile at five locations. From bottom: A: recorded water depths; B: measured longshore current velocities; C: measured shear component of the radiation stress. After Whitford and Thornton (1988).

can be caused by irregularities in the coastal profile along the coast.

One of the best known examples is the phenomenon of rip channels. On a barred profile the wave breaking on the bar will induce a wave set-up, causing an increase in the water level inshore of the bar. However, a bar will in many cases be interrupted by holes - rip channels - found at more or less regular intervals. The wave-breaking is less intensive in the rip channels due to the larger depth and because wave refraction may concentrate the wave energy on the bars at the sides of the channel.

The shoreward decrease of the radiation stress is therefore not the same over the bar and in the rip channels, and a balance between the radiation stress and the pressure of a constant water level in the trough inshore of the bar cannot be obtained at all cross sections along the coast. An offshore-directed flow is therefore driven out through the rip channels due to the wave set-up behind the bar. The out-going water flux is compensated by an onshore net flow across the bar.

In this way the irregular beach profile causes a wave-driven circulation current: shorewards over the bar, seawards in the rip channels, and alongshore in the trough towards the rip channels, Fig. 5.16.

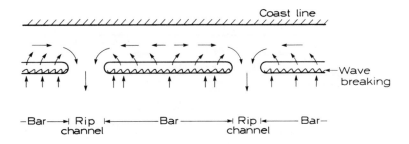

Figure 5.16 The circulation current generated by normally incident waves on a barred coast with rip channels.

If the waves approach the coast obliquely, the pattern becomes even more complex. The waves that break on the bar will exert a longshore driving force similar to the conditions on a uniform coast; a longshore current will be generated on the bar. The longshore current velocity profile may, however, be strongly modified by the shoreward flux over the bar, and some of its longshore momentum will be transferred to the flow in the trough. In the trough the flow is feeding the seaward flow in the rip channels, and the flow in the troughs may locally be stronger or weaker than the longshore flow on the bar, and it may locally go against the direction corresponding to the incoming wave direction. An irregular beach topography thus contributes many complications in the modelling of flow and sediment transport in the coastal zone when planning field measurements or interpreting observations.

Example 5.5:   Circulation driven by normally incident waves

In order to illustrate the order of magnitude of the circulation currents generated by normally incident waves, a very simplified model is considered. The geometry is illustrated in Fig. 5.17. The length of the bar is 2 L. The width of the rip channel is assumed to be so large that the flow behind two neighbouring bars does not interfere, and that the water level in the channel corresponds to the sea level. This flow situation was considered by Dalrymple (1978), who took the energy dissipation of the mean flow to be caused by bed friction. In the present example the energy loss is described as head losses where the flow crosses the bar, and where it flows from the trough into the rip channel.

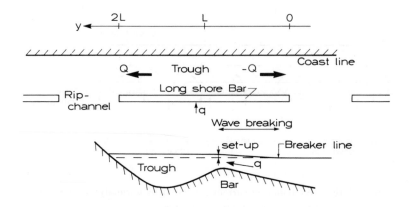

Figure 5.17 Definition sketch for the simplified model for wave-driven circulation.

The determination of the wave set-up, in the case of no cross-shore current across the bar, is described in Chapter 4. With no net current and uniform conditions, the set-up behind the bar is ΔD_0. An actual set-up ΔD, smaller than ΔD_0, will be associated with a net onshore cross current with a specific discharge q. The magnitude of the onshore current is estimated by assuming that the energy loss is equal to the velocity head of the net discharge over the bar

$$\frac{(q/D_c)^2}{2g} = \Delta D_0 - \Delta D \tag{5.56}$$

or

$$q = D_c\sqrt{2g(\Delta D_0 - \Delta D)} \tag{5.57}$$

where D_c is the water depth over the bar crest. In this description the bed shear stress on the bar has not been taken into account. Including the bed shear stress would give a small decrease of the calculated circulation current.

The longshore flow in the trough inshore of the bar is described by the momentum equation for non-uniform flow. The discharge is Q, and the cross-sectional area of the trough is A. The wave set-up is assumed to be small compared to the water depths. If the bed shear stress is neglected, the shore-parallel projection of the momentum equation for the flow in the trough reads, cf. Fig. 5.18

$$\rho Q(Q/A) + \rho g \Delta D A =$$

$$\rho\left(Q + \frac{dQ}{dy}dy\right)\left(Q + \frac{dQ}{dy}dy\right)/A + \rho g\left(\Delta D + \frac{d\Delta D}{dy}dy\right)A \qquad (5.58)$$

or

$$\frac{d}{dy}\left(\Delta D + \frac{Q^2}{gA^2}\right) = 0 \qquad (5.59)$$

In Eq. 5.58 the hydrostatic pressure, corresponding to a water surface at $\Delta D = 0$, has been omitted at both sides of the equation. Further, it should be noted that the cross flow over the bar carries no longshore momentum and is therefore not included directly in Eq. 5.59.

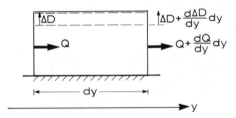

Figure 5.18 Sketch showing the terms in the longshore momentum equation

The continuity equation gives the relation between the longshore and the cross-shore current

$$\frac{dQ}{dy} = q \qquad (5.60)$$

The boundary conditions are that $\Delta D = 0$ at $y = 0$, corresponding to a head loss of $Q^2/(2gA^2)$ at the outflow from the trough and $Q = 0$ at $y = L$ due to symmetry.

The discharge is found by solving Eqs. 5.57, 5.59 and 5.60. The solution is

$$\frac{Q}{A\sqrt{g\Delta D_0}} = \sinh\left(\sqrt{2}\frac{D_c y}{A}\right) - \tanh\left(\sqrt{2}\frac{D_c L}{A}\right)\cosh\left(\sqrt{2}\frac{D_c y}{A}\right) \qquad (5.61)$$

The variation of the longshore velocity $V = Q/A$ is shown in Fig. 5.19 as functions of y and L.

An impression of the velocity magnitude predicted by this very simplified model can be obtained from Fig. 5.20, which shows the velocity scale $V_0 = \sqrt{g\Delta D_0}$ as a function of the deep water wave steepness H_0/L_0 and the still water depth over the bar crest, $D_{0c} = D_c - \Delta D$. The magnitude of the velocity scale is illustrated

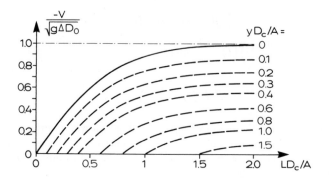

Figure 5.19 The longshore current velocity V as a function of y and L.

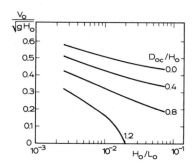

Figure 5.20 The scale for the longshore current velocity V_0 as a function of
the depth at the crest of the bar and the deep water steepness.

by taking a typical value of $V_0/\sqrt{gH_0}$ equal to 0.3. A wave height of $H_0 = 1$ m
then gives $V_0 = 0.9$ m/s and $H_0 = 3$ m gives $V_0 = 1.6$ m/s.

It is interesting to note from Fig. 5.19 that the outflow discharge at $y = 0$ is
practically constant for LD_c/A larger than 1 - 1.5. This sets an upper limit for the
length of a bar that can be expected to be stable. Consider a bar with existing rip
channels being maintained by the flushing effect from the flow generated by the
waves. If a section of the bar had a length corresponding to $LD_c/A = 3$ or longer,
then a new rip channel formed in the middle of the bar would draw a rip current
with practically the same strength as in the existing channels, without noticeably
reducing these. The new channel would therefore be maintained as effectively as
the existing ones. According to this very simplified model the maximum spacing

of rip channels can be expected to be of the order 2 - 3 times A/D_c, or a few times the distance from the shore to the bar.

Example 5.6:   Longshore current with a cross flow

As illustrated in Example 5.5, circulation currents will be generated on a non-uniform beach topography by normally incident waves. In a bar and rip channel system, the flow is directed onshore over the bars and offshore through the channels. If the waves approach obliquely, a longshore current will be generated on the bars, but the cross flow has an impact on the dynamics, which must be taken into account.

In order to illustrate the effect, an infinitely long uniform bar profile, Fig. 5.21 is considered. Over the bar there is an onshore directed flow with the discharge q. For steady state conditions, the depth-integrated flow equation in the longshore direction becomes

$$\int_0^D \rho \frac{dV_y}{dt} dz = \int_0^D \rho V_x \frac{dV_y}{dx} dz = \rho q \frac{dV_y}{dx} = -\frac{dS_{xy}}{dx} - \tau_{by} \qquad (5.62)$$

where V_x is the depth-integrated velocity in the x-direction, and the longshore velocity V_y is taken to be constant over the depth. S_{xy} is the shear radiation stress and τ_{by} is the shear stress in the y-direction.

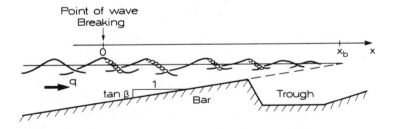

Figure 5.21 Wave-breaking on a bar with a cross flow.

In Eq. 5.62 the shear-induced momentum exchange is neglected, since the purpose of this example is to illustrate the effect of the convective acceleration term at the left-hand-side of the equation. The following simplifications are introduced:

- The wave angle is small, $\cos\alpha \approx 1$.
- The ratio K between the wave height and the water depth is constant.
- The bed slope $\tan\beta$ is constant.
- Shallow water wave theory is used.
- A linear bed shear stress formula is used:

$$\tau_{by} = \rho K_1 V_y \tag{5.63}$$

With these assumptions the driving force from the radiation stress is found (cf. Eq. 5.15) as

$$-\frac{dS_{xy}}{dx} = -\frac{5}{16}\rho\, K^2(gD)^{3/2}\,\frac{dD}{dx}\frac{\sin(\alpha_0)}{C_0} \tag{5.64}$$

If there is no cross current, the longshore current at the breaker line is

$$V_0 = \frac{5}{16}\frac{K^2(gD_b)^{3/2}}{K_1}\tan\beta\,\frac{\sin(\alpha_0)}{c_0} \tag{5.65}$$

where D_b is the water depth at the point of wave-breaking. By using V_0 as a velocity scale and x_b as a length scale, Eq. 5.62 is made dimensionless (x_b is defined as $D_b/\tan\beta$, cf. Fig. 5.21)

$$\frac{d(V_y/V_0)}{d(x/x_b)} = \frac{x_b K_1}{q}\left[\left(\frac{D}{D_b}\right)^{3/2} - \frac{V_y}{V_0}\right] \tag{5.66}$$

or

$$\frac{d(V_y/V_0)}{d(x/x_b)} = \frac{x_b K_1}{q}\left[\left(1 - \frac{x}{x_b}\right)^{3/2} - \frac{V_y}{V_0}\right] \tag{5.67}$$

The boundary condition to the differential equation is

$$V_y/V_0 = 0 \text{ for } x/x_b = 0 \tag{5.68}$$

which reflects that the water moving into the surf zone has no longshore velocity and must be accelerated from zero by the radiation stress gradient.

The solution to Eq. 5.67 is shown in Fig. 5.22 for different values of the coefficient $x_b K_1/q$. The mathematical solution is given for $0 \le x/x_b < 1$, but the solution is only valid until the crest of the bar. It is seen that for small values of $K_1 x_b/q$ (that is for a strong cross flow) the velocity profile becomes very flat. This is because each water particle moves so fast through the surf zone, where the longshore driving force is acting, that the particle is not accelerated to a speed comparable to V_0. With increasing values of $K_1 x_b/q$, the velocity profile gradually approaches the equilibrium profile, shown by the dashed line, where bed shear stress locally balances the radiation stress gradient. For small values of $K_1 x_b/q$, the total bed shear stress on the bar is much smaller than the driving force and

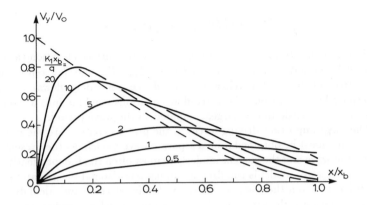

Figure 5.22 Longshore current velocity distribution, for uniform conditions with a cross flow.

a considerable part of the longshore momentum, delivered by the radiation stress in the surf zone, is convected to the trough inshore of the bar. The amount of longshore momentum which is convected is $\rho q V_{y1}$, where V_{y1} is the longshore velocity at the crest.

Modelling of wave-driven currents over a complex topography

In the general case of obliquely incident waves on a complex coastal topography, it is not possible to use analytical methods to describe the flow pattern. It is then necessary to use numerical models to solve the general, depth-integrated flow equations for two horizontal dimensions. The flow equations can be written

momentum in the x-direction:

$$\rho D \left(\frac{\partial V_x}{\partial t} + V_x \frac{\partial V_x}{\partial x} + V_y \frac{\partial V_x}{\partial y} \right) + \frac{\partial S_{xx}}{\partial x} + \frac{\partial S_{xy}}{\partial y} - \frac{\partial T_{xx}}{\partial x} - \frac{\partial T_{xy}}{\partial y}$$

$$+ \tau_{bx} + \rho g D \frac{\partial \eta}{\partial x} = 0 \qquad (5.69)$$

momentum in the y-direction:

$$\rho D \left(\frac{\partial V_y}{\partial t} + V_y \frac{\partial V_y}{\partial y} + V_x \frac{\partial V_y}{\partial x} \right) + \frac{\partial S_{yy}}{\partial y} + \frac{\partial S_{xy}}{\partial x} - \frac{\partial T_{yy}}{\partial y} - \frac{\partial T_{xy}}{\partial x}$$

$$+ \tau_{by} + \rho g D \frac{\partial \eta}{\partial y} = 0 \qquad (5.70)$$

continuity:

$$\frac{\partial \eta}{\partial t} + \frac{\partial}{\partial x}\left(DV_x\right) + \frac{\partial}{\partial y}\left(DV_y\right) = 0 \qquad (5.71)$$

Here D is the total water depth, η is the water surface elevation (averaged over the wave period) relative to a fixed level, T_{xx}, T_{yy} and T_{xy} represent the Reynolds-stresses due to the velocities (turbulence, secondary flows) which are not described by the model. The three flow equations combine into a model for the flow on a time scale that is large, compared to the wave period.

The most important driving forces in the present context are the radiation stresses, which must be obtained from a separate wave model. At the present state of the art, the wave and the current models are most often run independently. This means that the wave model is run first, assuming a horizontal mean water surface and no current. From the wave model the radiation stresses are calculated. The radiation stresses are then included in the current model as driving forces. The current model then produces the current velocities and the mean water level in the area. By this simplification, the effect of wave set-up and current refraction is not taken into account in the wave model.

Example 5.7:   Circulation current on a barred coast with rip channels

The current velocities over a complex topography can be illustrated by examining the model by Zyserman (1989). Zyserman considers a schematized bar profile with regularly spaced rip channels.

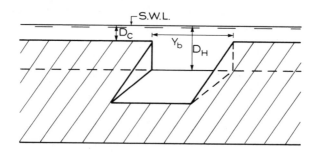

Figure 5.23 The configuration of rip channels in the model by Zyserman (1989).

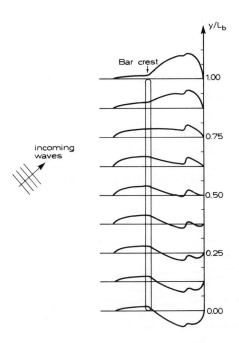

Figure 5.24 Longshore current on a coast with a bar and rip currents. L_b is
the distance between the rip channels, from Zyserman (1989).

The bar profile is composed of two parts with constant slopes, and a discontinuity at the bar crest. The rip channels have rectangular cross sections through the bar with the depth D_H and width Y_b, Fig. 5.23. Zyserman introduced a number of simplifications in order to develop an analytical-numerical model. For instance the cross-shore current over the bar crest is determined by the head loss, cf. Eq. 5.57. The important mechanism of cross-shore convection of momentum on the bar and in the trough is included in the description. Fig. 5.24 shows the profiles of the shore-parallel flow for a case with a bar length of $L_b = 180$ m and a width of the rip channel of $Y_b = 50$ m, the distance from the shore to the bar was 90 m. The deep water wave height and angle of approach was 1.2 m and 45° and the period was 7.5 s. It can be seen how the current in the trough becomes stronger than the longshore component on the bar, and that the flow in the trough is directed against the longshore current over a distance of about 1/3 of the spacing between the rip channels.

It is seen that the cross-shore flow over the bar, which is stronger near the rip channels than at the centre of the bar, has an impact on the longshore velocity on the bar, cf. Example 5.6. The effect of the rip channels on the longshore sediment transport is treated in Chapter 12.

Example 5.8: Detailed modelling of wave-driven currents at an actual site

The site in this example is Pesaro at the Adriatic coast of Italy, at the important tourist area between Arcona and Rimini. The topography in the area of interest is given in Fig. 5.25. The present example is based on a study carried out for Aquater, Italy, by the Danish Hydraulic Institute, as reported by Madsen et al. (1987).

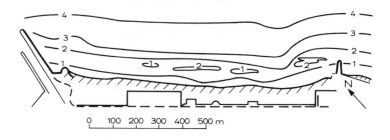

Figure 5.25 Topography of Pesaro site.

The waves are described by a refraction model which was found to be adequate, because the effect of wave diffraction is minor in this case. Fig. 5.26 shows an example of the refraction calculation, with waves coming from the east (height 2.3 m, period 8.4 s). The wave orthogonals show a rather smooth wave pattern, but with some concentration of incoming wave energy at certain locations along the beach. The wave-breaking and the wave height variation after breaking have been described by the empirical wave height index of Andersen and Fredsøe (1983), Eq. 4.3. The refraction model thus gives the details of wave height and direction over the entire model area. This is the basis for the calculation of the radiation stresses and their gradients, included as driving forces in the current model. The current is calculated by the model S21 of the Danish Hydraulic Institute, which solves the depth-integrated momentum and continuity equations, Eqs. 5.69-5.71 on a rectangular grid. The numerical techniques used in this modelling system are described by: Abbott et al. (1973), Abbott (1979) and Abbott et al. (1981).

The model area is represented by a rectangular grid, with a local bed level at each grid point. In the present case the model area of 1700 × 750 m is resolved

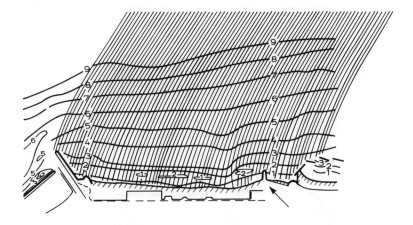

Figure 5.26 Computed wave refraction pattern at Pesaro. Waves from the
east. From Madsen et al. (1987).

in a mesh size of 25 m along the coast and 10 m perpendicular to the coast. The
flow equations are discretized into finite difference equations in this grid, and the
equations are solved from one time step to the next, using an implicit numerical
scheme. In this case a time step of 2.5 s is used, about 1000 s are required to
obtain steady state conditions.

Fig. 5.27 shows the calculated steady current pattern, driven by the waves
of Fig. 5.26. Fig. 5.27A represents the existing situation, while Fig. 5.27B shows
a situation with submerged, detached breakwaters, as indicated on the figure. The
top of the breakwaters is 0.8 m below the undisturbed still water level. It can be
seen how the breakwaters cause the location of maximum velocities to be shifted
offshore, because the waves now break on the breakwaters. A system of vortices
is formed in the lee of the breakwaters. These vortices are related to the gaps
between the breakwaters.

5.4   Low frequency oscillations

Until now, the effect of irregular waves on the wave-driven currents has
been taken into account by considering only the statistical distribution of the
wave heights and averaging the contributions from the different wave heights.
This approach will give a steady wave-driven current because the sequence of the
incoming wave heights is neglected. In a train of irregular wind-generated waves

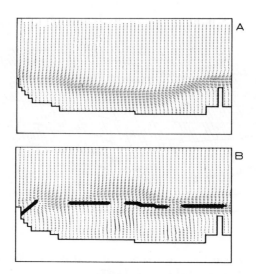

Figure 5.27 Computed wave-induced current pattern, at Pesaro. Wave situation as in Fig. 5.26. A: Present situation. B: With submerged, detached breakwaters.

the high and low waves will normally appear in groups, with typically 4–8 waves
in each group. Fig. 5.28 shows an example of a wave train where the wave groups
are illustrated by the dashed curve which is the envelope of the wind waves. The
wave groups will propagate in the same direction as the waves but with the group
velocity c_g, which is half the wave celerity for deep water waves and is equal to
the wave celerity for linear shallow water waves (cf. Chapter 1).

Infragravity waves, edge waves, surf beat

The existence of the wave groups gives rise to second order waves with
periods corresponding to the period of the groups. Due to their long periods they
are termed infragravity waves. Some of the mechanisms behind the generation of
infragravity waves are outlined in the following.

The radiation stress is a function of the wave height and is therefore varying
along the wave train in phase with the wave groups. The gradients in the radiation
stress must be balanced by variations in the mean (wave-averaged) water surface
and by accelerations of the water, resulting in a low mean water surface under
groups of high waves where the radiation stress is high and a high mean water
surface under the small waves where the radiation stress is low. This mechanism
is analogous to the wave set-down caused by wave shoaling as described in Section

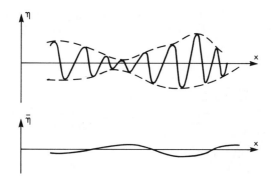

Figure 5.28 The wave groups and the associated mean water level.

4.2.1. Therefore a train of second order waves travels with the irregular waves with a celerity corresponding to the group velocity of the irregular waves and with a wave length of the same order as the length of the wave groups. These low-period waves are called bound long waves. As the wave train approaches the shore, the height of the bound long waves will drastically increase as the group velocity approaches the phase velocity of the short period waves in the shallow water. The significance of the bound long waves is therefore increased as the waves progress into shallower water (Longuet-Higgins and Stewart, 1962). The existence of infragravity waves was first noticed in the nearshore area, where slow oscillations or 'surf beat' was observed (Munk, 1949 and Tucker, 1950).

When the short waves come into the surf zone and loose height and energy due to the wave-breaking, they can no longer balance the bound long waves that have been following them, and the long waves are then released from the short waves. If the short waves have approached the coast perpendicularly, the long waves will now normally be reflected from the coast and propagate in the offshore direction as free long waves. If the short waves approach the coast at an angle, the long waves may still be reflected and propagate into deeper water. In many cases, however, the large difference between the wave period of the long waves and that of the short ones makes it impossible for the reflected long waves to reach deep water. When the bound long waves are released and reflected near the surf zone, they will be refracted as they move into deeper water. This refraction is determined by their own long period and it may turn them around to move in the shore-parallel direction before they reach very deep water, as sketched in Fig. 5.29. In this case the long waves are unable to leave the coast; this is called the 'trapped mode'. If the reflected long waves can propagate into deep water, it is termed the 'leaky mode'. The reason for the long waves being trapped, even if they have approached from deep water, is that during the approach they were bound to the short waves and were thus following their refraction determined by the short wave periods. The trapped waves will then progress along the coast in the form of three-dimensional edge waves, with a wave height decreasing with the distance from the shoreline.

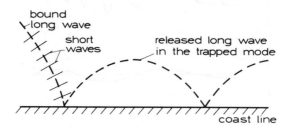

Figure 5.29 Long waves released in the trapped mode.

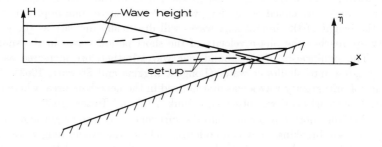

Figure 5.30 The wave height and mean water surface (set-up) for large (fully drawn lines) and small (dashed lines) waves, equilibrium conditions.

Another mechanism generating low-period waves in the trapped or the leaky mode was first considered by Symonds et al. (1982). It is related to the wave-breaking. When the irregular waves break, the force associated with the radiation stress gradient will vary in strength with the height of the breaking waves. In a quasi-steady situation this will result in a larger set-up in the surf zone when the large waves break than when the smaller waves break, cf. Fig. 5.30. In reality, the typical period of the wave groups will be so short that the situation in the surf zone is dynamic, with low frequency waves being generated by the time-varying forcing from the breaking of the irregular waves. Fig. 5.31 shows an example of a

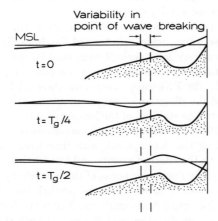

Figure 5.31 Calculated infragravity wave over a barred profile, over one half of the period of the wave groups, T_g. After Symonds and Bowen (1984).

low frequency wave generated by the time-varying location of the point of wave-breaking on a coastal profile with one bar, Symonds and Bowen (1984). A model which combines the effect of a time-varying break point position and the effect of the incident long waves was presented by Schäffer (1990). The model describes oblique incidence with the possibility of trapping as well as normal incidence. It is found that the two effects are of the same order of magnitude.

In many situations the low frequency waves may be expected to play an important role in the coastal morphology (cf. Hunt and Johns, 1963, Holman and Bowen, 1982), but until now no quantitative models describing their significance in terms of sediment transport rates have been developed, and much of the support for their importance has been indirect. For example it has been suggested that the distance from the shoreline to a longshore bar corresponds to the wave length of a standing long wave, Bowen (1980).

Far infragravity waves - instability of longshore currents

All the long waves described above are gravity waves, which means that the important forces in their dynamics are pressure, inertia, and gravity. It has therefore been possible to determine the dispersion relation, which is the relation between the wave length and the period, for these waves. Recently field measurements have detected waves that progressed in the longshore direction, but did not follow the dispersion relation for any known type of gravity wave, Oltman-Shay et al. (1989). The wave periods were of the order 10^2s which is in the low frequency domain, but the wave lengths were much too small (of the order 10^2m) to comply with any known theory for gravity-driven waves.

It has been proposed (Bowen and Holman, 1989) that an explanation for the occurrence of these waves, which are termed Far Infra Gravity Waves (FIG-waves) because of their very long periods compared to the wave length, might be an instability mechanism in the wave-driven longshore current. In the study by Oltman-Shay et al. the FIG-waves were only observed in connection with a longshore current, and the direction of propagation was always the same as the direction of the current. The stability of the longshore current was investigated by Bowen and Holman (1989) by adding a small perturbation to a steady longshore current and then finding the development with time from the depth-integrated flow equations neglecting the effect of gravity and bed friction. The flow equations were linearized so that only first order terms of the perturbation were maintained. The initial exponential growth or decay of the perturbation was then predicted by the solution to the linearized flow equations. The perturbations were taken to be periodic in the longshore direction and the wave length that gave the fastest growth rate was interpreted as representing the FIG-wave that would emerge in a given situation. The instability mechanism that gives rise to the growth of the perturbations is connected to the velocity distribution of the longshore current, and it is found that the instability only exists because the vorticity of the longshore current velocity field has a maximum and the FIG-waves predicted by this model can therefore be described as shear waves on the longshore current.

The fastest growing perturbations had a typical longshore wave length of two times the width of the surf zone and a celerity of 1/3 of the maximum longshore current velocity.

REFERENCES

Abbott, M.B. (1979): Computational Hydraulics; Elements of the Theory of Free Surface Flows. London, Pitman, 326 pp.

Abbott, M.B., McCowan, A. and Warren, I.R. (1981): Numerical modelling of free-surface flows that are two-dimensional in plan. In: Transport Models for Inland and Coastal Waters, Proc. Symp. Predictive Ability, Berkeley, CA 1980, Ed. by H.B. Fischer, London Academic Press, pp. 222-283.

Abbott, M.B., Damsgaard, A. and Rodenhuis, G.S. (1973): System 21, Jupiter, A design system for two-dimensional nearly horizontal flows. J. Hydr. Res., 11:1-28.

Andersen, O.H. and Fredsøe, J. (1983): Transport of suspended sediment along the coast. Progress Report No. 59, Inst. of Hydrodynamics and Hydraulic Engineeing, ISVA, Techn. Univ. Denmark, pp. 33-46.

Battjes, J.A. (1974): Computation of set-up, longshore currents, run-up and over-topping due to wind-generated waves. Delft University of Technology, Communications on hydraulics, 74-2.

Battjes, J.A. (1975): Modelling of turbulence in the surf zone. Modelling 75, Proc. 2. Symp. Modelling Techniques, ASCE, San Francisco, 2:1050-1061.

Battjes, J.A. (1983): Surf zone turbulence. Proc. 20. IAHR Congress, Moscow 1983, Vol. 7, Seminars, pp. 137-140.

Battjes, J.A. and Janssen, J.P.F.M. (1978): Energy loss and set-up due to breaking in random waves. Proc. 16th Coastal Eng. Conf., Hamburg, 1:569-587.

Bowen, A.J. (1969): The generation of longshore currents on a plane beach. J. Mar. Res., 27:206-215.

Bowen, A.J. (1980): Simple models of nearshore sedimentation: Beach profiles and longshore bars, in 'The Coastline of Canada'. Ed. by McCann, S.B. Paper 80-10, Geological Survey of Canada, pp. 1-11.

Bowen, A.J. and Holman, R.A. (1989): Shear instabilities of the mean longshore current, 1, Theory, J. Geophys. Res., 94:18,023-18,030.

Dalrymple, R.A. (1978): Rip currents and their causes. Proc. 16th Coastal Eng. Conf., Hamburg, 2:1414-1427.

Dalrymple, R.A. (1980): Longshore currents with wave-current interaction. J. Waterway, Port, Coastal and Offshore Div., ASCE, 106:414-420.

Deigaard, R., Fredsøe, J. and Hedegaard, I.B. (1986): Suspended sediment in the surf zone. J. Waterway, Port, Coastal and Ocean Eng., ASCE, 112(1):115-128.

Deigaard, R., Fredsøe, J. and Hedegaard, I.B. (1986a): Mathematical model for littoral drift. J. Waterway, Port, Coastal and Ocean Eng., ASCE, 112(3):351-369.

Dingemans, M.W., Radder, A.C. and de Vriend, H.J. (1987): Computation of the driving forces of wave-induced currents. Coastal Eng., 11:539-563.

Engelund, F. (1981): A simple theory of weak hydraulic jumps. Progress Report No. 54, Inst. of Hydrodynamics and Hydraulic Engineering, ISVA, Techn. Univ. Denmark, pp. 29-32.

Holman, R.A. and Bowen, A.J. (1982): Bars, bumps and holes: Models for the generation of complex beach topography. J. Geophys. Res., 87:457-468.

Hunt, J.N. and Johns, B. (1963): Currents produced by tides and gravity waves. Tellus, 15:343-351.

Jonsson, I.G., Skovgaard, O. and Jacobsen, T.S. (1974): Computation of longshore currents. Proc. 14th Coastal Eng. Conf., Copenhagen, 2:699-714.

Longuet-Higgins, M.S. (1970): Longshore currents generated by obliquely incident sea waves 1. J. Geophys. Res., 75:6778-6789.

Longuet-Higgins, M.S. (1970a): Longshore currents generated by obliquely incident waves 2. J. Geophys. Res., 75:6790-6801.

Longuet-Higgins, M.S. and Stewart, R.W. (1962): Radiation stress and mass transport in gravity waves, with application to surf beats. J. Fluid Mech., 13:481-504.

Madsen, P., Deigaard, R. and Hebsgaard, M. (1987): A coastal protection study based on physical and mathematical modelling. Proc. 2nd Int. Conf. on Coastal and Port Eng. in Developing Countries, Beijing, 1:223-239.

Munk, W.H. (1949): Surf beats. Trans. Am. Geophys. Union, 30:849-854.

Oltman-Shay, J., Howd, P.A. and Birkemeier, W.A. (1989): Shear instabilities of the mean longshore current, 2, Field observations. J. Geophys. Res., 94:18,031-18,042.

Schäffer, H.A. (1990): Infragravity water waves induced by short wave groups. Series paper No. 50, Inst. of Hydrodynamics and Hydraulic Engineering, ISVA, Techn. Univ. Denmark, 168 pp.

Southgate, H.N. (1989): A nearshore profile model of wave and tidal current interaction. Coastal Eng., 13:219-246.

Stive, M.J.F. and Wind, H.G. (1986): Cross-shore mean flow in the surf zone. Coastal Eng., 10(4):325-340.

Symonds, G., Huntley, D.A. and Bowen, A.J. (1982): Two-dimensional surf beat: long wave generation by a time-varying break point. J. Geophys. Res., 87(C1):492-498.

Symonds, G. and Bowen, A.J. (1984): Interaction of nearshore bars with incoming wave groups. J. Geophys. Res., 89(C2):1953-1959.

Thornton, E.B. (1970): Variation of longshore current across the surf zone. Proc. 12th Coastal Eng. Conf., Washington D.C., 1:291-308.

Tucker, M.J. (1950): Surf beats: sea waves of 1 to 5 min. period. Proc. Roy. Soc. London, A202:565-573.

Whitford, D.J. and Thornton, E.B. (1988): Longshore current forcing at a barred beach. Proc. 21st Coastal Eng. Conf., Malaga, 1:71-90.

Zyserman, J. (1989): Characteristics of stable rip-current systems on a coast with a longshore bar. Series paper No. 46, Inst. of Hydrodynamics and Hydraulic Engineering, ISVA, Techn. Univ. Denmark, 178 pp.

Chapter 6. Current velocity distribution in the surf zone

In Chapter 5 wave-driven currents were considered with the emphasis on the depth-integrated velocities. In this chapter their detailed three-dimensional structure will be analyzed. This concerns not only the velocity distribution of the depth-integrated wave-driven currents, which is of importance for the magnitude of the suspended sediment transport, but also the secondary currents normal to the mean flow direction, which become zero when integrated over the depth. Because of these secondary currents, the velocity vectors will deviate from the shore-parallel direction of the depth-integrated longshore current. In the surf zone, the velocity near the surface has an onshore component, while the near-bed velocity has an offshore component. The secondary circulation currents are therefore important for the sediment transport normal to the coastline and for the development of the coastal profile.

In hydraulics the discharge of a pipe or a conduit can be calculated from only the driving pressure gradient and the friction factor. However, if the details of the velocity distribution are required, it is necessary to determine the shear stress distribution over the cross section of the pipe. Similarly, the first step for determining the velocity distribution of wave-driven currents is to analyse the shear stress distribution.

6.1   Normally incident waves, shear stress distribution

As a starting point, the situation with normally incident waves on a uniform coast is considered. This equates to the situation represented in a wave flume. It was first noted by Dyhr-Nielsen and Sørensen (1970) that shear stresses and currents could be caused by the variation in the wave height across the surf zone. Their argument can be briefly summarized as follows:

The two most important forces integrated over a wave period in the surf zone are the radiation stress gradient and the pressure gradient due to the set-up, cf. Section 4.2.1, where the wave set-up is calculated for the condition where the two forces balance each other.

This balance is, however, not perfect. If the two forces do not have the same vertical distribution, shear stresses must be introduced in order to obtain a balance, and these shear stresses will then be associated with a mean current velocity profile.

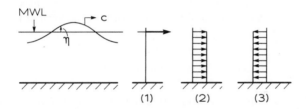

Figure 6.1 The vertical distribution of the horizontal forces in the surf
zone. (1) The pressure part of the radiation stress. (2) The
momentum part of the radiation stress. (3) The pressure gradient due to the set-up.

The distribution of the two forces is illustrated in Fig. 6.1 for the case of linear shallow water waves. The pressure gradient due to the wave set-up is evenly distributed over the vertical

$$\frac{\partial p}{\partial x} = -\rho g S \qquad (6.1)$$

The momentum part of the radiation stress is also evenly distributed over the vertical, as the horizontal orbital velocity is constant over the depth. The pressure part of the radiation stress is, however, located at the surface, between the wave top and wave trough levels. This can be seen from the time-averaged pressure, which under trough level is equal to the hydrostatic pressure under an undisturbed water surface at mean water level. The radiation stress due to the pressure must therefore be located above the wave trough level.

In this way there is an indication that the waves in the surf zone induce a shear stress distribution and a mean circulation current. A mathematical model using these principles was developed by Dally and Dean (1984). Svendsen (1984A) introduced the effect of the surface rollers in the model, and Stive and Wind (1986) have made careful experimental investigations of the different terms in the momentum balance. Deigaard and Fredsøe (1989) made a modification of the theoretical shear stress distribution by including an important contribution from the vertical flux of the horizontal momentum.

In the following, the vertical distribution of the shear stress in the surf zone is derived. It is assumed that the energy dissipation due to the wave-breaking is totally dominant over the dissipation in the near-bed wave boundary layer.

The water motion

The water motion in the waves is described according to the model outlined in Section 4.2. The pressure is assumed to be hydrostatic, and the horizontal orbital velocity u is constant over the depth. The wave profile is described as a cosine function. This is a simplification, but other wave profiles may be introduced by use of Fourier series, and the final result, in the form of the shear stress distribution, is not affected by the wave profile.

$$\eta = \frac{H}{2} \cos(kx - \omega t) \tag{6.2}$$

where η is the water surface elevation, H is the wave height, k is the wave number, and ω is the cyclic frequency. Here and in the following only the lowest non-zero term in (H/D) is maintained. In addition to the wave motion, surface rollers, each with the cross-sectional area A, follow each wave front, travelling with the phase velocity c of the waves.

When calculating the orbital velocity u it is important to include the effect of the variation in the wave height. Due to the energy dissipation the wave height decreases towards the shore. Over a short distance a linear approximation can be applied

$$H = H_0 + xH'_x \tag{6.3}$$

where $x = 0$ at the location considered, and $H = H_0$ at $x = 0$. The continuity equation integrated over the depth reads

$$D\frac{\partial u}{\partial x} = -\frac{\partial \eta}{\partial t} \tag{6.4}$$

Eqs. 6.2, 6.3 and 6.4 give the following expression for u

$$u = \frac{Hc}{2D(1 + (H_x'/kH)^2)} \left(\cos(kx - \omega t) - \frac{H_x'}{kH} \sin(kx - \omega t) \right)$$

$$\approx \frac{Hc}{2D} \left(\cos(kx - \omega t) - \frac{H_x'}{kH} \sin(kx - \omega t) \right) \tag{6.5}$$

Eq. 6.5 can easily be verified by insertion in Eq. 6.4. It has been assumed that the variation in the wave height is weak

$$\frac{H_x'}{kH} \ll 1 \tag{6.6}$$

Eq. 6.5 shows that there is a small phase difference between u and η because of the non-uniform wave conditions: when the wave height decreases towards the coast, water is transferred from the wave tops to the wave troughs, and u cannot be exactly zero, where η is zero. It should be noted that Eqs. 6.2 and 6.5 do not fulfill the flow equation

$$\frac{\partial u}{\partial t} = -g \frac{\partial \eta}{\partial x} \tag{6.7}$$

because the pressure from the surface rollers should be included in the flow equation. It is the pressure from the surface rollers that extracts the energy from the waves and thus gives the gradient in the wave height.

As u is constant over the vertical, the continuity equation gives a vertical wave-orbital velocity which increases linearly with the distance from the bed

$$w = \frac{z}{D} \frac{\partial \eta}{\partial t} = \frac{z\omega H}{2D} \sin(kx - \omega t) \tag{6.8}$$

In addition to the wave motion there is a small slope on the mean water surface: the wave set-up, cf. Section 4.2.1. The actual wave surface elevation is therefore

$$\eta = \frac{H}{2} \cos(kx - \omega t) + Sx \tag{6.9}$$

where S is the set-up, which is of the order $O((H/D)^2)$.

The local thickness of the surface roller is η^+, and the cross-sectional area of the roller is given by

$$A = c \int_0^T \eta^+ \, dt = cT \, \overline{\eta^+} \tag{6.10}$$

The shear stress distribution

From the presented description of wave kinematics the shear stress distribution can be found by using the momentum equation on the control surface shown in Fig. 6.2. The control surface is fixed in space and has horizontal and vertical sides. Its width is dx, and it extends from the level z into the air above the wave.

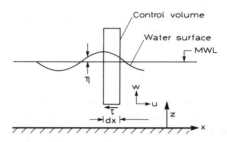

Figure 6.2 Control volume (Eulerian description) to which the momentum equation is applied.

The momentum equation is a vector equation which contains integrals over the volume and the surface area of the control surface

$$\int_V \rho \frac{\partial \vec{u}}{\partial t} dV =$$

$$-\int_A \rho \vec{u} \left(\vec{u} \cdot d\vec{A} \right) - \int_A p d\vec{A} + \int_V \rho \vec{g} dV + \int_A d\vec{T} \qquad (6.11)$$

where $d\vec{A}$ is the area vector of a surface element, directed out of the surface, and $d\vec{T}$ is the force from the shear stress acting on a surface element. Eq. 6.11 states that the acceleration of the mass in the volume of the control surface is equal to the sum of the momentum flux through the surface, the pressure force, the gravity, and the shear stress force acting on the control volume.

In the following, only the horizontal projection of Eq. 6.11 is used, and the equation is averaged over one wave period. Only the lowest order terms in (H/D) are maintained. It should be noted that the control surface extends into the air, and that all terms in Eq. 6.11 become zero above the instantaneous level of the water surface. Each of the elements in Eq. 6.11 can now be evaluated:

The left-hand-side becomes zero after time-averaging, because the waves are periodic, i.e. no momentum is added to the system during one wave period.

The time-averaged momentum flux can be written

$$\vec{e} \cdot \overline{\int_A \rho \vec{u} \, (\vec{u} \cdot d\vec{A})} = - \overline{\rho u u (D - z + \eta)} - \overline{\rho c c \eta^+}$$

$$+ \overline{\rho \left(u + \frac{\partial u}{\partial x} dx\right)^2 \left(D - z + \eta + \frac{\partial \eta}{\partial x} dx\right)} + \overline{\rho c c \eta^+} + \overline{\rho \frac{\partial}{\partial x}(c c \eta^+) dx} - \overline{\rho u w dx} \approx$$

$$\rho \left\{ \frac{d}{dx}(\overline{u^2}) \, (D - z) + \frac{d}{dx}\left(\frac{\overline{c^2 A}}{cT}\right) - \overline{uw} \right\} dx \approx$$

$$\rho \left\{ \frac{d}{dx}\left(\overline{\left(\left(\frac{Hc}{2D} \cos(kx - \omega t)\right)^2\right)}\right)(D - z) + \frac{d}{dx}\left(\frac{cA}{T}\right) \right.$$

$$\left. - \overline{\frac{Hc}{2D}\left(\cos(kx - \omega t) - \frac{H_x'}{kH}\sin(kx - \omega t)\right)\frac{z \omega H}{2D}\sin(kx - \omega t)} \right\} dx =$$

$$\rho \left\{ \frac{2H \, H_x' g}{4D}\frac{1}{2}(D - z) + \frac{d}{dx}\left(\frac{cA}{T}\right) + \frac{c H_x'}{2Dk}\frac{z \omega H}{2D}\frac{1}{2} \right\} dx =$$

$$\left\{ \rho \frac{d}{dx}\left(\frac{cA}{T}\right) + \rho g H H_x'\left(\frac{1}{4} - \frac{1}{8}\frac{z}{D}\right) \right\} dx \tag{6.12}$$

where $\vec{e}$ is a horizontal unit vector.

The pressure is found from the hydrostatic pressure distribution

$$\vec{e} \cdot \overline{\int_A p \, d\vec{A}} = -\frac{1}{2} \, \overline{\rho g \left(\eta^+ + \eta + D - z\right)^2}$$

$$+ \frac{1}{2}\rho g \overline{\left(\eta^+ + \frac{\partial \eta^+}{\partial x} dx + \eta + \frac{\partial \eta}{\partial x} dx + D - z\right)^2} \approx$$

$$\rho g \overline{\left(\frac{\partial \eta^+}{\partial x} + \frac{\partial \eta}{\partial x}\right)\left(\eta^+ + \eta + D - z\right)} dx =$$

$$\rho g \overline{\left(\frac{\partial \eta}{\partial x}\eta + \left(\frac{\partial \eta^+}{\partial x} + \frac{\partial \eta}{\partial x}\right)(D - z) + \frac{\partial \eta^+}{\partial x}\eta^+ + \frac{\partial}{\partial x}(\eta \eta^+)\right)} dx \tag{6.13}$$

The two last terms describe the gradient in the pressure force due to the shoreward change in the surface roller geometry, and have been demonstrated to be small by Svendsen (1984). The rollers are important because of their large velocity, but their volume is small compared to the volume of the wave profile. Using the expression for the wave profile, Eq. 6.2, the pressure force is calculated

$$\vec{e} \cdot \int_A p d\vec{A} = \rho g \overline{\left(\eta \frac{\partial \eta}{\partial x} + \left(\frac{\partial \eta}{\partial x} + \frac{\partial \eta^+}{\partial x} \right)(D - z) \right)} dx =$$

$$\rho g \overline{\Big(\big(Sx + \frac{H}{2} \cos(kx - \omega t) \big) \big(S + \frac{H'_x}{2} \cos(kx - \omega t) - \frac{Hk}{2} \sin(kx - \omega t) \big)}$$

$$+ S(D - z) \Big) dx \approx$$

$$\rho g \left(\frac{H H'_x}{8} + S(D - z) \right) dx \tag{6.14}$$

The gravity force becomes zero as a horizontal projection is used. The only shear stress force which is non-zero is due to the shear stress acting on the horizontal bottom of the control surface

$$\vec{e} \cdot \overline{\int_A d\vec{T}} = -\,\overline{\tau} dx \tag{6.15}$$

where $\overline{\tau}$ is the shear stress being described.

The momentum equation can now be written

$$0 = -\rho \frac{d}{dx}\left(\frac{cA}{T}\right) dx - \rho g H H'_x \left(\frac{1}{4} - \frac{1}{8}\frac{z}{D}\right) dx$$

$$-\rho g \left(\frac{H H'_x}{8} + S(D - z) \right) dx - \overline{\tau}\, dx \tag{6.16}$$

or

$$\overline{\tau} = -\frac{\rho g}{8} \frac{d(H^2)}{dx} \left(1 + \frac{D - z}{2D} \right) - \frac{\rho}{T}\frac{d(cA)}{dx} - S\rho g D \frac{D - z}{D} =$$

$$-\frac{1}{c}\frac{dE_{fw}}{dx}\left(1 + \frac{D - z}{2D} \right) - \frac{\rho}{T}\frac{d(cA)}{dx} - S\rho g D \frac{D - z}{D} \tag{6.17}$$

where E_{fw} is the energy flux of the wave motion.

The mean shear stress in the surf zone is thus composed of three contributions:

The first is due to the pressure and momentum fluxes associated with the decaying wave motion in the surf zone. It has a trapezoidal shape.

The second is due to the change of momentum in the surface rollers, and is constant over the depth.

The third is due to the slope S of the mean water surface, the wave set-up. It has the triangular distribution which is well known from uniform channel flow in hydraulics.

The three contributions are illustrated in Fig. 6.3 for the special case where the wave set-up exactly balances the radiation stress gradient, giving a mean bed shear stress of zero. The calculations have been made for a wave height $H/D = 0.6$ and a wave period $T^2 g/D = 300$. The surface roller volume has been calculated according to the model by Engelund (1981), cf. Example 4.2.

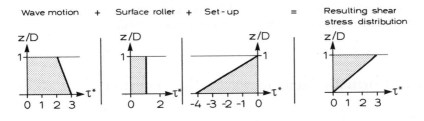

Figure 6.3 Variation in $\tau^* = -\overline{\tau} \Big/ \left(\frac{1}{16}\rho g \frac{d(H^2)}{dx}\right)$ with depth.

It is clear from Fig. 6.3 that with a non-zero near-surface shear stress it is impossible to find a value for the wave set-up which balances the radiation stress gradient without causing shear stresses over the vertical. These shear stresses will in turn drive a circulation current.

Near the mean water surface (at the wave trough level), $z \approx D$ the shear stress becomes

$$\overline{\tau}_s = -\frac{1}{c}\frac{dE_{fw}}{dx} - \frac{\rho}{T}\frac{d(cA)}{dx} =$$
$$\frac{1}{c}\overline{D} - \frac{\rho}{T}\frac{d(cA)}{dx} \qquad (6.18)$$

where $\overline{D}$ is the rate of loss of wave energy per unit bed area.

In Eq. 6.18 the first contribution is due to the pressure and momentum flux of the wave motion, and the second, which was first recognized by Svendsen (1984), is due to a change in the momentum of the surface rollers. This expression is generally applicable for deep water or shallow water waves, and for a non-uniform steady wave condition as considered here or for an unsteady wave condition, e.g. waves decaying after a storm. For a uniform unsteady wave situation the near-surface mean shear stress can be written as (cf. Phillips, 1977; Longuet-Higgins, 1969; Liu and Davis, 1977)

$$\overline{\tau}_s = -\frac{1}{c}\frac{dE}{dt} - \frac{\rho}{L}\frac{d(cA)}{dt} \qquad (6.19)$$

where E is the wave energy per unit bed area (cf. Eqs. 1.35 and 1.37)

$$E = \frac{1}{8}\rho g H^2 \tag{6.20}$$

It must, however, be kept in mind that these expressions (Eqs. 6.18 and 6.19) are only valid for waves where the energy dissipation occurs near the water surface. If the energy dissipation is located at another level, the shear stress distribution will be different, as discussed in Example 6.1. For instance, in the case of energy dissipation in a near-bed wave boundary layer, it is in fact possible to find a value of the wave set-up which does not give any shear stresses over the entire water depth outside the wave boundary layer itself.

6.2 Normally incident waves, the undertow

Consider a steady situation with regular waves breaking on a coast. For normally incident waves the vertical distribution of the mean shear stress in the broken waves was determined in the previous section. In the surf zone the waves cause an onshore directed flow due to two mechanisms, the wave drift, described in Example 1.2, and the surface rollers carrying water shorewards. In a strictly two-dimensional situation the shore-normal discharge is zero. Therefore the shoreward discharge caused by the waves must be compensated by a current in the offshore direction. While the shoreward discharge lies near the mean water surface (rollers and Eulerian drift), the return current has its maximum near the bed. This circulating current with its offshore-directed flow near the bed is called the undertow. A detailed determination of its velocity distribution can be made by solving the flow equations taking the mean shear stress distribution into account. The undertow is limited to the surf zone, partly because outside the surf zone no surface rollers transport water towards the coast, and partly because the energy dissipation occurring mainly in the near-bed wave boundary layer is weak and does not cause shear stresses outside the wave boundary layer. The situation with energy dissipation in a boundary layer is treated in Example 6.1 and in Chapter 11. In the following the distribution of the undertow is determined.

If the higher order convective terms are neglected and linear shallow water wave theory is used, the horizontal flow equation can be written as

$$\frac{\partial u}{\partial t} = -\frac{1}{\rho}\frac{\partial \tilde{p}}{\partial x} + \frac{1}{\rho}\frac{\partial}{\partial z}(\tau - \overline{\tau}) \tag{6.21}$$

where u is the horizontal flow velocity, $\tilde{p}$ is the periodic pressure variation due to the wave motion, and τ is the turbulent shear stress. $\overline{\tau}$ is the time-averaged shear stress, which is given by Eq. 6.17. The periodic pressure variation $\tilde{p}$ can be found from the wave-orbital velocity u_0 outside the wave boundary layer

$$-\frac{1}{\rho}\frac{\partial \tilde{p}}{\partial x} = \frac{\partial u_0}{\partial t} \qquad (6.22)$$

It is seen from Eq. 6.22 that the time average of $\tilde{p}$ is zero. That is because the mean pressure gradient associated with the mean water slope is included in the expressions for the mean shear stress (cf. Eq. 6.17).

The instantaneous shear stress is determined through the eddy viscosity

$$\frac{\tau}{\rho} = \nu_t \frac{\partial u}{\partial z} \qquad (6.23)$$

Thus Equation 6.21 can be written as

$$\frac{\partial(u - u_0)}{\partial t} = \frac{\partial}{\partial z}\left(\nu_t \frac{\partial u}{\partial z}\right) - \frac{1}{\rho}\frac{\partial \overline{\tau}}{\partial z} \qquad (6.24)$$

It is seen that, except for the term $\overline{\tau}$, Eq. 6.24 is identical to the equation used for the calculation of the wave boundary layer, Eq. 3.55, and that it is similar to the equation used for analysing the combined wave and current boundary layer, where $\overline{\tau}$ is caused only by mean water slope. Equation 6.24 can thus be used over the entire water column, including the oscillatory boundary layer near the bed. For a hydraulically rough bed with a roughness k_N the boundary condition is

$$u = 0 \quad \text{for} \quad z = k_N/30 \qquad (6.25)$$

and the near-surface boundary condition is

$$\overline{\nu_t \frac{\partial u}{\partial z}}\bigg|_s = \overline{\tau_s} \qquad (6.26)$$

where the subscript s indicates a near-surface value.

By taking the time-average of Eq. 6.24 it is seen that the turbulent stresses balance the driving forces

$$\overline{\nu_t \frac{\partial u}{\partial z}} = \frac{1}{\rho}\overline{\tau} \qquad (6.27)$$

In some cases Equation 6.24 is used to determine the near-bed current velocity profile assuming the eddy viscosity and the flow velocity are independent

$$\frac{\partial \overline{u}}{\partial z} = \frac{1}{\rho \overline{\nu_t}}\overline{\tau} \qquad (6.28)$$

This equation can, however, only be used as a rather crude first approximation, because the eddy viscosity and flow velocity will be correlated in the near-bed wave boundary layer, and a complete solution can only be obtained through Eq. 6.24.

Determining the set-up

The solution of Eq. 6.24 will give a velocity distribution of the mean current for a given distribution of the driving force $\bar{\tau}$. The distribution of $\bar{\tau}$ depends on the slope of the mean water surface, the wave set-up S. In the present case S is an unknown parameter which must be determined by considering the depth-integrated continuity equation. The net discharge is calculated from the mean flow velocity integrated over the depth

$$q = \int_0^D \bar{u}\,dz \qquad (6.29)$$

The net discharge will vary with $\bar{\tau}$ and is thus a function of S. The actual value of the mean water slope S is the one that gives the correct net cross-shore discharge. In the strictly two-dimensional case the net cross-shore discharge is zero, which is the situation represented by a wave flume. Zero cross-shore discharge does not imply that $\bar{q}$ is also equal to zero, because the wave motion itself gives a discharge which must be compensated by the mean current.

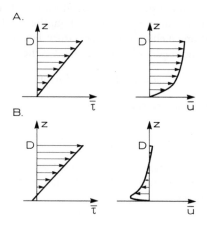

Figure 6.4 The mean shear stress and mean velocity distribution. A: Exact balance between set-up and radiation stress gradient. B: Increased wave set-up.

The wave motion is associated with a mass drift. For linear shallow water waves the drift can (cf. Example 1.2) be expressed by

$$q_{\text{drift}} = D\overline{u^2}/c = cBH^2/D \qquad (6.30)$$

where B is the shape factor of the wave profile, cf. Section 4.2. In addition to the mass drift, water is transported in the surface rollers, as found by Svendsen (1984), giving a mean discharge of

$$q_{\text{roller}} = A/T \qquad (6.31)$$

In order to obtain a zero discharge the circulation must balance these two contributions

$$q = -\left(q_{\text{drift}} + q_{\text{roller}}\right) \qquad (6.32)$$

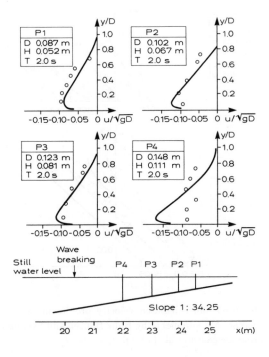

Figure 6.5 Measured and calculated mean current, zero net flow. Measurements by Svendsen et al. (1987). From Deigaard et al. (1991).

How this is achieved is illustrated in Fig. 6.4. First the shear stress distribution shown in Fig. 6.3 is considered, where the set-up exactly balances the radiation stress gradient. In this case the time-averaged shear stress is zero at the bed and positive over the entire vertical. The mean current velocity is also positive (on-shore) over the vertical, and q is positive. Equation 6.32 is therefore not satisfied with this mean shear stress distribution. By increasing the set-up the near-bed shear stresses become negative, and a circulation current is generated with an offshore-directed near-bed velocity. The equilibrium set-up then makes the circulation current fulfill the continuity equation (6.32). The equilibrium set-up is only slightly different (of the order 10%) from the set-up which balances the radiation stress gradient. The traditional method for calculating the mean water level, which is based on this balance, can therefore be used as a good approximation, as long as no information on the velocity distribution is required.

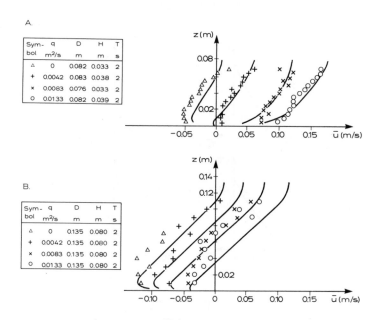

Figure 6.6 Measured and calculated velocity profiles in case of a mean
            current. Measurements by Buhr Hansen and Svendsen (1986).
            From Deigaard et al. (1991).

Fig. 6.5 shows a comparison between measured undertow velocities and calculated velocity profiles. The measurements have been made by Svendsen et al. (1987) in a closed wave flume with zero mean flow. The beach profile had a constant slope of 0.029. The experimental set-up is also sketched in Fig. 6.5.

The mean shear stress is determined by the method described in Section 6.1, with the small contribution from the streaming in the wave boundary layer included in the mean shear stress, cf. Section 2.4. The volumes of the surface rollers are determined by Engelund's (1981) model, cf. Example 4.2. The eddy viscosity is determined by the one-equation model described in Section 4.3.2 with a coupled modelling of the turbulence generated by the broken waves and in the near-bed wave boundary layer. The flow has been calculated by the unsteady flow equation 6.24, and the mean current is found by time-averaging the velocities. The wave profile has been assumed to be described by the saw-toothed profile, corresponding to a B-value of 1/12, cf. Section 4.2. It is seen from Fig. 6.5 that the agreement between the measured and calculated velocity profiles is good for the three inner cross sections, while it is less good for the cross section P4 closest to the breaking point.

Experiments with a net current have been carried out by Buhr Hansen and Svendsen (1986) in a wave flume with a bar profile. Water was pumped from the trough inshore of the bar to an offshore point near the wave generator, creating an onshore-directed net current. Tests have been carried out for four different discharges : $q = 0.0042$ m^2/s, $q = 0.0083$ m^2/s, $q = 0.0133$ m^2/s and q equal zero. The comparison has been made for two points, A: at the crest of the bar with zero bed slope, and B: 3.6 m offshore of the crest, with a bed slope of 0.028. Generally the agreement between measured and calculated profiles is good, especially at the crest, where the net discharge is relatively more significant because of the smaller water depth.

In these examples it has been assumed that on a time scale larger than the wave period the water motion is steady. No model has yet been developed to describe the vertical distribution of the velocities in an unsteady motion, as for instance in the case of low period oscillations associated with irregular incoming waves. The results derived in this section can, however, still be used in connection with an unsteady model to give the variation in time and space of the driving forces due to the energy dissipation in the broken waves.

Example 6.1: Energy dissipation in the wave boundary layer

In Section 6.1 it was stated that the calculated mean shear stress distribution is only valid for situations where the energy dissipation takes place near the mean water surface. This is the case for broken waves where the near-surface shear stress is directly related to the energy loss, cf. Eq. 6.18. In this example the opposite situation is discussed: waves with energy dissipation taking place only in the near-bed wave boundary layer.

The mean shear stress can be determined by the same principles for both cases, but there are significant differences in the water motion, which change the vertical transfer of horizontal momentum. As described in Section 2.4 the non-uniformity of the wave boundary layer causes a vertical displacement of the fluid above it.

The displacement gives an additional vertical velocity outside the wave boundary layer of (cf. Eq. 2.94)

$$w_\infty = -\frac{\tau_b}{\rho c} \tag{6.33}$$

where τ_b is the instantaneous bed shear stress. For linear shallow water waves this additional velocity is constant over the vertical outside the wave boundary layer and gives a deformation of the wave surface. Inside the wave boundary layer the displacement decreases becoming zero at the bed. By combining w_∞ with the horizontal orbital velocity an extra contribution to the $\overline{uw}$ term in the momentum equation is found

$$\overline{uw_\infty} = -\frac{\overline{\tau_b u}}{\rho c} \tag{6.34}$$

The numerator in Eq. 6.34 is, however, equal to the energy dissipation in the wave boundary layer, cf. Chapter 2, and can therefore be expressed by the wave energy flux

$$\overline{\tau_b u} = -\frac{dE_{fw}}{dx} \tag{6.35}$$

By including this additional $\overline{uw}$ term in the momentum equation, the shear stress distribution, given by Eq. 6.17, is modified to

$$\overline{\tau} = -S\rho g D\left(1 - \frac{z}{D}\right) - \frac{1}{2c}\frac{dE_{fw}}{dx}\left(1 - \frac{z}{D}\right) \tag{6.36}$$

with the surface roller contribution also being omitted, as only non-broken waves are considered.

In this situation, with energy dissipation near the bed, it can be seen that the mean shear stress becomes zero near the mean water surface, and that the set-up can now balance the momentum flux due to the wave height decay. For a set-up given by

$$S = -\frac{1}{2c\rho g D}\frac{dE_{fw}}{dx} = -\frac{1}{8}\frac{H}{D}\frac{dH}{dx} \tag{6.37}$$

the shear stress becomes exactly zero over the entire water depth outside the wave boundary layer. Inside the wave boundary layer the mean shear stress increases towards the bed due to the streaming, as described in Section 2.4.

It may be noted that the set-up given by Eq. 6.37 is only one third of the set-up required to balance the radiation stress gradient, as described in Section 4.2.1. The remaining two thirds of the radiation stress gradient are balanced by the mean bed shear stress of the streaming.

Example 6.2:   Three-dimensional shear stress distribution

Until now only the two-dimensional situation, which can be represented in a wave flume, has been treated. The same principles can, however, be used to determine the mean shear stress distribution in a three-dimensional situation. In the following the simplest three-dimensional shear stress distribution is discussed, but not all details of the calculations are worked out.

A uniform coastal profile with waves approaching at an angle α is now considered. At the location in question, the bed is horizontal, and the waves are broken as they progress through the section of interest. The applied coordinate systems are shown in Fig. 6.7. The x-axis is perpendicular to the coastline, with the positive direction onshore. The y-axis is shore-parallel, and the z-axis is vertical with origin at the bed.

The waves approach the coast with an angle α between the wave crests and the y-axis. The x', y' coordinate system is aligned with the waves, the x'-axis being in the direction of wave propagation and the y'-axis parallel to the wave crests.

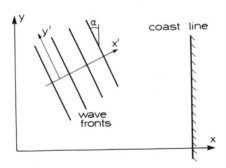

Figure 6.7 The coordinate systems used for the three-dimensional analysis.

The water motion

The water motion in the waves is analysed with the same assumptions as used in the two-dimensional case. The horizontal orbital velocity in the direction of propagation $u_{x'}$ is therefore similar to Eq. 6.5

$$u_{x'} = \frac{Hc}{2D}\left(\cos(kx' - \omega t) - \frac{\partial H}{\partial x'}\frac{1}{kH}\sin(kx' - \omega t)\right) \qquad (6.38)$$

The vertical orbital velocity is given by

$$w = \frac{\partial \eta}{\partial t} \frac{z}{D} = \frac{H\omega}{2} \frac{z}{D} \sin(kx' - \omega t) \tag{6.39}$$

The two orbital velocity components are thus similar to the results obtained from the two-dimensional analysis, but the three-dimensionality gives a small, but significant correction to the velocity in the y'-direction. In the two-dimensional case the velocity perpendicular to the direction of propagation is of course zero. In the present case, the wave height is decreasing in the shoreward, x-direction, and as the conditions are uniform along the coast, the wave height H is constant for a given value of x. Because of the angle between the wave crests and the y-axis, this means that the wave height varies along the crests of the waves. The water surface slope along the wave crests gives a velocity component, which can be determined by the flow equation

$$\frac{\partial v_{y'}}{\partial t} = -g \frac{\partial \eta}{\partial y'} = -g \frac{\eta}{H} \frac{\partial H}{\partial y'} =$$

$$-\frac{g}{2} \frac{\partial H}{\partial y'} \cos(kx' - \omega t) = \frac{g}{2} H'_x \sin \alpha \cos(kx' - \omega t) \tag{6.40}$$

giving

$$v_{y'} = -\frac{g H'_x}{2\omega} \sin \alpha \sin(kx' - \omega t) \tag{6.41}$$

assuming that H is only a function of x, and that

$$\frac{\partial y'}{\partial x} = -\sin \alpha \tag{6.42}$$

The wave-orbital motion is thus seen to form ellipses in the horizontal plane due to the wave height decay towards the coast. The orbital velocities in the x- and y-direction are found by projection of $u_{x'}$ and $v_{y'}$

$$u = u_{x'} \cos \alpha - v_{y'} \sin \alpha =$$

$$\frac{Hc}{2D}\left(\cos(kx' - \omega t) - \frac{H'_x}{kH} \cos \alpha \sin(kx' - \omega t)\right) \cos \alpha$$

$$+ \frac{g H'_x}{2\omega} \sin^2 \alpha \sin(kx' - \omega t) \tag{6.43}$$

$$v = u_{x'} \sin \alpha + v_{y'} \cos \alpha =$$

$$\frac{Hc}{2D}\left(\cos(kx' - \omega t) - \frac{H'_x}{kH} \cos \alpha \sin(kx' - \omega t)\right) \sin \alpha$$

$$- \frac{g H'_x}{2\omega} \sin \alpha \cos \alpha \sin(kx' - \omega t) \tag{6.44}$$

In addition to the wave-orbital motion the rollers carry water with each wave crest. The water in the rollers has a velocity of c in the x'-direction and a mean specific discharge of A/T.

The shear stress distribution

The shear stresses are determined by establishing the momentum equation for the box-shaped control volume shown in Fig. 6.8.

The horizontal shear stresses, τ_{zx} and τ_{zy}, are determined by projecting the momentum equation on to the x- and y-axis. The new terms in the momentum balance relative to the two-dimensional analysis are the momentum fluxes through the control surface described by the term $\overline{uv}$. It is assumed that the shear stresses acting on the vertical sides of the control surface are negligible compared to τ_{zx} and τ_{zy}. This assumption is justifiable over large parts of the surf zone.

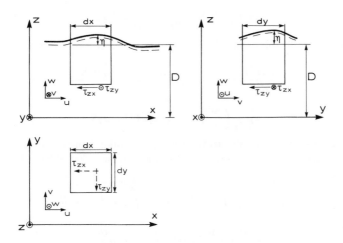

Figure 6.8 The control surface used to determine the shear stresses.

By using the approximations introduced in the two-dimensional analysis, the projection of the momentum equation on the x-direction can be written

$$\tau_{zx} = -\rho\frac{\partial}{\partial x}\overline{\int_z^D u^2 dz} - \rho\frac{\partial}{\partial y}\overline{\int_z^D uv\,dz} + \rho\overline{uw}$$

$$-\rho\frac{\partial}{\partial x}\left(\frac{Ac}{T}\right)\cos^2\alpha - \rho g\eta\overline{\frac{\partial\eta}{\partial x}} - \rho g(D-z)S \qquad (6.45)$$

Each term in Eq. 6.45 can be evaluated by the expressions for water motion. The first term on the right-hand-side is calculated in a similar manner to the two-dimensional case

$$\rho\frac{\partial}{\partial x}\overline{\int_z^D u^2 dz} = \left(1-\frac{z}{D}\right)\frac{\rho g}{8}2HH_x'\cos^2\alpha \qquad (6.46)$$

The second term is zero, due to the assumption of uniform conditions along the coast. The third term is calculated from Eqs. 6.39 and 6.43

$$\rho \, \overline{uw} = \rho \, \frac{H\omega}{2} \frac{z}{D} \left(-\frac{Hc}{2D} \frac{H'_x}{kH} \cos^2 \alpha + \frac{gH'_x}{2\omega} \sin^2 \alpha \right) \frac{1}{2} =$$

$$\frac{z}{D} \frac{\rho g}{16} 2HH'_x \left(-\cos^2 \alpha + \sin^2 \alpha \right) \tag{6.47}$$

The fourth term is not manipulated further, and the fifth, again similar to the two-dimensional case, gives

$$\rho g \eta \frac{\partial \eta}{\partial x} = \frac{\rho g}{2} \frac{\partial (\eta^2)}{\partial x} = \frac{\rho g}{16} 2HH'_x \tag{6.48}$$

Combining all the terms gives the following expression for τ_{zx}

$$\tau_{zx} = -\left(1 - \frac{z}{D} \right) \frac{\rho g}{8} \frac{d(H^2)}{dx} \cos^2 \alpha - \frac{\rho g}{16} \frac{d(H^2)}{dx} + \frac{z}{D} \frac{\rho g}{16} \frac{d(H^2)}{dx} \left(-\cos^2 \alpha + \sin^2 \alpha \right)$$

$$-\rho \frac{d}{dx} \left(\frac{Ac}{T} \right) \cos^2 \alpha - \rho g(D - z)S \tag{6.49}$$

which can be seen to be similar to the two-dimensional expression for $\alpha = 0$.

The shore-parallel component of the shear stress is found from the momentum equation

$$\tau_{zy} = -\rho \frac{\partial}{\partial y} \int_z^D v^2 dz - \rho \frac{\partial}{\partial x} \int_z^D \overline{uv} dz + \rho \overline{vw}$$

$$-\frac{\partial}{\partial x} \left(\rho \frac{Ac}{T} \right) \cos \alpha \sin \alpha - \rho g \eta \frac{\partial \eta}{\partial y} \tag{6.50}$$

Due to the uniform conditions the first and the fifth term in Eq. 6.50 are zero. The second and the third term are calculated as follows

$$\rho \frac{\partial}{\partial x} \int_z^D \overline{uv} dz \approx \rho(D - z) \frac{\partial}{\partial x} \left(\left(\frac{Hc}{2D} \right)^2 \frac{1}{2} \cos \alpha \sin \alpha \right) =$$

$$\left(1 - \frac{z}{D} \right) \frac{\rho g}{8} 2HH'_x \cos \alpha \sin \alpha \tag{6.51}$$

and

$$\rho \overline{vw} = \rho \frac{H\omega}{2} \frac{z}{D} \frac{1}{2} \left(-\frac{Hc}{2D} \frac{H'_x}{kH} - \frac{gH'_x}{2\omega} \right) \cos \alpha \sin \alpha =$$

$$-\frac{z}{D} \frac{\rho g}{8} 2HH'_x \cos \alpha \sin \alpha \tag{6.52}$$

giving a constant τ_{zy} over the vertical

$$\tau_{zy} = -\frac{\rho g}{8} \frac{d(H^2)}{dx} \cos \alpha \sin \alpha - \rho \frac{d}{dx} \left(\frac{Ac}{T} \right) \cos \alpha \sin \alpha \tag{6.53}$$

The result of this analysis is a shear stress τ_{zx} that is qualitatively similar to the shear stress distribution found for the two-dimensional case, and a longshore shear stress τ_{zy} which is constant over the vertical. This field of driving forces gives a combination of a longshore current and an (often weaker) cross-shore circulation. This flow situation has been treated by Svendsen and Lorenz (1989), who used a slightly different shear stress distribution.

The velocity profile of the longshore flow can be determined from the assumption of a constant driving force over the vertical, cf. Section 5.2.2, and the circulation flow is rather similar to the two-dimensional undertow described in Section 6.2. The bed shear stress given by Eqs. 6.49 and 6.53 is similar to the total driving forces determined from the radiation stress gradients, cf. Section 5.1. The near-surface shear stress is given by

$$
\vec{\tau}_s = \left\{ \begin{array}{c} \overline{\tau_{sx}} \\ \overline{\tau_{sy}} \end{array} \right\} = \left\{ \begin{array}{c} -\dfrac{\rho g}{16} \dfrac{d(H^2)}{dx}(1 + \cos^2\alpha - \sin^2\alpha) - \rho \dfrac{d}{dx}\left(\dfrac{Ac}{T}\right)\cos^2\alpha \\[3mm] -\dfrac{\rho g}{8} \dfrac{d(H^2)}{dx}\cos\alpha\sin\alpha - \rho \dfrac{d}{dx}\left(\dfrac{Ac}{T}\right)\cos\alpha\sin\alpha \end{array} \right\}
$$

$$
= \left(-\dfrac{\rho g}{8} \dfrac{d(H^2)}{dx}\cos\alpha - \rho \dfrac{d}{dx}\left(\dfrac{Ac}{T}\right)\cos\alpha \right) \left\{ \begin{array}{c} \cos\alpha \\ \sin\alpha \end{array} \right\} =
$$

$$
\left(-\dfrac{\rho g}{8} \dfrac{d(H^2)}{dx'} - \rho \dfrac{d}{dx'}\left(\dfrac{Ac}{T}\right) \right) \left\{ \begin{array}{c} \cos\alpha \\ \sin\alpha \end{array} \right\} \tag{6.54}
$$

where $\vec{\tau}_s$ is the mean surface shear stress vector and $\overline{\tau_{sx}}$ and $\overline{\tau_{sy}}$ is the surface shear stress in the x- and y-direction, respectively.

When comparing Eq. 6.54 to Eq. 6.18 it is seen that the near-surface shear stress has the same magnitude for both cases, and is in the direction of wave propagation. The magnitude of the surface shear stress is related to the loss of wave energy $\overline{\mathcal{D}}$ and the change of momentum in the surface rollers, as given by Eq. 6.18.

Example 6.3: The magnitude of the surface shear stress

It is remarkable that the magnitude of the near-surface shear stress is found to be the same in the two- and three-dimensional analysis, and it is of interest to investigate if this relationship (Eq. 6.18) is general. The cases treated so far are the simplest possible, with a horizontal bed, and even here the calculations are rather extensive. The work would become much more complicated if the effects of shoaling and refraction were to be included even on a long uniform coast. It is

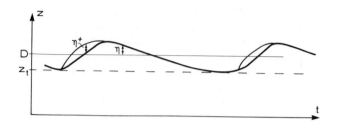

Figure 6.9 The surface along which the shear stress is determined.

therefore not feasible to extend the calculations with the fixed control surface to the general case. Instead a simple, heuristic approach is adopted.

Consider the variation with time of a water column under the action of the wave shown in Fig. 6.9. The near-surface shear stress is evaluated at the wave trough level by the momentum equation. As only the lowest order non-zero term in (H/D) is of interest, the shear stress could, at any instance, just as well be evaluated at the surface of the wave motion: $z = D + \eta$. Similarly, the difference between the horizontal shear stress at the surface and the shear stress following the tangent of the surface is negligible.

At the order considered it is therefore just as good to take the time-averaged shear stress along the wave surface, the thick line at Fig. 6.9, as at the wave trough level z_t. Over the main part of the wave the surface shear stress is zero as it is only in contact with air. But under the surface rollers the waves are subject to a shear stress due to the weight of the surface rollers.

The shear stress at the boundary between the roller and the wave can be determined by assuming hydrostatic pressure in the surface roller, as shown in Fig. 6.10. The pressure at the interface is

$$p_r^+ = \eta^+ \rho g \qquad (6.55)$$

and the shear stress is calculated to be

$$\tau_s dx = -p_r^+ \frac{\partial \eta}{\partial x} dx + \frac{1}{2} \rho g \eta^{+2} - \frac{1}{2} \rho g \left(\eta^+ + \frac{\partial \eta^+}{\partial x} dx \right)^2 \qquad (6.56)$$

giving

$$\tau_s = -\rho g \eta^+ \left(\frac{\partial \eta^+}{\partial x} + \frac{\partial \eta}{\partial x} \right) \qquad (6.57)$$

The mean surface shear stress is then found by averaging τ_s over the wave period.

In the present model of the broken wave, the surface roller is treated as a separate body of water and a shear layer is thus found between the surface

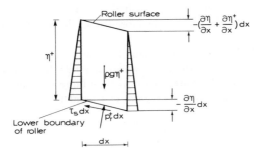

Figure 6.10 Forces acting on a vertical column of the roller.

roller and the water in the wave motion, as sketched in Fig. 6.11. The energy dissipation in a shear layer is equal to the product of the shear stress at the centerline multiplied by the velocity difference across the shear layer. As only the lowest order terms in (H/D) are considered, the velocity difference across the shear layer can be taken as the wave celerity c. The dissipation therefore becomes

$$\mathcal{D} = \tau_s c \tag{6.58}$$

This gives a mean surface shear stress of

$$\overline{\tau_s} = \frac{\overline{\mathcal{D}}}{c} = -\frac{1}{c}\frac{dE_{fw}}{dx} \tag{6.59}$$

which is identical to Eqs. 6.18 and 6.54, except that variation in the momentum of the surface rollers has not been taken into account in this simple analysis.

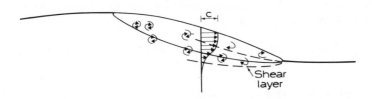

Figure 6.11 The shear layer under the surface roller.

If no surface rollers are present, i.e. no energy dissipation near the surface, the mean surface shear stress is zero according to Eq. 6.59. This is in agreement with the analysis of wave dissipation in the near-bed boundary layer made in

Example 6.1. The same is found when considering wave shoaling, which can be described by potential flow theory and therefore cannot be associated with shear stresses.

The simple analysis made in this example is more general than the calculations using the momentum equation. The waves may be refracting and shoaling in addition to the breaking, the water depth may be large or small compared to the wave length, and the waves may be regular or irregular.

REFERENCES

Buhr Hansen, J. and Svendsen, I.A. (1986): Experimental investigation of the wave and current motion over a longshore bar. Proc. 20th Coastal Eng. Conf., Taipei, pp. 1166-1179.

Dally, W.R. and Dean, R.G. (1984): Suspended sediment transport and beach evolution. ASCE, J. Waterw. Harbors Coastal Ocean Eng. Div., 110(1):15-33.

Deigaard, R. and Fredsøe, J. (1989): Shear stress distribution in dissipative water waves. Coastal Eng., 13:357-378.

Deigaard, R., Justesen, P. and Fredsøe, J. (1991): Modelling of undertow by a one-equation turbulence model. Coastal Eng., 15:431-458.

Dyhr-Nielsen, M. and Sørensen, T. (1970): Sand transport phenomena on coasts with bars. Proc. 12th Coastal Engineering Conf., pp. 855-866.

Engelund, F. (1981): A simple theory of weak hydraulic jumps. Progress Report No. 54, Inst. of Hydrodynamics and Hydraulic Engineering, ISVA, Techn. Univ. Denmark, pp. 29-32.

Liu, A.K. and Davis, S.H. (1977): Viscous attenuation of mean drift in water waves. J. Fluid Mech., 81(1):63-84.

Longuet-Higgins, M.S. (1969): A nonlinear mechanism for the generation of sea waves. Proc. R. Soc. London, Ser. A, 311:371-309.

Phillips, O.M. (1977): The Dynamics of the Upper Ocean. Cambridge Univ. Press, Cambridge, 336 pp.

Stive, M.J.F. and Wind, H.G. (1986): Cross-shore mean flow in the surf zone. Coastal Eng., 10(4):325-340.

Svendsen, I.A. (1984): Mass flux and undertow in a surf zone. Coastal Eng., 8:347-366.

Svendsen, I.A., Schäffer, H.A. and Buhr Hansen, J. (1987): The interaction between the undertow and the boundary layer flow on a beach. J. Geophys. Res., 92(C11):11845-11856.

Svendsen, I.A. and Lorenz, R.S. (1989): Velocities in combined undertow and longshore currents. Coastal Eng., 13:55-80.

Chapter 7. Basic concepts of sediment transport

7.1 Transport modes

This chapter gives a short introduction to the basic concepts of sediment transport mechanisms. The application of these concepts to describe the sediment transport in the surf zone is outlined in the next chapters.

It is common to split the sediment transport modes up into three parts:

- Bed load
- Suspended load
- Wash load

The wash load consists of very fine particles which are transported by the water and which normally are not represented in the bed. Therefore, the knowledge of bed material composition does not permit any prediction of the rate of wash load transport. Hence, when the term "total sediment discharge" is applied, the wash load is neglected.

Of the total sediment load a distinction between two categories is made, the *bed load* and the *suspended load*. No precise definitions of these terms have been given so far, but the basic idea of splitting up the total sediment load in two parts is that, roughly speaking, two different mechanisms are effective during the transport.

The bed load is defined as the part of the total load that is in more or less continuous contact with the bed during the transport. It primarily includes grains that roll, slide, or jump along the bed. Thus the bed load must be determined almost exclusively by the effective bed shear acting directly on the sand surface.

The suspended load is the part of the total load that is moving without continuous contact with the bed as a result of the agitation of fluid turbulence.

7.2   Sediment properties

The sediment transported in the coastal zone usually contains particles ranging from gravel or sand down to very small particles classified as silt or clay. The very fine fractions are carried as wash load.

The subject of this and the following chapters is to consider non-cohesive sediment. This covers usually the particles in the range from sand (0.06 mm to 2 mm) to gravel (2 mm to 20 mm), and the following brief account of sediment properties is devoted to these fractions exclusively. From a hydraulic point of view the most important sediment properties are related to size, shape, and specific gravity.

7.2.1   Particle size characteristics

The most usual and convenient method for the analysis of particle size distribution is the sieve analysis, which is applicable for particle sizes not smaller than 0.06 mm. An adequate number of representative sediment samples is analyzed, and the result is presented as a frequency curve (Fig. 7.1) or as a cumulative frequency curve (distribution curve, Fig. 7.2).

In the frequency or probability density function curve in Fig. 7.1 the abscissa represents the sieve diameter d_s, and the ordinate the concentration of the total sample contained in the corresponding intervals of the d_s. Very often the distribution curve of sediments approaches the normal probability curve when plotted as in Fig.7.1, so that the distribution function is log-normal and given by

$$f(d) = \frac{1}{\sqrt{2\pi}\,d\ln(\sigma_g)} \exp\left(-\frac{1}{2}\left(\frac{\ln(d/d_{50})}{\ln(\sigma_g)}\right)^2\right) \tag{7.1}$$

in which σ_g is the geometric standard deviation given by

$$\sigma_g = \sqrt{d_{84}/d_{16}} \tag{7.2}$$

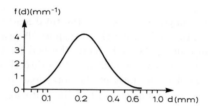

Figure 7.1 Frequency or probability density function of a log-normal grain size distribution. $d_{50} = 0.25$ mm. $\sigma_g = 1.5$.

In Eq. 7.1 d_{50} corresponds to the median, i.e. 50 p.c. by weight being finer (and 50 p.c. coarser). Similarly, in Eq. 7.2, d_{16} and d_{84} correspond to the diameter of which, respectively, 16 and 84 p.c. are finer.

In the cumulative (or grain-size) curve in Fig. 7.2 the ordinate indicates how much per cent (by weight) of the total sample is finer than the diameter d_s of the abscissa.

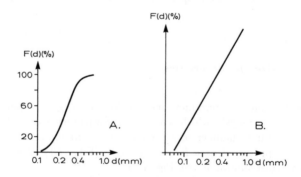

Figure 7.2 Grain-size distribution curve for a log-normal grain size distribution for $d_{50} = 0.25$ mm and $\sigma_g = 1.5$. A: Semi-logarithmic plot. B: Plot on probability paper.

The grain-size curve, shown in Fig. 7.2, is obtained from Eq. 7.1 by integration to be given by

$$F(d) = P\{d' < d\}$$

$$= \int_{\infty}^{d} f(d')dd'$$

$$= \frac{1}{2}\left[1 + \text{Erf}\left(\frac{1}{\sqrt{2}}\frac{\ln(d/d_{50})}{\ln(\sigma_g)}\right)\right] \tag{7.3}$$

in which Erf is the error function, defined by

$$\text{Erf}(y) = \frac{2}{\sqrt{\pi}}\int_{0}^{y} e^{-x^2}dx \tag{7.4}$$

On probability paper, Fig. 7.2B, Eq. 7.3 forms a straight line.

Table 7.1 Values of d_s, d_v, d_f and w_s for typical sand fractions (from Engelund and Hansen, 1972).

d_s mm	d_v mm	d_f mm	$w_s(10°\text{C})$ m/s	$w_s(20°\text{C})$ m/s
0.089	0.10	0.10	0.005	0.008
0.126	0.14	0.14	0.010	0.013
0.147	0.17	0.16	0.013	0.016
0.208	0.22	0.22	0.023	0.028
0.25	0.25	0.25	0.028	0.033
0.29	0.30	0.29	0.033	0.039
0.42	0.46	0.40	0.050	0.058
0.59	0.64	0.55	0.077	0.084
0.76	0.80	0.70	0.10	0.11
1.25	1.4	1.0	0.15	0.16
1.8	1.9	1.2	0.17	0.17

Another measure of particle size different from the sieve diameter is the spherical diameter d_v, defined as the diameter of a sphere having the same volume as the given particle. In practice d_v is determined by weighing a counted number of particles from a certain fraction of the sample.

This measure does not take any account of the shape of the sediment grains, for which reason the so-called fall diameter d_f is a more satisfying parameter.

The fall diameter of a particle is defined as the diameter of a sphere having the same settling velocity in water at 24°C. For a fixed particle volume d_f will be greater for angular grains than for rounded grains, so that this size measure takes some account of the shape.

Table 7.1 lists a number of simultaneous values of d_s, d_v and d_f for some typical fractions of natural sand. The ratio between d_v and d_s is nearly constant, which is very natural when the shape of the grains is not too much dependent on the grain size. The fall diameter d_f, however, becomes significantly smaller than d_s and d_v for the larger grains.

7.2.2 Specific gravity

The specific gravity γ_s of the grains is the parameter which exhibits the smallest variation under natural conditions.

The ratio

$$s = \gamma_s/\gamma ,$$
(7.5)

in which γ denotes the specific gravity of water at 4°C is called the relative density. For natural sediments, s is usual very close to 2.65.

7.2.3 Settling velocity

The settling or fall velocity w_s of a grain is defined as the terminal velocity attained when the grain is settling in an extended fluid under the action of gravity.

The fall velocity w_s depends on several parameters, the most important are grain size, specific gravity, shape, and the dynamic viscosity of the fluid.

The drag force F on a submerged body is given by the general expression

$$F = \frac{1}{2}c_D\rho V^2 A ,$$
(7.6)

in which c_D is the drag coefficient, ρ the density of the fluid, V the relative velocity, and A the area of the projection of the body upon a plane normal to the flow direction.

Consider now the settling of a single spherical particle of diameter d. The combined action of gravity and buoyancy gives the force

$$(\gamma_s - \gamma)\frac{\pi}{6}d^3$$
(7.7)

which under equilibrium conditions must be balanced by the drag, so that we obtain the following equation

$$(\gamma_s - \gamma)\frac{\pi}{6}d^3 = c_D\frac{1}{2}\rho w_s^2\frac{\pi}{4}d^2 \qquad (7.8)$$

from which

$$w_s = \sqrt{\frac{4(s-1)gd}{3c_D}} \qquad (7.9)$$

For a single spherical particle in an extended fluid the value of c_D depends on the grain Reynolds number

$$\mathcal{R} = \frac{w_s d}{\nu} \qquad (7.10)$$

exclusively.

In Fig. 7.3 the value of c_D for a number of typical sand fractions is plotted against the grain Reynolds number $\mathcal{R}$. For the sake of simplicity c_D is defined by the expression

$$F = \frac{1}{2}c_D\rho w_s^2\frac{\pi}{4}d^2 \qquad (7.11)$$

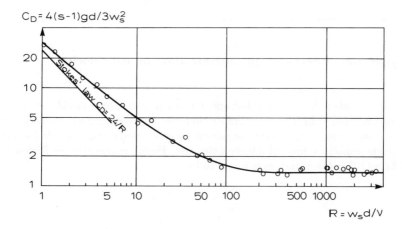

Figure 7.3 Variation in drag coefficient with Reynolds number for natural sand.

The relationship depicted in Fig. 7.3 can be expressed by

$$c_D = 1.4 + 36/\mathcal{R} \qquad (7.12)$$

from which an analytical expression for w_s can be obtained.

Example 7.1:   Fall velocity of fine sediment

For very small values of $\mathcal{R}$ Stokes' law will apply:

$$F = 3\pi\mu d w_s \ , \tag{7.13}$$

corresponding to the expression

$$c_D = 24/\mathcal{R} \tag{7.14}$$

Under these circumstances Eq. 7.9 becomes

$$w_s = \frac{(s-1)gd^2}{18\nu} \tag{7.15}$$

Example 7.2:   Effect of high concentrations on fall velocity

If the sphere is not single but one of many particles settling simultaneously, the observed fall velocity w_s is smaller than the above expression indicates, the ratio w_s/w_{so} being a function of the volume concentration c. w_{so} is the fall velocity of a single grain in a fluid of infinite extension. From experiments by Richardson and Zaki (1954) the variation is found to be

$$\frac{w_s}{w_{so}} = (1-c)^n \tag{7.16}$$

where
$$
\begin{array}{ll}
n = 4.35\mathcal{R}^{-0.03} & 0.2 < \mathcal{R} < 1 \\
n = 4.45\mathcal{R}^{-0.10} & 1 < \mathcal{R} < 500 \\
n = 2.39 & 500 < \mathcal{R}
\end{array}
$$

Usually, c is much smaller than 1, so Eq. 7.16 can be written as

$$w_s = w_{so}(1 - nc) \tag{7.17}$$

Fig. 7.4 shows this effect graphically.

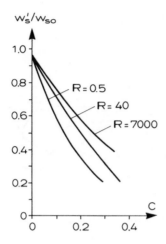

Figure 7.4 Reduction in fall velocity at large concentrations.

7.3   Critical bed shear stress

In the following, steady flow over a bed composed of cohesionless grains is considered. These grains will not move at very small flow velocities, but when the flow velocity becomes large enough, the driving forces on the sediment particles will exceed the stabilizing forces, and the sediment will start to move. This flow velocity is called the critical flow velocity.

A now classical solution to the problem was offered by Shields (1936). His analysis was based on dimensional analysis. The threshold of particle motion is supposed to be attained for a given ratio between driving and stabilizing forces.

The driving forces on a sediment particle resting on other particles on an originally plane horizontal bed are the tractive stress τ_o (horizontal) and the lift force, see Fig. 7.5.

The horizontal drag F_D, created by the flow, consists of skin friction acting on the surface of the grain and form drag due to a pressure difference on the up- and downstream sides of the grain because of flow separation. From the elementary theory of drag it is known that

$$F_D = \frac{1}{2}\rho c_D \frac{\pi}{4} d^2 U^2 \tag{7.18}$$

where U is a characteristic velocity near the bed, d is the grain diameter, and c_D is the drag coefficient which is known to depend on the local Reynolds number.

In the following it is assumed that the shape effects are accounted for sufficiently well by using the fall diameter.

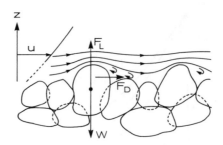

Figure 7.5 Forces acting on grains resting on the bed.

Generally, a lifting force F_L in excess of the natural buoyancy is also created by the flow. This lift is partly due to the curvature of the streamlines which locally will decrease the pressure to be lower than the hydrostatic pressure at the top of the grains. Further, the flow separation also involves a positive lift force on the grains. The lift force is given by a similar expression as Eq. 7.18, so the driving forces can be given in the following form

$$F_D = \frac{1}{2}\rho c_D \frac{\pi}{4}d^2\left(\alpha U_f'\right)^2 \tag{7.19}$$

in which U_f' is the friction velocity due to the friction directly acting on the bed surface (skin friction), while c_D and α are non-dimensional coefficients. $\alpha U_f'$ is the flow velocity at a distance of the order of magnitude d from the bed. Assuming the validity of the ordinary velocity distribution in rough channels (cf. Example 2.2), α must be of the order of 10. The factor c_D stands for a drag (and lift) coefficient.

The stabilizing forces can be modelled as frictional forces acting on a particle. For a non-moving particle resting in the bed this can not exceed

$$F_s = \rho g(s-1)\frac{\pi}{6}d^3\mu_s = W\mu_s \tag{7.20}$$

in which W is the submerged weight of the particle, and μ_s is the measure for the maximum friction between the grain and the surrounding grains. This can be taken as equal to

$$\mu_s = \tan(\phi_s) \tag{7.21}$$

where ϕ_s is the static friction angle (angle of repose) for the sediment.

The particle can remain resting on the bed without moving as long as the driving forces are smaller than the maximum retarding force, given by Eq. 7.20.

This means that the particle will not move as long as U'_f is smaller than U_{fc}, where U_{fc} is determined by

$$\frac{1}{2}\rho c_D \frac{\pi}{4} d^2 (\alpha U_{fc})^2 = \frac{\pi}{6} d^3 (s-1)\rho g \mu_s \tag{7.22}$$

or

$$\frac{U_{fc}^2}{(s-1)gd} = \frac{\mu_s}{c_D} \frac{4}{3\alpha^2} \tag{7.23}$$

The parameter on the left-hand-side is called the critical Shields parameter, and is defined by

$$\theta_c = \frac{U_{fc}^2}{(s-1)gd} \tag{7.24}$$

The term on the right-hand-side of Eq. 7.23 is, despite a weak function of the Reynolds number, a constant quantity which Shields found to be of the order 0.05 for sand placed smoothly on a horizontal bed. Fig. 7.6 shows some of Shields' experimental observations which relate θ_c to the Reynolds number $\mathcal{R}e$ (for the grain) defined as

$$\mathcal{R}e = \frac{U'_f d}{\nu} \tag{7.25}$$

For large values of $\mathcal{R}e$ the critical value increases to about 0.06.

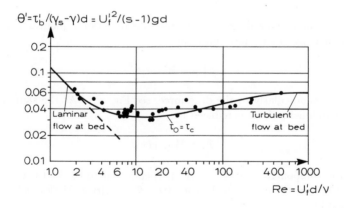

Figure 7.6 The Shields diagram giving the threshold value θ_c as a function of $\mathcal{R}e$.

Example 7.3:   Effect of a transverse bed slope on the critical Shields parameter

If sand grains are placed on a bed with a transverse slope, it is easier for them to move. So the critical Shields parameter will be reduced.

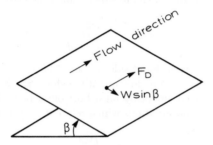

Figure 7.7 Driving forces on a particle placed on a bed with transverse slope.

While the driving drag forces in the flow direction remain unchanged, an additional driving gravity force given by

$$F_T = W \sin \beta \tag{7.26}$$

now acts perpendicularly to the drag force, see Fig. 7.7. At the same time, the maximum stabilizing force is reduced to

$$F_s = W \mu_s \cos \beta \tag{7.27}$$

because the component of gravity force perpendicular to the bed is reduced by a factor $\cos \beta$. Now the critical bed shear stress θ_c or critical friction velocity U_{fc} is determined by

$$\sqrt{F_{Dc}^2 + F_T^2} = F_s \tag{7.28}$$

where F_{Dc} is the critical drag force. By use of Eqs. 7.18, 7.26 and 7.27, Eq. 7.28 reads

$$\left[\frac{1}{2}\rho c_D \frac{\pi}{4} d^2 (\alpha U_{fc})^2\right]^2 + W^2 \sin^2 \beta = W^2 \mu_s^2 \cos^2 \beta \qquad (7.29)$$

or

$$\left(\frac{c_D}{\mu_s} \frac{3\alpha^2}{4}\right)^2 \left(\frac{U_{fc}^2}{(s-1)gd}\right)^2 + \frac{\sin^2 \beta}{\mu_s^2} = \cos^2 \beta \qquad (7.30)$$

From Eqs. 7.22 and 7.23 it is seen that

$$\frac{c_D}{\mu_s} \frac{3\alpha^2}{4} = \frac{1}{\theta_{co}} \qquad (7.31)$$

where the index *co* indicates the critical Shields parameter on a bed with no slope. Inserting Eq. 7.31 together with Eq. 7.20 into Eq. 7.30 finally gives

$$\theta_c = \theta_{co} \cos \beta \sqrt{1 - \frac{\tan^2 \beta}{\tan^2(\phi_s)}} \qquad (7.32)$$

Example 7.4:   Effect of a longitudinal slope on the critical Shields parameter

The principles outlined in the foregoing section are easily transferred to the case where the bed has a slope γ in the flow direction. Now, the critical bed shear stress is determined by

$$F_{Dc} + F_T = F_s \qquad (7.33)$$

or

$$\frac{1}{2}\rho c_D \frac{\pi}{4} d^2 (\alpha U_{fc})^2 + W \sin \gamma = W \mu_s \cos \gamma \qquad (7.34)$$

which, similarly to the former example, can be rearranged to

$$\theta_c = \theta_{co} \cos \gamma \left[1 - \frac{\tan \gamma}{\tan(\phi_s)}\right] \qquad (7.35)$$

7.4   Bed load transportation

One of the first theoretical approaches to the problem of predicting the rate of bed load transport was presented by H.A. Einstein (1950). One of the most important innovations in his analysis was the application of the theory of probability to account for the statistical variation of the agitating forces on bed particles caused by turbulence.

If the magnitude of the instantaneous agitating forces on a certain bed particle exceeds the stabilizing forces on the particle, the particle will begin to jump, roll, or slide along the bed until it becomes deposited downstream at a location where the magnitude of the instantaneous forces is smaller than the stabilizing forces. Based on experimental observations, Einstein assumed that the mean distance, travelled by a sand particle between erosion and subsequent deposition, is simply proportional to the grain diameter and independent of the hydraulic conditions and the amount of sediment in motion.

The principle in Einstein's analysis is as follows: the number of particles, deposited in a unit area, depends on the number of particles in motion and on the probability that the dynamical forces permit the particles to deposit. The number of particles eroded from the same unit area depends on the number of particles within the area and on the probability that the hydrodynamic forces on these grains are sufficiently strong to move them. For equilibrium conditions the number of grains deposited must equal the number of particles eroded. In this way, a functional relation (bed load function) is derived between the two non-dimensional quantities.

$$\Phi_B = \frac{q_B}{\sqrt{(s-1)gd^3}} \tag{7.36}$$

and

$$\theta' = \frac{\tau_b'}{\rho g(s-1)d} = \frac{U_f'^2}{(s-1)gd} \tag{7.37}$$

where q_b is the rate of bed load transport in volume of material per unit time and width. Hence, Φ is a non-dimensional form of bed load discharge, while θ' is the non-dimensional tractive stress (the Shields parameter) due to skin friction.

R.A. Bagnold (1954) pointed out one of the shortcomings in Einstein's formulation by stating the following paradox. Consider the ideal case of fluid flow over a bed of uniform, perfectly piled spheres in a plane bed, so that all particles are equally exposed. Statistical variations due to turbulence are neglected. When the tractive stress exceeds the critical value, all particles in the upper layer are peeled off simultaneously and are dispersed. Hence the next layer of particles is exposed to the flow and should consequently also be peeled off. The result is that all the subsequent underlying layers are also eroded, so that a stable bed could not exist at all when the shear stress exceeds the critical value.

Bagnold explained the paradox by assuming that in a water-sediment mixture the total shear stress τ' would be separated in two parts

$$\tau' = \tau_F + \tau_G \tag{7.38}$$

where τ_F is the shear stress transmitted by the intergranular fluid, while τ_G is the shear stress transmitted because of the interchange of momentum caused by the encounters of solid particles, i.e. a tangential dispersive stress. The existence of such dispersive stresses was confirmed by his experiments.

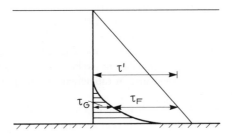

Figure 7.8 Distribution of fluid shear stress and grain shear stress in uniform open channel flow.

Bagnold argues that when a layer of spheres is peeled off, some of the spheres may go into suspension, while others will be transported as bed load. Thus a dispersive pressure on the next layer of spheres will develop and act as a stabilizing agency. Hence, a certain part of the total bed shear stress is transmitted as a grain shear stress τ_G, and a correspondingly minor part as fluid stress $(\tau_F = \tau' - \tau_G)$. Continuing this argumentation, it is understood that exactly so many layers of spheres will be eroded that the residual fluid stress τ_F on the first immovable layer is equal to (or smaller than) the critical tractive stress τ_c. The mechanism in transmission of a tractive shear stress τ greater than the critical is then the following: τ_c is transferred directly from the fluid to the immovable bed, while the residual stress $\tau' - \tau_c$ is transferred to the moving particles and further from these to the fixed bed as a dispersive stress.

Example 7.5: Bagnold's experiments

Bagnold (1954) measured the dispersive stress in a mixture of fluid and sediment particles (spheres with a density equal that of the water in order to avoid disturbing centrifugal forces). The mixture was placed between two co-axial cylinders of which the one was fixed while the other (outer cylinder) could rotate. The spacing between the two cylinders was small compared to their radii. As a result a nearly uniform distribution of the shear stress was obtained in the mixture. The shear stress as well as the normal stress acting on the cylinder wall were measured. The experiments were carried out for the same rotation speed of the cylinder with and without sediment in the fluid. Therefore, the dispersive tangential and normal force due to the presence of sediment was obtained as the difference between the two above mentioned measured quantities.

The dispersive stresses originate from exchange of momentum between the particles because of collisions resulting in movement of the particles transversely to the mean flow direction.

Bagnold suggested that the dispersive stresses could be described mathematically by a relationship between two non-dimensional parameters, namely

$$N = \frac{\sqrt{\lambda} s d^2}{\nu} \frac{du}{dz} \tag{7.39}$$

and

$$G = \frac{d}{\nu}\sqrt{\frac{\sigma}{\lambda}\frac{s}{\rho}} \tag{7.40}$$

In both equations λ is the so-called linear concentration related to the real volumetric concentration c by

$$c = c_o/(1 + 1/\lambda)^3 \tag{7.41}$$

in which c_o is the maximum value for the volumetric concentration. For natural sand, this number is around 0.65. Geometrically, d/λ is a measure for the average distance between two grains approaching zero for the most dense packing.

The parameter given by Eq. 7.40 is a kind of Reynolds number for the sediment, because the square root of the stresses (like the usual friction velocity) defines a velocity scale. N appearing in Eq. 7.39 is a measure for the shear in the flow.

The measured relations between N and G are depicted in Fig. 7.9. Two different values for G are given, namely G_τ and G_σ, where G_τ represents the dispersive shear stress, while G_σ is the dispersive normal stress.

If the dispersive shear stresses first is considered, it is seen that at large values of N the relation between G^2 and N on double logarithmic scale approaches

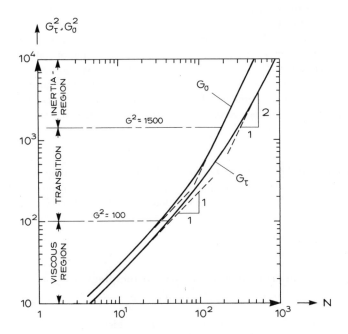

Figure 7.9 Variation in dimensionless grain shear stresses G with N.

a straight line with the slope 2:1. Here, the relation between N and G can be rewritten as

$$\tau_G = 0.013\rho s(\lambda d)^2 \left(\frac{du}{dz}\right)^2 \quad \text{for} \quad N > 450 \tag{7.42}$$

At low values of N, the relation between G^2 and N approaches another straight line with a smaller slope, namely 1:1. Here, the corresponding expression similar to Eq. 7.42 becomes

$$\tau_G = 2.2\lambda^{3/2}\rho\nu\frac{du}{dz} \quad \text{for} \quad N < 40 \tag{7.43}$$

Savage and McKeown (1983) later suggested that this last expression be modified to

$$\tau_G = 1.2\lambda^2\rho\nu\frac{du}{dz} \tag{7.43a}$$

Eq. 7.42 corresponds to the so-called inertia-region in which the intergranular collisions occur so often and with such a strength that the inertia of the grains will dominate the transfer of dispersive stresses. In this region, the fluid viscosity is of no importance.

Eq. 7.43, on the other hand, corresponds to the viscous region where the fluid viscosity becomes important. Here the concentration of sediment is so weak that the velocity of the grains is mainly determined by the interaction with the fluid rather than by the momentum transferred by intergranular collisions.

In Fig. 7.9 the dimensionless dispersive normal stress G_σ is plotted. The relation between the dispersive shear stresses and normal stresses can be written as

$$\tau_G = \sigma_G \tan(\phi_d) \qquad (7.44)$$

in which ϕ_d can be viewed as a dynamic friction angle. In Bagnold's experiments it was found that $\tan(\phi_d)$ was around 0.75 in the viscous region and decreased to about 0.32 in the inertia-dominated region (see also Example 7.8 for a discussion of ϕ_d).

7.4.1   Bed load transportation close to incipient motion

Fernández Luque (1974), (see also Fernández Luque and Beek, 1976) has argued against some of Bagnold's ideas and has developed a consistent theory for the transport of bed load on a plane bed considering the motion of individual particles. His theory is supported by a series of careful experimental observations.

One of his issues is that close to incipient particle motion (small transport rates) only the topmost grains will be eroded, and the bed load will not effectively reduce the fluid part of the turbulent bed shear stress. This can hardly be expected, either, since in these conditions the moving bed load particles cover only a small portion of the bed surface. According to Luque's model, the bed load particles reduce the maximum fluid shear stress at the bed surface to the critical value τ_c by exerting an average reaction force on the surrounding fluid. Hence, the bed load forms a kind of protective shield at higher bed load concentrations, which control the erosion rate.

From these considerations it is possible to construct a bed load formula in a relatively simple way (Engelund and Fredsøe, 1976). The model is, as the considerations on the critical Shields' parameter Section 7.3, a kind of model equation containing time-averaged quantities rather than an exact description of the forces, which is of fluctuating character.

The bed load particles are transported with a mean transport velocity U_B, when they are moving. Hereby, the agitating forces (drag and lift) are given by

$$F_D = \frac{1}{2}\rho c_D \left[\alpha\, U_f' - U_B\right]^2 \frac{\pi}{4}d^2 \qquad (7.45)$$

This expression is similar to Eq. 7.19, and the symbols are the same. The stabilizing forces are like Eq. 7.20 given by

$$F_s = W\mu_d \tag{7.46}$$

in which μ_d represents the dynamic friction, given by

$$\mu_d = \tan(\phi_d) \tag{7.47}$$

where ϕ_d is the dynamic friction angle for the bed load sediment.

The model equation then expresses the equilibrium of agitating and stabilizing forces

$$\frac{1}{2}\rho c_D \frac{\pi}{4} d^2 \left[\alpha \, U_f' - U_B\right]^2 = \frac{\pi}{6}\rho g(s-1)d^3 \mu_d \tag{7.48}$$

which gives

$$U_B = \alpha U_f' \left[1 - \sqrt{\frac{\theta_0}{\theta'}}\right] \tag{7.49}$$

in which

$$\theta_0 = \frac{4\mu_d}{3\alpha^2 c_D} \tag{7.50}$$

and θ' is the Shields parameter, Eq. 7.37. While in Shields' experiments it was intended to place the sediment regularly in the bed, naturally placed sand will have particles which are more exposed to the flow than others.

As a particle lying on the bed is easier to move than a particle located in the bed, it must be expected that $\theta_0 < \theta_c$. From his experiments Luque found θ_0 to be $\frac{1}{2}\theta_c$, so that Eq. 7.48 may be written

$$\frac{U_B}{U_f'} = 10\left[1 - 0.7\sqrt{\frac{\theta_c}{\theta'}}\right] \tag{7.51}$$

taking $\alpha = 10$, as suggested earlier. Eq. 7.51 is compared with measurements by Luque (1974) and by Meland and Normann (1966) in Fig. 7.10.

From the knowledge of mean particle velocity, an expression for the rate of bed load transport q_b can now be derived under the assumption that the bed load is the transport of a certain fraction p (= probability) of the particles that may be in one single layer. As the total number of surface grains per unit area is $1/d^2$, we get

$$q_B = \frac{\pi}{6}d^3 \frac{p}{d^2} U_B \tag{7.52}$$

or, after insertion of Eq. 7.51

$$q_B = 10\frac{\pi}{6}dpU_f' \left[1 - 0.7\sqrt{\theta_c/\theta'}\right] \tag{7.53}$$

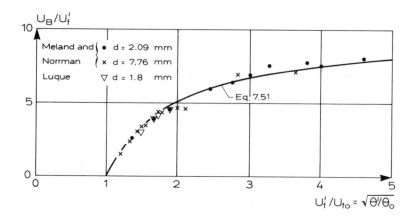

Figure 7.10 Experiments on the transport velocity U_B of bed load particles.

This is made non-dimensional by the divisor $\sqrt{(s-1)gd^3}$

$$\Phi_B \simeq 5p\left[\sqrt{\theta'} - 0.7\sqrt{\theta_c}\right] \qquad (7.54)$$

An estimate on p can be obtained based on the assumption that only the part τ_c of the total shear stress τ is transferred directly to the immobile bed as skin friction, while the residual part $\tau - \tau_c$ is carried as drag on the moving bed particles and indirectly transferred to the bed by occasional encounters. This idea leads to the equation

$$\tau = \tau_c + nF_D \qquad (7.55)$$

where F_D is the average drag on a single moving bed particle, while n is the number of moving particles per unit area. If this expression is divided by $\rho g(s-1)d$, and F_D is estimated as

$$F_D \simeq \rho g(s-1)\frac{\pi}{6}d^3\mu_d \qquad (7.56)$$

the result becomes

$$\theta' = \theta_c + \frac{\pi}{6}\mu_d(nd^2) = \theta_c + \frac{\pi}{6}\mu_d p \qquad (7.57)$$

Luque's experiments give empirical information about p, as the measurements comprise Φ_B, θ' and θ_c, so p can be evaluated from Eq. 7.54. A comparison between Eq. 7.57 (with $\mu_d = 0.8$ and $\theta_c = 0.045$) and the measured values of p is shown in Fig. 7.11.

Besides the measured values of p by Luque some additional data, obtained from Guy et al. (1966), (the Fort Collins data), are included. From Fig. 7.11 it is seen that Eq. 7.57 shows good agreement with measured values for small values of the Shields' parameter. At larger values of θ' however, Eq. 7.57 falls below the Fort Collins data. This seems reasonable, because Eq. 7.57 for larger values of θ' becomes larger than unity, which is the upper limit of p, corresponding to the situation where all particles in the bed load layer are moving.

In the Fort Collins series, four runs (corresponding to "standing waves") are marked by triangles in Fig. 7.11. In these runs the transport rate was large but still largely occurring as bed load. The fact that they all gave values of p close to unity is an experimental support for the idea that p approaches unity for increasing values of θ'.

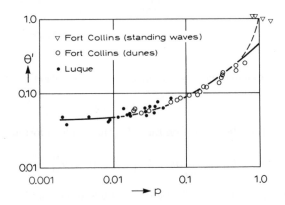

Figure 7.11 Probability p versus non-dimensional effective shear stress θ'.
Solid line: Eq. 7.57. Dashed line: Eq. 7.58.

By accepting a limiting value of $p = 1$, the expression for p has to be modified, for instance to the following expression

$$p = \left[1 + \left(\frac{\frac{\pi}{6}\mu_d}{\theta' - \theta_c} \right)^4 \right]^{-\frac{1}{4}} \tag{7.58}$$

which is about equal to Eq. 7.57 for θ close to θ_c and approaches unity for large values of θ'.

If Eq. 7.58 is adapted, the force balance given by Eq. 7.55 can not be fulfilled at large values of θ' because of the limiting value of n. For this reason, other mechanisms must be included in the transfer of shear stress to the bed at higher values of θ'. In the coming sections, two different mechanisms will be investigated, namely (i): The occurrence of bed load in several layers (sheet flow), and (ii): The influence of dispersive stresses from suspended sediment.

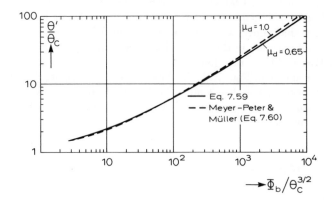

Figure 7.12 Comparison between Eq. 7.59 and Meyer-Peter's formula, Eq. 7.60.

Keeping p from the expression Eq. 7.57, the bed load formula Eq. 7.54 now reads

$$\Phi_B = \frac{30}{\pi \mu_d}(\theta' - \theta_c)\left(\sqrt{\theta'} - 0.7\sqrt{\theta_c}\right) \tag{7.59}$$

This formula is depicted in Fig. 7.12 for $\mu_d = 1.0$ and $\mu_d = 0.65$, as shown in Fig. 7.12. For $\mu_d = 1.0$ the formula, given by Eq. 7.59 becomes close to the widely used semi-empirical formula of Meyer-Peter and Müller (1948)

$$\Phi_B = 8(\theta' - \theta_c)^{\frac{3}{2}} \tag{7.60}$$

However, both formulae overestimate the bed load transport at high shear stresses.

Example 7.6: Bed load transport on a (small) transverse slope

A bed load particle moving along a bed with a transverse slope β will move in a different direction than that of the flow, see Engelund and Fredsøe (1982).

The angle between the direction of particle movement and the flow direction is called ψ, see Fig. 7.13. This angle can be found from the following force balance considerations: In the longitudinal direction the force balance is described by the equation

$$F_D \cos(\psi_1) = W \mu_d \cos \beta \tag{7.61}$$

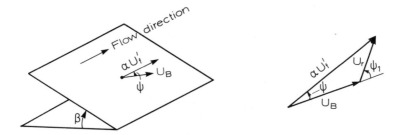

Figure 7.13 Particle motion on a transverse slope

Here ψ_1 is the angle between the particle path and the drag. As shown in Fig. 7.13, the drag force acts in the direction determined by the relative velocity between flow velocity and particle velocity. This velocity is called U_r and by simple geometry (cf. Fig. 7.13) is

$$U_r \cos(\psi_1) = \alpha U_f' \cos \psi - U_B$$

$$= \alpha U_f' - U_B \tag{7.62}$$

as the angle ψ has to be small. Inserting Eq. 7.62 into Eq. 7.61 and taking

$$F_D = \frac{1}{2} \rho c_D \frac{\pi}{4} d^2 U_r^2 \tag{7.63}$$

give similarly to the derivation of Eq. 7.48

$$1 - \frac{U_B}{\alpha U_f'} = \sqrt{\frac{\theta_0}{\theta'}} \tag{7.64}$$

in which θ_0 still is given by Eq. 7.50. In the present derivation the transverse slope β and ψ_1 are assumed small, so $\cos \beta$ and $\cos(\psi_1) \approx 1$. ψ_1 is from geometrical considerations (see Fig. 7.13) given by

$$\tan(\psi_1) = \frac{\alpha U_f' \sin \psi}{\alpha U_f' \cos \psi - U_B} = \sqrt{\frac{\theta'}{\theta_0}} \tan \psi \tag{7.65}$$

Next, the transverse force balance is considered. Because the dynamic friction is directed opposite to the particle path, the transverse balance is given by

$$W \sin \beta = F_D \sin(\psi_1) \tag{7.66}$$

By use of Eq. 7.61, this can be written as

$$\mu_d \tan(\psi_1) = \tan \beta \qquad (7.67)$$

Hereby, Eq. 7.65 finally gives

$$\tan \psi = \sqrt{\frac{\theta_0}{\theta'}} \frac{1}{\mu_d} \tan \beta \qquad (7.68)$$

Eq. 7.67 can be written as

$$\tan \psi \approx \psi = \frac{\tan \beta}{\tan \phi} \qquad (7.69)$$

in which ϕ is a measure for the friction angle, given by

$$\tan \phi = \sqrt{\frac{\theta'}{\theta_0}} \tan(\phi_d) \qquad (7.70)$$

As seen from Fig. 7.14, experiments indicate that the constants appearing in Eq. 7.70 should be calibrated to give

$$\tan \phi = 1.6 \sqrt{\theta'} \qquad (7.71)$$

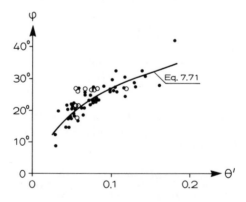

Figure 7.14 The friction angle ϕ versus θ'. Experiments: filled circles: Zimmermann and Kennedy (1978), open circles: Wan (1981).

7.4.2   Bed load transport and suspended bottom concentration at high shear stresses

In Section 7.1 we defined the bed load as that part of the load which is more or less in continuous contact with the bed during the transport. At small transport rates, this transport occurs in one single layer of particles moving over the fixed bed, as modelled in the previous section and sketched in Fig. 7.15A. At larger transport rates, however, some of the particles either go into suspension, Fig. 7.15C, or the particles move as bed load in several layers, Fig. 7.15B.

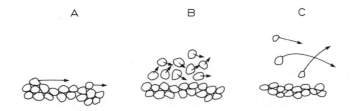

Figure 7.15 Different modes of sediment transport: A: bed load at small shear stresses. B: sheet flow. C: suspended sediment.

The distinction between bed load and suspended load can now be made as follows: the bed load is that part of the load which is travelling immediately above the bed and supported by intergranular collisions rather than by fluid turbulence (Wilson, 1966). The suspended sediment, on the other hand, must then be that part of the transported particles which mainly is supported by the fluid turbulence.

A complete description of this complex behaviour still awaits to be made. In the following, two simplified approaches to solve the problem are described:

Influence of suspended load

A very simplified extension of the model, described in Section 7.4.1 for weak bed load transport, was suggested by Engelund and Fredsøe (1976). They included the dispersive stress from the suspended load into Eq. 7.55, which now becomes

$$\tau' = \tau_c + nF_D + \tau_G \tag{7.72}$$

in which τ_G is the dispersive stress given by Eq. 7.42. In Eq. 7.72, the bed load is assumed only to consist of one single layer of particles. The remaining load is suspended load, where the concentration decays very fast away from the bed.

Hence, the velocity gradient can, as a first approximation, be assumed to be nearly unaffected by the presence of sediment, so

$$\frac{du}{dz} \simeq \frac{U_f'}{\kappa z} \tag{7.73}$$

The dispersive stress acting on the bed must depend on this velocity gradient calculated for a value of z in order of one particle diameter d

$$z = \alpha_1 d \tag{7.74}$$

where $\alpha_1 = O(1)$. Inserting Eqs. 7.42, 7.73 and 7.74 into Eq. 7.72 gives

$$\tau' = \tau_c + nF_D + \frac{0.013}{\kappa^2 \alpha_1^2} \rho s \left(\lambda_b U_f' \right)^2 \tag{7.75}$$

where λ_b is the linear concentration of suspended sediment at bed level, cf. Eq. 7.41. In non-dimensional form, Eq. 7.75 becomes

$$\theta' = \theta_c + \frac{\pi}{6} \mu_d p + \frac{0.013}{\kappa^2 \alpha_1^2} s\theta' \lambda_b^2 \tag{7.76}$$

after Eq. 7.56 is introduced.

Hence this model provides a method for calculation of c_b from the requirements of momentum transfer to the immobile sand surface if p is known.

When θ' becomes very large, corresponding to large suspended transport rates, we assume p to be unity and find that

$$\lambda_b = \kappa \alpha_1 \sqrt{\frac{\theta' - 0.3}{0.013 s\theta'}} \rightarrow \frac{8.77 \kappa \alpha_1}{\sqrt{s}} \tag{7.77}$$

For ordinary sand with $s = 2.65$ and $\kappa \simeq 0.4$ we get

$$\lambda_b = 2.16 \alpha_1 \tag{7.78}$$

For $\alpha_1 = 2$ (i.e. the velocity gradient in Eq. 7.73 is taken $2d$ above the bed), λ_b becomes 4.32, which corresponds to the volumetric bed concentration $c_b = 0.35$. This is estimated to be a reasonable maximum value for suspended sediment in motion. Theoretically, c_b can be as large as 0.65, but this corresponds to firm packing and does not allow free motion of the particles.

If the variation in p given by Eq. 7.58 is adapted, c_b can be found from Eq. 7.76 for all θ'-values. Fig. 7.16 shows an example on such a relation. For $\theta' < 0.1, c_b$ becomes extremely small, while c_b approaches 0.35 for large values of θ'.

Fig. 7.17 shows the sensitivity in the predicted bed concentration (by Eqs. 7.41, 7.58, and 7.76) with μ_d. It is seen that at small θ'-values, the predicted c_b-value strongly depends on μ_d.

It is still a little under debate which value of μ_d will give the best agreement between theory and data. Garcia and Parker (1991) compared different theoretical and empirical expressions for the bed concentration with laboratory data and found that the present suggested method with $\mu_d = 1.0$ gave too small a bed concentration at small θ'-values. Hence, a value of μ_d around 0.50 to 0.65 seems more appropriate.

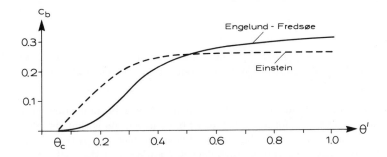

Figure 7.16 Bed concentration c_b versus θ', assuming $\theta_c = 0.05, s = 2.65$, and $\mu_d = 0.50$.

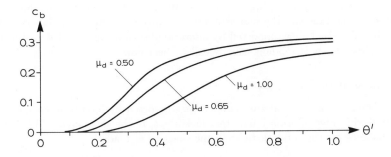

Figure 7.17 Sensitivity in predicted bed concentration to μ_d.

Example 7.7: Einstein's approach

The above presented dynamic approach is quite different from one of the first suggestions on how to determine c_b, namely the geometrical one by Einstein (1950). He simply suggested that the bed concentration was proportional to the concentration of bed load particles and hence given by

$$c_b = \frac{q_B}{U_B} \bigg/ 2d \qquad (7.79)$$

Here q_B/U_B represents the volume of bed load particles per unit area, while $2d$ is the thickness of the bed layer, which Einstein took to be equal to two grain diameters. Now, applying Eq. 7.52 gives

$$c_b = \frac{\pi}{12} p \tag{7.80}$$

where p is given by Eq. 7.58. Eq. 7.80 is also plotted in Fig. 7.16. It is interesting to see that this and the former quite different approach give quite similar results.

The above mentioned models may be insufficient in connection with coarse sediment: one criterion for a particle to go into suspension is that its fall velocity w_s is not so big compared to the near wall velocity fluctuations that it cannot be kept in suspension. The vertical turbulent fluctuation near the wall is in order of U_f', so usually the criterion for a particle to be moved in suspension is

$$\frac{w_s}{U_f'} < 0.8 - 1 \tag{7.81}$$

For coarse sediment, this restriction can not be fulfilled even for quite high values of θ', and instead the sediment will be transported in several layers as bed load, the particles being supported by intergranular collisions, as sketched in Fig. 7.15B. This transport mode is usually called sheet flow.

Thickness of the sheet layer

The mechanism that the sediment is supported by intergranular collisions can be used to give information on the thickness δ_s of the sheet layer. The vertical force balance for the solid particles in the sheet layer reads

$$\frac{d\sigma_G}{dz} = -\rho g(s-1)c \tag{7.82}$$

in which the left-hand-side represents the vertical gradient in the dispersive normal stresses, and the right-hand-side represents the submerged weight of the sediment. Integrating Eq. 7.82 across the sheet layer gives

$$\sigma_{G,b} - \sigma_{G,\delta} = \int_0^{\delta_s} \rho g(s-1)c\,dz \tag{7.83}$$

where index b and δ represent the value at the bed and at the sheet layer surface, respectively. If the dispersive stress at the top of the sheet layer is neglected, and if the relation

$$\tau_b' = \sigma_{G,b} \tan(\phi_d) \tag{7.84}$$

is used for the relation between bed shear stress and normal bed stresses, Eq. 7.83 becomes

$$\int_0^{\delta_s} c\,dz = \frac{\tau_b'}{\rho g(s-1)\tan(\phi_d)} \tag{7.85}$$

Wilson (1987) and Hanes and Bowen (1985) assumed the vertical distribution of sediment in the sheet layer to vary linearly as

$$c = c_0 - \frac{z}{\delta_s}(c_0 - c_\delta) \qquad (7.86)$$

in which c_δ is the concentration at the top of the sheet layer and c_0 the maximum concentration. If the influence of suspended sediment above the sheet layer is neglected, the upper concentration c_δ can be put equal to zero, and Eq. 7.85 gives

$$\delta_s = \frac{2\tau_b'}{\rho g(s-1)c_0 \tan(\phi_d)} \qquad (7.87)$$

or

$$\frac{\delta_s}{d} = \frac{2\theta'}{c_0 \tan(\phi_d)} \qquad (7.88)$$

from which it is seen that δ_s increases proportional with θ', (Wilson, 1987). Hanes and Bowen (1985) obtained the more general expression

$$\frac{\delta_s}{d} = \frac{2\theta' \Delta(\phi_d)}{(c_0 + c_\delta)} \qquad (7.89)$$

by using Eq. 7.86 and allowing a variation in ϕ through the sheet layer (see Example 7.7). In Eq. 7.89, $\Delta(\phi_d)$ is defined by

$$\Delta(\phi_d) = \frac{1}{\tan(\phi_b)} - \frac{1}{\tan(\phi_\delta)} \qquad (7.90)$$

where ϕ_b and ϕ_δ are the value of the dynamical friction angle at the bed and the sheet layer surface.

Velocity distribution inside the sheet layer

One of the main difficulties in establishing a sediment transport formula for the load in the sheet flow layer is to calculate the velocity of fluid and grains in the sheet flow layer.

Only very few experimental data are available today, and most of the theoretical work is therefore developed on a quite speculative background. A few of the suggestions are briefly described below.

A: Wilson's model

Wilson (1987) calculated the velocity profile in the sheet layer by application of a mixing length theory, taking

$$\sqrt{\frac{\tau_F}{\rho}} = l\frac{du}{dz} \qquad , \quad l = \kappa z \qquad (7.91)$$

cf. Eq. 2.68. κ, which for clear water flow is 0.4, may be different in the sheet flow. τ_F can be calculated as

$$\tau_F = \tau' - \tau_G \tag{7.92}$$

The τ_G-variation is found by

$$\tau_G = \tan(\phi_d) \int_z^{\delta_s} \rho g(s-1)c\,dz \tag{7.93}$$

whereby the fluid shear stress is estimated to vary like

$$\tau_F = \tau_b \left(2\left(\frac{z}{\delta_s}\right) - \left(\frac{z}{\delta_s}\right)^2 \right) \tag{7.94}$$

Eqs. 7.91 and 7.94 now give

$$U_f' \sqrt{2\frac{z}{\delta_s} - \left(\frac{z}{\delta_s}\right)^2} = \kappa z \frac{dU}{dz} \tag{7.95}$$

Close to the bed, this equation has the asymptotical solution

$$U = \frac{U_f'}{\kappa} 2 \sqrt{\frac{2z}{\delta_s}} \tag{7.96}$$

which combined with the concentration profile Eq. 7.86 (with $c_\delta = 0$) suggests a transport rate given by

$$q_B = \int_0^{\delta_s} U c\,dz = \frac{8\sqrt{2}}{15} \frac{U_f' c_0 \delta_s}{\kappa} \tag{7.97}$$

or

$$q_B = \frac{1.51}{\kappa g(s-1)\tan\phi} \left(\frac{\tau_b'}{\rho}\right)^{3/2} \tag{7.98}$$

In dimensionless form this gives

$$\Phi_B = \frac{1.51}{\kappa \tan(\phi_d)} (\theta')^{3/2} \approx 12(\theta')^{3/2} \tag{7.99}$$

taking $\tan(\phi_d) \approx 0.32$.

However, it is still open to debate whether a mixing length approach based on the fluid shear stresses, like that presented above, describes the real nature of the fluid flow in the sheet layer, because the fluid to a large extent is controlled by the moving solid particles in the sheet layer. Further, the mixing length might rather be correlated to the vertical exchange of particles between different layers in the sheet layer, which in turn results in a similar opposite exchange of fluid.

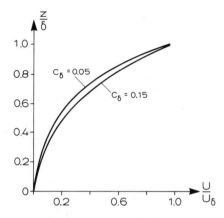

Figure 7.18 Velocity profiles in the sheet layer, after Hanes and Bowen (1985).

B: The model of Hanes and Bowen

Hanes and Bowen used a different approach to determine the velocity distribution in the sheet layer. They applied Bagnold's expression

$$\tau_G = 0.013 s \rho \left(\lambda d \frac{dU}{dz} \right)^2 \tag{7.100}$$

to relate the grain shear stresses and the velocity gradient. With a known value of τ_G it is possible to calculate the velocity profile.

Fig. 7.18 shows examples on the calculated dimensionless velocity profiles, obtained from Eqs. 7.77 and 7.100 for two different values of the bed concentration c_δ of suspended sediment at the top of the sheet layer. The velocity gradient is smallest near the stationary bed where the concentration is highest. As the concentration decreases away from the bed, the velocity gradient increases in order to maintain a constant shear stress. The flow velocity U_δ at the top of the shear layer is obtained from Eqs. 7.86 and 7.100, and from the calculations it is found that

$$\delta_s \sim U_f'^{\;3} \tag{7.101}$$

The transport in the sheet layer is found as the product of concentration and velocity integrated over the sheet layer thickness δ_s. This gives

$$\Phi \sim (\theta')^{5/2} \tag{7.102}$$

The absolute values of the parameters in Eqs. 7.101 and 7.102 are not given, as they, in a complex way, depend on the conditions in the suspension or saltation layer above the sheet layer. This description is also included in the Hanes-Bowen model.

It can finally be mentioned that the approach by Hanes-Bowen differs from the Engelund-Fredsøe model for suspended sediment, described earlier in this section, in the following sense: In the EF-model the bed concentration of suspended sediment is determined by requiring $\tau_G = \tau_b - \tau_c \simeq \tau_b$ at large τ-values and the velocity gradient being determined by the usual logarithmic shape. Thus the bed concentration λ_b can be determined.

In the Hanes-Bowen model, on the other hand, the concentration is instead assumed to be given in advance, whereby Eq. 7.94 gives the required velocity gradient.

In more complete models, such assumptions can be avoided, but the complexity of the models increases very much at the same time. An example of a more complete model for the sediment-fluid mixture based on the constitutive equations for two-phase flow is that by Kobayashi and Seo (1985).

Example 7.8: Variation in the ratio between normal and tangential stresses in the sheet flow

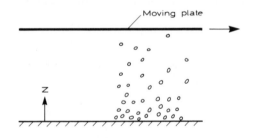

Figure 7.19 Couette-flow of solid particles.

Savage and Sayed (1984) and Hanes and Inman (1985) pointed out that the angle ϕ which determines the ratio between normal and tangential stresses by

$$\tau' = \sigma \tan(\phi_d) \tag{7.103}$$

probably is not constant through the sheet flow layer. A simple example can illustrate this: consider a Couette-flow between two parallel plates, the lower being in rest and the upper moving with a certain constant speed in its own plane. Only grains are present between the two plates, so all stresses are due to intergranular collisions. (This flow can be started by moving the plate over grains initially being in rest. As the flow develops, the upper plate must be allowed to move slightly up, due to the normal forces from the moving grains).

In such a Couette-flow, the tangential shear stress is constant over the depth, while the normal stresses decrease in the upward direction because the weight above decreases away from the bed. In this example it is easy to see from Eq. 7.103 that ϕ_d cannot be constant over the vertical.

REFERENCES

Bagnold, R.A. (1954): Experiments on a gravity-free dispersion of large solid spheres in a Newtonian fluid under shear. Proc. Roy. Soc. London (A) 225:49-63

Einstein, H.A. (1950): The bed-load function for sediment transportation in open channel flows. U.S. Dept. of Agriculture, Techn. Bulletin No. 1026.

Engelund, F. and Hansen, E. (1972): A Monograph on Sediment Transport in Alluvial Streams. 3.ed. Technical Press, Copenhagen.

Engelund, F. and Fredsøe, J. (1976): A sediment transport model for straight alluvial channels. Nordic Hydrology 7:293-306.

Engelund, F. and Fredsøe, J. (1982): Hydraulic theory of alluvial rivers. Advances in Hydroscience, 13:187-215.

Garcia, M. and Parker, G. (1991): Entrainment of bed sediment into suspension. Proc. ASCE, J. Hydr. Div., 117(4):414-435.

Guy, H.P., Simons, D.B. and Richardson, E.V. (1966): Summary of alluvial channel data from flume experiments, 1956-1961. U.S. Geological Survey Professional Paper, 462-I.

Hanes, D.M. and Inman, D.L. (1985): Experimental evaluation of a dynamic yield criterion for granular fluid flows. J. Geophys. Res., 90(B5):3670-3674.

Hanes, D.M. and Bowen, A.J. (1985): A granular-fluid model for steady, intense bed load transport. J. Geophys. Res., 90(C5):9149-9158.

Kobayashi, N. and Seo, S.N. (1985): Fluid and sediment interaction over a plane bed. J. Hydr. Eng. ASCE, 111(6):903-921.

Luque, R.F. (1974): Erosion and transport of bed-load sediment. Delft University of Technology, Diss.

Luque, R.F. and van Beek, R. (1976): Erosion and transport of bed-load sediment. J. Hydr. Res., 14(2):127-144.

Meland, N. and Normann, J.O. (1966): Transport velocities of single particles in bed-load motion. Geografiska Annaler, Vol. 48, Serie A.(4):165-182.

Meyer-Peter, E. and Müller, R. (1948): Formulas for bed-load transport. Rep. 2nd Meet. Int. Assoc. Hydraul. Struct. Res., Stockholm 1948, pp. 39-64.

Richardson, J.F. and Zaki, W.N. (1954): Sedimentation and fluidisation, Part 1. Trans. Instn. Chem. Engrs., Vol. 32., pp 35-53.

Savage, S.B. and McKeown, S. (1983): Shear stresses developed during rapid shear of concentrated suspensions of large spherical particles between concentric cylinders, J. Fluid Mech., 127:453-472.

Savage, S.B. and Sayed, M. (1984): Stresses developed by dry cohesionless granular materials sheared in an annular shear cell, J. Fluid Mech., 142:391-430.

Shields, I.A. (1936): Anwendung der Aehnlichkeitsmechanik und der Turbulenz-forschung auf die Geschiebebewegung. Mitt. Preuss. Versuchsanstalt, Berlin, 26.

Wan, Z. (1981): The deformation of a small bed with a transverse slope. Progress Report No. 53, Inst. of Hydrodynamics and Hydraulic Engineering, ISVA, Techn. Univ. Denmark, pp. 9-14.

Wilson, K.C. (1966): Bed-load transport at high shear stress. Proc. ASCE, J. Hydr. Div., 92(HY6):49-59.

Wilson, K.C. (1987): Analysis of bed load motion at high shear stresses. J. Hydr. Eng., ASCE, 113(1):97-103.

Zimmermann, C. and Kennedy, J.F. (1978): Transverse bed slopes in curved alluvial streams. Proc. ASCE, J. Hydraul. Div., 104(HY1):33-48.

Chapter 8. Vertical distribution of suspended sediment in waves and current over a plane bed

This and the following chapter describe how the suspended sediment is distributed vertically by waves and current turbulence. The presence of bed forms like ripples and sandwaves complicates this description. However, these bed forms are generally washed out at large shear stresses (or, strictly speaking, at large Shields parameters), and the bed becomes almost plane. This plane bed case is very important because the main part of the sediment transport takes place at large Shields parameters. This chapter considers only the plane bed case, while the influence of bed forms is introduced in the following two chapters.

8.1 Vertical distribution of suspended sediment in a steady current

Consider a steady flow over a plane bed. In this case, the temporal mean values of the velocity are parallel with the bed, and the sediment is kept in suspension by turbulent fluctuations. The classical approach to calculating the vertical distribution of suspended sediment is to adapt the same mixing length concept as outlined in Example 2.8.

Consider uniform sands with a settling velocity w_s. In a turbulent flow, these sands are assumed to settle relative to the surrounding water by their fall velocity. Now, following the concepts of the mixing length theory, fluid and sand will be transported from a lower level I where the (volumetric) concentration of suspended sediment is $c - \frac{1}{2}ldc/dz$ up to a higher level II where the concentration is $c + \frac{1}{2}ldc/dz$.

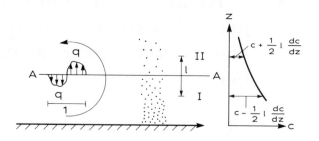

Figure 8.1 Suspended sediment in turbulent flow.

This exchange is caused by an upward fluid discharge q through A-A, which is a cross section parallel with the bed and located between level I and II. The fluid will transport the following amount of sediment up through A-A

$$q_u = (w' - w_s)\left(c - \frac{1}{2}l\frac{dc}{dz}\right) \tag{8.1}$$

in which w' is the vertical velocity fluctuation of the fluid.

The upward transport is compensated by a corresponding downward transport of fluid and sediment. Analogous to Eq. 8.1, the downward sediment transport is given by

$$q_d = (w' + w_s)\left(c + \frac{1}{2}l\frac{dc}{dz}\right) \tag{8.2}$$

In the case of a steady situation, q_u and q_d must be equal, which gives

$$cw_s + \frac{1}{2}w'l\frac{dc}{dz} = 0 \tag{8.3}$$

Here

$$\frac{1}{2}w'l \simeq ql \tag{8.4}$$

The term ql is the same as that evaluated in mixing length theory, see Example 2.8. Eq. 2.73 yields

$$ql = \frac{\tau}{\rho} \bigg/ \frac{du}{dz} \tag{8.5}$$

In steady channel flow the vertical shear stress distribution is given by

$$\frac{\tau}{\rho} = \frac{\tau_b}{\rho}\left(1 - \frac{z}{D}\right) \tag{8.6}$$

and the velocity gradient by

$$\frac{du}{dz} = \frac{U_f}{\kappa z} \tag{8.7}$$

Due to the presence of suspended sediment, a small deviation in the logarithmic shape of the velocity may occur, so Eq. 8.7 is - especially for large concentrations - an approximation as described in Example 8.3. From Eqs. 8.3 and 8.7, the following differential equation in c is obtained

$$cw_s + \kappa U_f z\left(1 - \frac{z}{D}\right)\frac{dc}{dz} = 0 \tag{8.8}$$

which by integration gives

$$c = c_b\left(\frac{D - z}{z}\frac{b}{D - b}\right)^{\frac{w_s}{\kappa U_f}} \tag{8.9}$$

This distribution is usually called the Vanoni-distribution, and the parameter $w_s/(\kappa U_f)$ the Rouse parameter. c_b is the reference concentration of suspended sediment a distance b above the bed. Usually b can be taken to be $2d$, see Section 7.4.2.

Example 8.1: Solution of the vertical distribution by the eddy viscosity concept

From Eqs. 2.61 and 2.73 the mixing length and eddy viscosity are connected by

$$\nu_T = ql = \kappa U_f z\left(1 - \frac{z}{D}\right) \tag{8.10}$$

which is the generally accepted distribution of eddy viscosity for steady channel flow. Inserting Eq. 8.10, Eq. 8.8 can be written

$$w_s c + \nu_T \frac{dc}{dz} = 0 \tag{8.11}$$

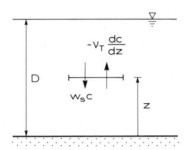

Figure 8.2 Settling and diffusion of sediment

The two terms in Eq. 8.11 are easy to interpret: the left term $w_s c$ represents the settling of the suspended grains through a unit area parallel to the bed, see Fig. 8.2. The other term $\nu_T dc/dz$ represents the diffusion of suspended sediment.

The diffusion is characterized by spreading of matter proportional to a diffusion coefficient and the negative gradient of the concentration. From Eq. 8.11 it is seen that the eddy viscosity can be taken as the mixing, or diffusion coefficient. However, it can be argued that the mixing of sediment is not completely analogous to the mixing of water. For this reason, it is convenient to introduce the following relationship

$$\varepsilon_s = \beta \nu_T \qquad (8.12)$$

where ε_s is the mixing coefficient for solid material and β is a factor which must depend on the sand size and the level of turbulence. One important reason for β to deviate from unity might be caused by the sediment settling out of the surrounding water before the water loses its earlier composition by mixing, as discussed in Example 8.2. Another important contribution may occur from the centrifugal forces having a larger effect on the sediment grains with their larger density than on the fluid particles. Based on measurements by Coleman (1981), van Rijn (1984) suggested that β should be given by

$$\beta = 1 + 2\left[\frac{w_s}{U_f}\right]^2 \quad , \quad 0.1 < \frac{w_s}{U_f} < 1 \qquad (8.13)$$

However, the variation in β is still not totally clear (Van de Graaff, 1988).

For high concentrations, Eq. 8.11 must be modified to take into account that the sediment grains occupy a certain fraction of the total volume. This implies that when a certain volume of sediment $w_s c$ settles through a unit area, this volume must be replaced from below by water and sediment. Here the concentration is also approximately c, so the volume of solid matter transported up through the unit area is $c(cw_s)$. Hence Eq. 8.11 becomes

$$w_s c - w_s c^2 + \nu_T \frac{dc}{dz} = 0 \qquad (8.14)$$

which corresponds to a decrease in the fall velocity w_s equal to a factor $(1-c)$. This partly explains the measured reduction in w_s at high concentrations as described in Example 7.2.

Example 8.2: Some considerations on the relation between the turbulent diffusion coefficient and the Rouse number

In this example, two mechanisms are presented which may explain the deviation of β appearing in Eq. 8.12 to be different from unity (Deigaard, 1991).

A. Turbulent exchange of sediment

By using mixing length theory the derivation of the vertical concentration profile can be extended as follows:

The water in the vertical exchange is moving with a velocity w which is approximately equal to the vertical turbulent velocity fluctuations U_f.

$$w = \alpha U_f \qquad (8.15)$$

where α is a coefficient of order one.

A single event of vertical exchange is considered as a simplified picture of an eddy. The volume X of the exchanged water per event at the level z is

$$X = \alpha_1 l^3 \qquad (8.16)$$

in which α_1 is a coefficient and l is the mixing length. It takes the time $l/(2\alpha U_f)$ to travel from the level $z - l/2$ to z (see Fig. 8.1). The sediment concentration at the start is given by $c - \frac{1}{2} l \frac{dc}{dz}$; cf. Eq. 8.1. As the water volume travels from $z - l/2$ to z it loses the sediment that settles out from it. A characteristic horizontal area of the volume is $\alpha_2 l^2$, and the amount of sediment lost is

$$\alpha_2 l^2 \frac{l}{2\alpha U_f} w_s (c - \Delta c) \qquad (8.17)$$

where Δc is defined as $\frac{dc}{dz} l/2$.

At the same time the water volume gains the sediment that settles into it. The gain is related to the average concentration $\bar{c}$ of sediment in the water surrounding the water volume of exchange

$$\alpha_2 l^2 \frac{l}{2\alpha U_f} w_s \bar{c} =$$

$$\alpha_2 l^2 \frac{l}{2\alpha U_f} w_s(c - \Delta c/2) \tag{8.18}$$

The sediment concentration at level z of a water volume which has travelled upwards is thus

$$c_{up} = \frac{1}{\alpha_1 l^3} \left\{ \alpha_1 l^3 (c - \Delta c) - \alpha_2 l^2 \frac{l}{2\alpha U_f} w_s(c - \Delta c) \right.$$

$$\left. + \alpha_2 l^2 \frac{l}{2\alpha U_f} w_s(c - \Delta c/2) \right\} =$$

$$c - \Delta c + \frac{\alpha_2 l^3}{\alpha_1 l^3} \frac{w_s \Delta c}{4\alpha U_f} = c - \left(1 - \frac{\alpha_2}{4\alpha_1 \alpha} \frac{w_s}{U_f} \right) \Delta c \tag{8.19}$$

Correspondingly, the sediment concentration at level z of a water volume after travelling downward is

$$c_{down} = c + \left(1 - \frac{\alpha_2}{4\alpha_1 \alpha} \frac{w_s}{U_f} \right) \Delta c \tag{8.20}$$

This gives a vertical net flux of sediment

$$\overline{w'c'} = q \left(c - \left(1 - \frac{\alpha_2}{4\alpha_1 \alpha} \frac{w_s}{U_f} \right) \Delta c \right)$$

$$-q \left(c + \left(1 - \frac{\alpha_2}{4\alpha_1 \alpha} \frac{w_s}{U_f} \right) \Delta c \right) =$$

$$-q \left(1 - \frac{\alpha_2}{4\alpha_1 \alpha} \frac{w_s}{U_f} \right) 2\Delta c = -ql \left(1 - \frac{\alpha_2}{4\alpha_1 \alpha} \frac{w_s}{U_f} \right) \frac{dc}{dz} =$$

$$-\varepsilon_s \frac{dc}{dz} \tag{8.21}$$

giving a sediment exchange factor of

$$\varepsilon_s = \left(1 - \frac{\alpha_2}{4\alpha_1 \alpha} \frac{w_s}{U_f} \right) ql = \left(1 - \beta_1 \frac{w_s}{U_f} \right) ql \tag{8.22}$$

It can be seen that these simple kinematic considerations lead to a sediment exchange factor which is always smaller than the momentum exchange coefficient (cf. Eqs. 8.3 and 8.4), with difference proportional to w_s/U_f. Mechanisms that give a ε_s which *increases* with w_s are considered in the following section.

B. Convective sediment exchange

Until now only exchange due to turbulent diffusion has been considered, where the upward flux of sediment is proportional to the gradient in the concentration. Instead the flux could be governed by a convective transport. In the following a very simple model for convective transport is considered. The model is not intended to be a quantitative representation of a real situation, but only an illustration. Situations where convection of sediment can be of importance are for instance:

A. *Wave ripples:* During the first half wave period sediment is transported both as bed load and in suspension in the boundary layer along the upstream side of each ripple. The sediment passes the ripple crest and is suspended in the vortex on the lee side. In the next half wave period the vortices with the sediment move away from the bed and are dissolved, and at the same time a new boundary layer is formed at the present upstream side of the ripples. The vertical flux of sediment carried in the vortices is clearly not a diffusion process (for a more detailed description, see Chapter 10).

B. *The bursting process:* In turbulent boundary layers a significant part of the momentum transfer is related to large coherent fluid motions, which can cause a vertical convection of suspended sediment (Sumer and Deigaard, 1981).

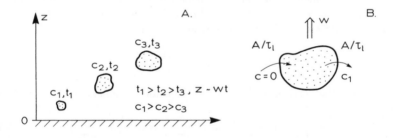

Figure 8.3   Exchange of sediment by convection. A: a mass of water rises with velocity w. B: exchange with surrounding water gives decreasing sediment concentration.

The main element of the simplified model is that fluid volumes are moving upward with a period T_c. In case of a bed with wave ripples T_c will be half of the wave period. The volume per unit bed area is A, and the sediment concentration in the volume is c_1, see Fig. 8.3. The total volume of the upward moving fluid is small, and c_1 is much larger than the concentration c of the surrounding fluid. The upward convective discharge is thus

$$q_{up} = \frac{A}{T_c} \tag{8.23}$$

As a fluid volume moves upward, it exchanges water (and sediment) with the surroundings. The time scale for the exchange is τ_e, giving an exchange rate of

$$\frac{A}{\tau_e} \qquad (8.24)$$

This means that the volume loses an amount of water A/τ_e per unit time with sediment concentration c_1 and receives the same amount of water with a much lower concentration, giving a variation of the concentration in the volume of

$$\frac{d(c_1 A)}{dt} = A\frac{dc_1}{dt} = -\frac{A}{\tau_e}c_1 \qquad (8.25)$$

It is assumed that the volume moves upward with the constant velocity w. Thus t in Eq. 8.25 can be substituted by z/w giving a variation with the level of

$$\frac{dc_1}{dz} = \frac{1}{w}\frac{dc_1}{dt} = -\frac{c_1}{w\tau_e} \qquad (8.26)$$

Eq. 8.26 gives

$$c_1 = c_{10}\exp\left(-\frac{z}{w\tau_e}\right) \qquad (8.27)$$

where c_{10} is the concentration in the volume when it starts its upward motion at $z = 0$. For a steady situation (averaged over T_c) the settling of sediment in the surrounding fluid must balance the vertical convection

$$w_s c = q_{up}c_1 = \frac{A}{T_c}c_1 \qquad (8.28)$$

Eq. 8.28 gives the vertical distribution of suspended sediment in the main water mass

$$c = \frac{A}{T_c w_s}c_1 = \frac{Ac_{10}}{T_c w_s}\exp\left(-\frac{z}{w\tau_e}\right) = c_0\exp\left(-\frac{z}{w\tau_e}\right) \qquad (8.29)$$

which determines the suspended sediment concentration profile.

If a concentration profile, given by Eq. 8.29, is measured, and then *analyzed* in order to determine the turbulent diffusion coefficient; that is, the concentration profile is assumed to be determined by the diffusion equation

$$\varepsilon_s\frac{dc}{dz} = -w_s c \qquad (8.30)$$

after which, inserting Eq. 8.29 in 8.30 gives an *estimated* turbulent diffusion coefficient of

$$\varepsilon_s' = -\frac{w_s c}{dc/dz} = -\frac{w_s c_0\exp(-z/w\tau_e)}{c_0(-1/w\tau_e)\exp(-z/w\tau_e)} = w_s w\tau_e \qquad (8.31)$$

which implies an apparent diffusion coefficient that increases with the settling velocity of the sediment.

Example 8.3:   Vertical distribution of graded sediment

The vertical distribution given by Eq. 8.9 is derived for uniform sand. For graded sediment, Eq. 8.9 is often used with a fall velocity based on the mean diameter of the sediment transported in suspension. This mean diameter can be found as follows: first a criterion for a particle able to go into suspension is needed. The simplest approach is that given by Eq. 7.81

$$w_c = 0.8 U_f'$$ (8.32)

where w_c is the critical fall velocity above which sediment is not able to go into suspension.

As an illustrative numerical example consider a flow along a plane bed consisting of sand with $d_{50} = 0.25$ mm and $\sigma_g = 1.5$. This corresponds to the grain distribution curves shown in Fig. 7.1 and 7.2. The friction velocity in the flow is assumed to be 0.04 m/s. Now the sediment must have a fall velocity smaller than 0.032 m/s in order to be able to go into suspension. At 20°C, Table 7.1, gives that the critical fall diameter is $d_{f,cr} = 0.24$ mm. From the particle-size distribution curve, Fig. 7.2, it is found that this corresponds to 48% being finer. The mean diameter of suspended sediment is then found from the 48/2 = 24% fractile, which gives $\overline{d_s} = 0.19$ mm.

A more refined method has been suggested by Engelund (1975). He divides the sediment with a fall velocity lower than w_c into equal fractions (by volume), and the relative vertical distribution of each fraction is calculated on the basis of the mean fall velocity of each fraction.

The bed concentration of each fraction is obtained from the following two requirements:

(i) the sum of the bed concentrations must be equal to that obtained using the mean diameter d_{50},

   and

(ii) only sediment which is moderately graded is considered. In this case it can be assumed that the particles in suspension will keep the original composition. Hence the integral

$$I = \int_0^D c\, dz$$ (8.33)

should be the same for all the chosen fractions.

In practice two or three fractions are sufficient.

Example 8.4:   Density effects at large concentrations

The presence of suspended sediment near the bed gives rise to a vertical gradient in density, which will slightly modify the velocity profile as pointed out by, among others, Einstein and Chien (1955) and Coleman (1981). Because the concentration of suspended sediment decreases away from the bed, the fluid becomes stabilized, and work must be spent on the fluid to mix it because it requires additional potential energy. The frictional effect of the turbulence is consequently slightly reduced. A discussion of the importance of this effect on the modification of the velocity profile is given by Soulsby and Wainwright (1987) who quantified the stabilizing effect by the Monin-Obukov parameter

$$M = \frac{\kappa g z \overline{\rho' w'}}{\overline{\rho} U_f^3} \tag{8.34}$$

Here, ρ' is the fluctuating part of the density around its time-mean value $\overline{\rho}$ (which as an approximation can be taken as ρ) and w' is the fluctuating component of the vertical velocity, so $\overline{\rho' w'}$ becomes the upward flux of density. The Monin-Obukov parameter is therefore a term describing the ratio of potential energy required to mix the density gradient to the turbulent kinetic energy supplied by the shear at a certain level.

The presence of this stabilizing parameter changes the velocity gradient by

$$\frac{dU}{d(lnz)} = \frac{U_f}{\kappa}(1 + 4.7M) \tag{8.35}$$

using an analogy to changes in the velocity profile of the atmosphere due to density changes caused by variations in temperature.

Soulsby and Wainwright suggested that the influence of density effects can be disregarded for

$$M < 0.03 \tag{8.36}$$

The parameter M can further be related to the vertical distribution of sediment by the following considerations: The fluctuation in density can be related to the fluctuation in volumetric concentration c' by

$$\overline{\rho' w'} = \rho(s - 1)\overline{c' w'} \tag{8.37}$$

In the case of equilibrium, the upward diffusive flux is balanced by the downward settling flux, so

$$\overline{c' w'} = cw_s = \varepsilon_s \frac{\partial c}{\partial z} \tag{8.38}$$

in which the eddy diffusion concept has been introduced. Taking the sediment diffusion coefficient to be equal to the eddy viscosity and

$$\nu_T = \kappa U_f z \tag{8.39}$$

the Monin-Obukov parameter becomes

$$M = \frac{\kappa g(s-1)w_s}{sU_f^3} bc_b\left(\frac{z}{b}\right)^{\left[1 - \frac{w_s}{\kappa U_f}\right]} \tag{8.40}$$

where c_b is the reference concentration at the level b above the bed.

From this equation the influence of density gradients can be evaluated for different values of vertical levels, grain diameters, and friction velocities. Soulsby and Wainwright carried out a detailed discussion on the variation in M with d and U_f, with the reference concentration based on the expression suggested by Smith and McLean (1977)

$$c_b = s\left(\frac{\theta' - \theta_c}{\theta_c}\right)\gamma_1 \tag{8.41}$$

in which γ_1 is a constant ($\sim 1.56 \times 10^{-3}$). The stratification effects are most pronounced for fine sediment combined with large shear stresses.

More important than the damping effect on the velocity profile is the damping effect on the diffusion coefficient for the sediment particles. Van Rijn (1984) expressed the diffusion coefficient by

$$\varepsilon_s = \beta \sigma \nu_T \tag{8.42}$$

in which Van Rijn suggested that β should be given by Eq. 8.13 based on calibrations with the data far away from the bed by Coleman (see also Example 8.2). Further, he suggested

$$\sigma = 1 + \left[\frac{c}{c_0}\right]^{0.8} - 2\left[\frac{c}{c_0}\right]^{0.4} \tag{8.43}$$

in which $c_0 = 0.65$ is the maximum bed concentration. This expression is empirically based on the measurements by Einstein and Chien (1955).

Fig. 8.4 shows how the incorporation of Eqs. 8.22, 8.24 and 8.25 into the diffusion equation, Eq. 8.11, changes the Vanoni profile as given by Eq. 8.9. The bed concentration is based on the Engelund-Fredsøe (1976) formula, Eqs. 7.58 and 7.76, which for the selected examples, gives quite high concentrations just above the bed and hence has a large influence on the ϕ-factor in Eq. 8.42. In Fig. 8.4 the fall velocity is given by

$$w_s = w_{so}(1 - c)^4 \tag{8.44}$$

as suggested by van Rijn (1984). This formula is quite like Eq. 7.16 and based on the same data. From Fig. 8.4 it can be seen that the calculations suggest significantly less sediment enters suspension due to density effects, the main contribution arising from the large near-bed concentrations. However, Eq. 8.42 is not verified at these large concentrations.

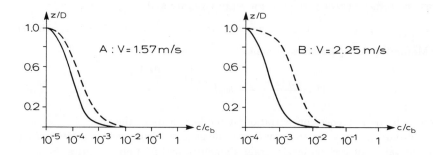

Figure 8.4 Calculated vertical profile of suspended sediment. Dashed line:
Vanoni profile (Eq. 8.9). Solid line: Modified Vanoni profile
due to density effects (van Rijn, 1984). Input data: $d = 0.15$
mm, $w_{so} = 0.018$ m/s, $D = 8$ m.

8.2   Distribution of suspended sediment in pure oscillatory flow

In a pure oscillatory flow, the turbulence is restricted to the thin oscillatory
boundary layer investigated in Chapter 2. Because the flow in this case is unsteady,
it is necessary to apply the complete continuity equation for suspended sediment
which for the plane case reads

$$\frac{dc}{dt} = w_s\frac{\partial c}{\partial z} + \frac{\partial}{\partial z}\left(\varepsilon_s\frac{\partial c}{\partial z}\right) + \frac{\partial}{\partial x}\left(\varepsilon_s\frac{\partial c}{\partial x}\right) \qquad (8.45)$$

where the left-hand-side represents the rate of change in the concentration of
suspended sediments, the first term on the right-hand-side represents the settling
of suspended sediments, and the two last terms the vertical and horizontal sediment
diffusion.

The last term on the right-hand-side can usually be neglected, because the
vertical gradient in c is much larger than the horizontal one. Further, the left-
hand-side of Eq. 8.45 can be approximated by

$$\frac{dc}{dt} = \frac{\partial c}{\partial t} + u\frac{\partial c}{\partial x} + w\frac{\partial c}{\partial z} \simeq \frac{\partial c}{\partial t} \qquad (8.46)$$

because the convective terms are higher order terms which can normally be neglected. Hence, the diffusion equation is reduced to

$$\frac{\partial c}{\partial t} = w_s\frac{\partial c}{\partial z} + \frac{\partial}{\partial z}\left(\varepsilon_s\frac{\partial c}{\partial z}\right) \qquad (8.47)$$

Example 8.5:   Solution of the time-averaged continuity equation for suspended sediment

Time-averaging of Eq. 8.47 gives

$$w_s \frac{\partial \bar{c}}{\partial z} + \frac{\partial}{\partial z}\left(\overline{\varepsilon_s \frac{\partial c}{\partial z}}\right) = 0 \qquad (8.48)$$

or

$$w_s \bar{c} + \overline{\varepsilon_s \frac{\partial c}{\partial z}} = 0 \qquad (8.49)$$

in which $\bar{c}$ is the time-averaged value of the concentration at a certain level z above the bed. Inserting a time-independent value for ε_s into Eq. 8.49 gives

$$w_s \bar{c} + \bar{\varepsilon}_s \frac{\partial \bar{c}}{\partial z} = 0 \qquad (8.50)$$

This equation can easily be solved with known vertical variations of $\bar{\varepsilon}_s$. Such an expression was suggested by Lundgren (1972), who constructed the expression

$$\varepsilon_s = \kappa U_{f,\max} z \left/ \left[1 + 1.34 \sqrt{\frac{1}{2} f_w} \frac{z}{\delta_1} \exp\left(\frac{z}{\delta_1}\right)\right]\right. \qquad (8.51)$$

where $U_{f,\max}$ = the maximum shear velocity during one wave period, f_w = the wave friction factor and δ_1 = the mean boundary layer thickness. Eq. 8.51 is empirically based on the measurements by Jonsson (1963). Since Eq. 8.51 is the mean averaged value over one wave period, Eq. 8.50 can easily be solved to give

$$\bar{c} = K z^{-\frac{w_s}{\kappa U_f}} \exp\left[-\frac{w_s}{\kappa U_f} 1.34 \sqrt{\frac{1}{2} f_w} \exp\left(\frac{z}{\delta_1}\right)\right] \qquad (8.52)$$

However, the assumption that the eddy viscosity is time-invariant is usually an oversimplification because

$$\overline{\varepsilon_s \frac{\partial c}{\partial z}} \neq \bar{\varepsilon}_s \frac{\partial \bar{c}}{\partial z} \qquad (8.53)$$

This can be illustrated by the following example:

A periodic motion is considered which during the first half-period $T/2$ has a constant flow velocity V_1 and during the second half-period has zero velocity. If

T is assumed to be so large that the time-scale for the settling of suspended sediment is much smaller than T, the correct distribution will be

$$\bar{c} = \frac{1}{2}c_{b1}\left(\frac{D-z}{z}\frac{b}{D-b}\right)^{\frac{w_s}{\kappa U_{f1}}} \qquad (8.54)$$

where U_{f1} is the friction velocity according to the mean flow velocity V_1, and c_{b1} the bed concentration determined from the conditions at flow velocity V_1. If, instead, the diffusion equation based on the mean eddy viscosity during the total period of motion is solved then the eddy viscosity will only be half the eddy viscosity for the flow velocity V_1, consequently Eq. 8.50 gives

$$\bar{c} = c_{b2}\left(\frac{D-z}{z}\frac{b}{D-b}\right)^{\frac{2w_s}{\kappa U_{f1}}} \qquad (8.55)$$

where c_{b2} is a nominal concentration different from c_{b1}. It can easily be seen that even by an appropriate choice of c_{a2}, Eqs. 8.54 and 8.55 cannot be identical.

Substitution of a time-averaged eddy viscosity may also lead to significantly incorrect predictions of the resulting sediment transport in combined wave-current motion.

For unsteady flow, the instantaneous amount of sediment in suspension is not determined by the instantaneous value of the bed shear stress, because the sediment takes some time to settle after being picked up from the bed. This means that the sediment transport must be determined by

$$q_s = \frac{1}{T}\int_0^T\int_0^D ucdzdt \qquad (8.56)$$

where u and c are the instantaneous vertical velocity distribution and the concentration distribution, respectively. Because of the lag between velocity and concentration, Eq. 8.56 differs from

$$q_1 = \int_0^D U\bar{c}dz \qquad (8.57)$$

where U is the average velocity and $\bar{c}$ the average concentration (over one wave period).

For the reasons given in the preceding example, the correct approach for solving the vertical distribution of suspended sediment in unsteady flow will be to solve the complete version of Eq. 8.47, calculating the time and space variations in c during one wave period. Using this approach it will be possible to take into account the time variation in the eddy viscosity and to introduce the boundary condition at the bed in a more rigorous manner.

Fig. 8.5 shows the result of such a calculation. The *flow description* is based on the momentum integral method for the wave boundary layer, see Section 2.2. In this description, the variations in bed shear stress and boundary layer thickness were calculated at each phase during a wave cycle. From this information, the instantaneous value of the eddy viscosity can be taken as

$$\nu_T = \kappa |U'_f| z (1 - z/\delta) \qquad (8.58)$$

where both U'_f and δ vary with time. Taking $\varepsilon_s = \nu_T$, Eq. 8.47 must now be solved numerically. The boundary conditions being

(i) The time variation in c must be periodic, so

$$c(t, z) = c(t + T, z) \qquad (8.59)$$

(ii) At the water surface, there must be no vertical flux, so

$$\nu_T \frac{\partial c}{\partial z} + w_s c = 0 \quad , z = D \qquad (8.60)$$

However, because of the very limited vertical extent of the wave boundary layer, Eq. 8.60 will in practice always degenerate into

$$c \to 0 \quad \text{for} \quad z \to \infty \qquad (8.61)$$

(iii) The last boundary condition is related to the bed concentration of suspended sediment. As it appears from Chapter 7, this problem still does not have a completely satisfactory solution, even for the more simple case of steady current. In the following, the Engelund-Fredsøe (1976) approach (see Section 7.4.2) for the bed concentration has been applied. It relates the bed concentration at $z = 2d$ to the Shields parameter

$$c_b = c_b(\theta') \qquad (8.62)$$

where this relationship is given by Eqs. 7.58 and 7.76. In the case of unsteady flow, Eq. 8.62 is used as the bed boundary condition for the instantaneous bed shear stress.

Fig. 8.5 shows the variation in c with time at different distances from the bed for a specific run. The bed roughness k_N has been set to $2.5d$, so $z/k_N = 0.8$ corresponds to $z = 2d$. The ordinate indicates the relative concentration $c/c_{b,\text{max}}$, where $c_{b,\text{max}}$ is the bed concentration for $U'_f = U'_{f,\text{max}}$.

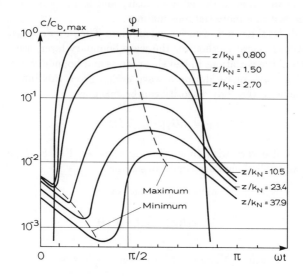

Figure 8.5 Variation in concentration with time at different levels from the bed for $a/k_N = 3916$, $\theta'_{max} = 1.45$ and $w_s/U_{1m} = 0.018$. (Fredsøe et al., 1985).

Just above the bed, the maximum concentration occurs at the same time as the maximum bed shear stress occurring, as described in Chapter 2, at a phase ϕ before the maximum wave velocity outside the boundary layer. Increasing distance from the bed causes the maximum concentration to lag more and more behind the maximum bed shear stress, the suspended sediment reacting to changes in the flow with a certain lag, because it takes some time for the sediment to settle after it has been picked up from the bed.

Furthermore, Fig. 8.5 shows that the variation in c follows an asymmetric pattern: the rise in concentration occurs much faster than the fall, for two reasons: First, the variation in U_f is asymmetric, see Chapter 2. However, this contribution is not very important, as can be seen from Fig. 8.5, as the variation in the bed concentration for $z/k_N \simeq 0.8$ is nearly symmetric. The reason is that the bed concentration is insensitive to changes in bed shear stress if θ' is sufficiently large (see Fig. 7.16).

The second contribution to the asymmetric shape arises because the rise in the concentration is very rapid when the sediment is brought into suspension and pushed away from the bed during periods of large bed shear stresses, while the fall in concentration simply occurs because the sediment falls towards the bed at the fall velocity w_s with almost no turbulence present.

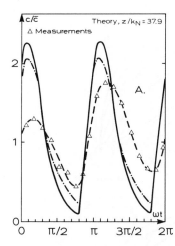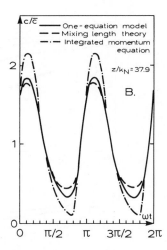

Figure 8.6 Comparison between measured and predicted variation in c during one wave period 1.8 cm above the bed. Fig. A: Experimental data: $d_{50} = 0.19$ mm, $a = 1.86$ m, $T = 9.1$ s (from Staub et al., 1984). (These data correspond to the non-dimensional data in Fig. 8.4). Dashed-dotted curve, Fig. A: Gradation of sediment incorporated. Fig. B: Comparison between three different hydrodynamic approaches to estimating ε_s. (Justesen and Fredsøe, 1985).

Fig. 8.6 shows a comparison between the results shown in Fig. 8.5 and the time variation in the concentration measured by Staub et al. (1984), who applied a mechanical suction system to measure the concentration. It is seen that both the asymmetric behaviour and the amplitude in the variation of the normalized concentration $c/\bar{c}$ during a wave cycle confirm the theoretical predictions.

As outlined in the last part of Example 8.3, the effect of including the sediment gradation in the theory is indicated by a dashed-dotted curve in Fig. 8.6A. The suspended sediment has been divided into three fractions.

Figs. 8.7A and B depict the average concentration over one wave period at different combinations of the dimensionless parameters a/k_N, $w_s/U_{f,max}$, and θ_{max}. From the figure it is noticed that changes in θ'_{max} are only important close to the bed. The fully-drawn curves are those predicted from theory setting $\theta'_{max} = 10$, while the dashed ones are based on $\theta'_{max} = 1$.

On the ordinate in Fig. 8.7, the value δ_1/k indicates a measure of the mean value of the turbulent boundary layer thickness, δ_1 being the boundary layer

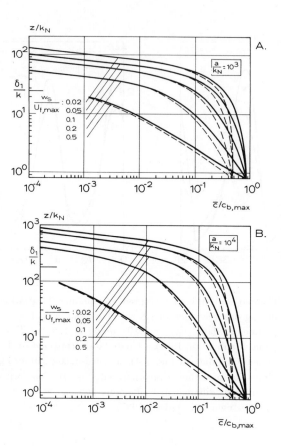

Figure 8.7 Theoretical vertical distribution of the average concentration $\bar{c}$ in pure wave motion. A: $a/k_N = 10^3$. B: $a/k_N = 10^4$. The fully drawn line corresponds to $\theta'_{max} = 10$, while the dashed line corresponds to $\theta'_{max} = 1$. (Fredsøe et al., 1985).

thickness at $\omega t = \pi/2$ (maximum wave-induced velocity outside the boundary layer). It is seen that a significant amount of sediment may be present above this mean boundary layer thickness.

Other hydrodynamic approaches

Instead of adopting the eddy viscosity variation given by Eq. 8.58 obtained from the momentum integral method, ν_T can be obtained from more refined models like those described in Section 2.3. In Fig. 8.6B three different models for ν_T have been used to solve the diffusion equation, Eq. 8.47, for the same data as those used in Fig. 8.6A. Besides the one mentioned above, the other approaches are: the eddy viscosity found from mixing length theory and from a one-equation turbulent model (k-equation). As seen from Fig. 8.6B, the latter two approaches predict slightly smaller variations in the concentration during a wave cycle, because the predicted time variation in the eddy viscosity is smaller, as described in Chapter 2.

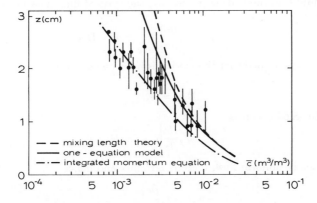

Figure 8.8 Predicted vertical mean concentration distribution $\bar{c}$. Experiments by Staub et al. (1984).

In Fig. 8.8 the predictions of the time-averaged vertical distribution of the concentration $\bar{c}$ from the same three different models are shown. These predictions are compared with measurements by Staub et al. (1984), showing the average concentration for the data at different distances from the bed. The line surrounding each experimental point indicates the uncertainties in the measurement. These uncertainties arise from the inexact determination of the bed level, which varies slightly with time, even through a single wave period.

It is seen that the more refined hydrodynamic models predict a slightly larger concentration of suspended sediment away from the bed. The experimental data base is, however, too limited to confirm which theory should be preferred.

Example 8.6: Discussion on the bed boundary condition

Some researchers are not in favour of the bed boundary condition given by Eq. 8.62 in the case of unsteady flow, and prefer instead the instantaneous bed concentration to be given by a so-called 'pick-up function', also called the gradient boundary condition. This is based on the following considerations (Parker, 1978): The vertical flux of sediment at the height z above the bed is given by

$$\text{Flux} = -\varepsilon_s \frac{\partial c}{\partial z} - w_s c \tag{8.63}$$

Just above the bed, this flux must be equal to the amount of sediment eroded from the bed into the fluid, E, minus the amount of sediment deposited from the fluid into the bed, Dep. Because the vertical flux must be continuous at the bed

$$-w_s c - \varepsilon_s \frac{\partial c}{\partial z}\bigg|_{z=2d} = E - \text{Dep} \tag{8.64}$$

Now, it is argued that the mechanism for bed deposition is the fall velocity, so

$$\text{Dep} = w_s c\big|_{z=2d} \tag{8.65}$$

whereby Eq. 8.64 becomes

$$-\varepsilon_s \frac{\partial c}{\partial z}\bigg|_{z=2d} = E \tag{8.66}$$

In steady uniform flow, there is zero flux and hence

$$-w_s c - \varepsilon_s \frac{\partial c}{\partial z}\bigg|_{z=2d} = 0 \tag{8.67}$$

In this case, Eqs. 8.66 and 8.67 give

$$E = w_s c_b \tag{8.68}$$

In unsteady (or non-uniform) flow, E must be evaluated as a function of flow conditions in order to solve Eq. 8.66. Suggestions concerning the function E for the wave case is for instance given by Nielsen (1979). However, no experimental verification of the time variation in the suggested variation in E is available today, and the bed concentration boundary condition, Eq. 8.62, can easily be translated into another suggestion for the appearance of the unknown function E.

The implications of the two boundary conditions can be illustrated by a specific example as suggested by Parker (1978): he considered an example where uniform steady flow in a channel has developed. If, for simplicity, the diffusion coefficient ε_s is assumed to be a constant quantity, i.e. independent of the height z, then the diffusion equation 8.47 reads

$$w_s \frac{\partial c}{\partial z} + \varepsilon_s \frac{\partial^2 c}{\partial z^2} = 0 \tag{8.69}$$

With the requirement of no flux through the water surface, Eq. 8.60, and with the bed boundary condition, Eq. 8.66, Eq. 8.69 has the solution.

$$c_o = \frac{E}{w} \exp\left(-\frac{w_s}{\varepsilon_s}(z - 2d)\right) \tag{8.70}$$

Parker imagined that this channel flow was subjected to a uniform rain of particles from above the water surface, so there would be a constant downward volumetric flux I through the water surface. The concentration can now be written as

$$c(z) = c_o(z) + e(z) \tag{8.71}$$

where c_o is given by Eq. 8.70. The overload concentration also satisfies the diffusion equation 8.47

$$w_s \frac{\partial e}{\partial z} + \varepsilon_s \frac{\partial^2 e}{\partial z^2} = 0 \tag{8.72}$$

and the boundary condition for e at the water surface becomes

$$\left[w_s e + \varepsilon_s \frac{\partial c}{\partial z} \right]_{z=D} = I \tag{8.73}$$

If the bed concentration boundary condition is used, c_b must remain unchanged in spite of the additional supply of sediment from above, because the bed shear stress is unchanged. Hence

$$e\Big|_{z=2d} = 0 \tag{8.74}$$

The solutions of Eqs. 8.72 - 8.74 give

$$e = \frac{I}{w_s} \left(1 - \exp\left(-\frac{w_s}{\varepsilon}(z - 2d) \right) \right) \tag{8.75}$$

If, on the other hand, the gradient boundary condition is used, the rate of erosion is unaffected by overloading, so

$$\frac{de}{dz}\Big|_{z=2d} = 0 \tag{8.76}$$

which together with Eqs. 8.72 and 8.73 gives

$$e = \frac{I}{w_s} \tag{8.77}$$

The results of these two imaginary cases are shown in Fig. 8.9. In Fig. 8.9A it is seen that the gradient condition (b.c. I) implies that overloading can be felt identically over the whole depth, while the bed concentration boundary condition (b.c. II) implies that overloading is mostly felt at the water surface.

If the supply is very heavy as sketched in Fig. 8.9C and D, the situation can occur where the concentration will decrease close to the bed if the bed concentration condition is used, see Fig. 8.9D. This seems intuitively wrong, but it must be mentioned that in the case of vanishing ε_s, both Eq. 8.75 and Eq. 8.77

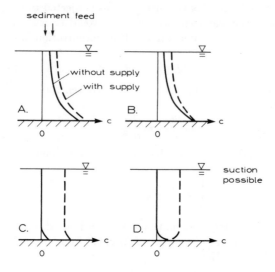

Figure 8.9 Solution for steady uniform overloading. A and C: gradient
boundary condition. B and D: bed concentration boundary
condition. Fully drawn line: concentration profile without supply from above. Dashed line: concentration profile with supply
from above.

describe a uniform distribution of sediment through the depth which is easily seen
to be the correct solution.

On the other hand, if a suction system at $z = D$ is considered instead
of a supply system, the gradient boundary condition implies that even the bed
concentration of suspended sediment will decrease, while the bed concentration
conditions give the more reasonable result that the concentration close to the bed
far and away from the suction system is nearly unaffected.

In unsteady flow, similar considerations can be made: If a hypothetical case
is considered in which the flow velocity is suddenly reduced to zero from a certain
steady value where $\theta = \theta_0$, then the bed concentration condition gives the result
that c_b drops instantaneously from $c_b(\theta_0)$ to zero. However, all the sediment in
suspension before the change in flow velocity occurs cannot settle faster than the
settling velocity w_s. This means that in the still water the time evolution of c_b at
the distance b above the bed is determined by

$$c_b(b, t) = c(b + \Delta z, t - \Delta t) \tag{8.78}$$

in which $\Delta z = w\Delta t$.

   This equation simply states that the suspended sediment settles without changing the shape of its profile.

   The previous example corresponds to the overloaded case. The underloaded case corresponds to the flow situation where the fluid velocity is suddenly increased from zero to a certain flow velocity where $\theta = \theta_0$: here it is possible for the bed concentration to obtain the new value $c_b(\theta_0)$ immediately because the suspended sediment only has to be transported a very small distance (of the order of one grain diameter) from the bed.

   From the considerations given above it seems reasonable to apply the bed concentration condition for waves provided that the concentration profile close to the bed is underloaded (negative vertical concentration gradient), while the bed concentration in the case of a positive concentration gradient should be determined by Eq. 8.78.

   Hence the resulting boundary conditions become

$$c_b = \max\{c_b(\theta'), \ c(b + w_s\Delta t, \ t - \Delta t)\} \tag{8.79}$$

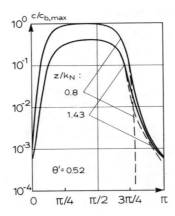

Figure 8.10 Effect of choice of boundary condition for the sediment con-
        centration. Solid lines indicate the results obtained using Eq.
        8.79, whereas the dashed lines show results by using the bed
        concentration condition (b.c. I). Input data: $a/k_N = 10^3$ and
        $w_s/U_{1m} = 0.038$. (From Justesen et al., 1986).

   The correct choice of the bed boundary condition is, however, in practice not very important: Fig. 8.10 shows the effect of the selected boundary condition: the dashed line in Fig. 8.10 is the result obtained by applying the bed concentration boundary condition, while the fully drawn line is obtained from Eq. 8.79. It is seen that the effect of the different boundary conditions is limited to a few grain sizes away from the bed.

8.3   Vertical distribution of suspended sediment in combined wave-current motion

The presence of a current together with the waves implies that the turbulence will now be present over the whole flow depth, cf. Chapter 3. It is quite easy to extend the approach from the preceding section to cover the general wave-current case, still solving the unsteady diffusion equation, Eq. 8.45, together with the boundary conditions. In the diffusion equation Eq. 8.45, the diffusion coefficient ε_s, which is taken equal (or proportional) to the eddy viscosity ν_T, must now be determined from the hydrodynamic approach given in Chapter 3.

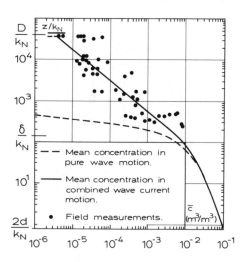

Figure 8.11 Calculated mean concentration profile in combined wave-current motion. $H = 1.7$ m. $T = 12$ s, $V = 0.20$ m/s. $d_{50} = 0.07$ mm and $D = 7$ m. The field experiments are taken from Kirkegaard and Sørensen (1972). The dashed line is the solution for the same parameters except that the current has been set equal to zero. (From Fredsøe et al., 1985).

Fig. 8.11 shows the result of such a calculation in which the eddy viscosity is evaluated from the integrated momentum equation approach, cf. Section 3.2. This is done as follows:

In the outer region, the eddy viscosity is assumed to be parabolic and independent of time

$$\varepsilon_c = \kappa z U_{fc}(1 - z/D) \tag{8.80}$$

where U_{fc} is the current friction velocity, which is related to the average value of the bed shear stress in the current direction by the relation

$$\overline{\tau_b} = \rho U_{fc}^2 \tag{8.81}$$

Inside the wave boundary layer, the eddy viscosity is assumed time dependent. Here, the eddy viscosity is taken to vary as

$$\varepsilon_s = \kappa z U_f' \left[1 - \frac{z}{\delta}\left(1 - \frac{U_{fc}}{U_f}\right) \right] \tag{8.82}$$

so ε_s has the same parabolic shape as Eq. 8.58 for the pure oscillating boundary layer, but equates to the outer eddy viscosity given by Eq. 8.75 at the top of the boundary layer.

The bed boundary condition Eq. 8.62 is based on the instantaneous value of τ_b calculated for the combined wave-current motion.

The influence of the current is easily vizualized from Fig. 8.11. The solid line is the theoretical solution for the time-averaged concentration $\overline{c}$, while the broken line shows a calculation based on the same data, except for the current which has been set equal to zero. It is easily seen that the presence of suspended sediment far away from the bed is due to the presence of the current, while close to the bed, the presence of a persistent current does not greatly change the predicted concentration profile because the turbulence in this layer is mainly caused by the wave motion.

Fig. 8.12 presents a more general diagram. In the combined wave-current motion, three additional parameters are necessary in order to define the flow situation compared to pure wave motion; the angle γ between the current direction and the direction of wave propagation, the dimensionless flow depth D/k_N and, finally, the strength of the current compared with the strength of the near-bed wave-induced flow velocities. The latter can, for instance, be represented by the dimensionless quantity U_{1m}/U_{fc}. If U_{1m}/U_{fc} approaches zero, the motion approaches the pure current situation, and the distribution of the sediment will be equal to the well-known Rouse-Vanoni distribution. On the other hand, if U_{1m}/U_{fc} approaches infinity, the situation will correspond to the case of pure wave motion, so the distribution will approach the one described in the previous section. This behaviour can be seen from Fig. 8.12 where the parameter a/k_N held at 100, while the parameter U_{1m}/U_{fc} is varied. The solid lines are those for $\gamma = 0$ (current direction is the same as the direction of wave propagation). It is interesting to note that for some intervals (around $U_{1m}/U_{fc} \sim 10$) the bed concentration decreases for a fixed value of $\theta_{\max}$. This is because the near-bed wave and current motions are of the same order in this interval, so when the wave-induced motion is

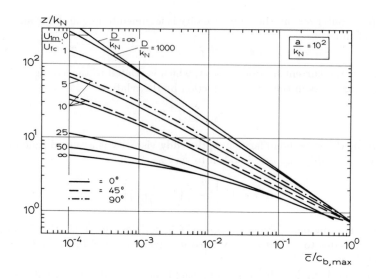

Figure 8.12 Theoretical vertical distribution of the average concentration $\bar{c}$
in combined wave-current motion. $a/k_N = 100$. $\theta'_{max} = 10$.
$w_s/U_{f,max} = 0.6$. (From Fredsøe et al., 1985).

opposite to the current motion, the bed shear stress becomes small during nearly half a wave period. If $\gamma = 90°$, this phenomenon does not occur.

For $U_{1m}/U_{fc} = 10$, Fig. 8.12 furthermore shows the variation in C with γ. The three-dimensional case clearly exhibits a greater sediment suspension.

From the model of suspended sediment together with a model for the flow velocity in combined wave-current it is easy to calculate the resulting sediment transport in combined wave-current motion by use of Eq. 8.56. Fig. 8.13 shows some examples of the results of such a calculation for different combinations of dimensionless parameters.

Example 8.7: Suspended sediment in irregular waves

In the case of irregular waves, representative values for the wave height and wave period are usually applied so mathematical modelling can be performed with regular waves.

When selecting the regular waves that best represent the irregular train, a possible criterion would be that the amount of sediment brought into suspension

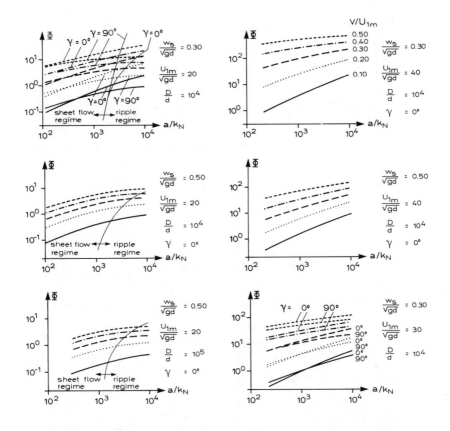

Figure 8.13 Transport of suspended sediment in the case of co-directional flow.

should be the same for the two conditions, i.e.

$$\left[\int_0^D \bar{c}\,dz\right]_{\text{regular}} = \left[\int_0^D \bar{c}\,dz\right]_{\text{irregular}} \tag{8.83}$$

where the overbar denotes time-averaging.

However, it is of greater engineering interest to require the same amount of transported sediment; i.e.

$$\left[\int_0^D \overline{cu}\,dz\right]_{\text{regular}} = \left[\int_0^D \overline{cu}\,dz\right]_{\text{irregular}} \tag{8.84}$$

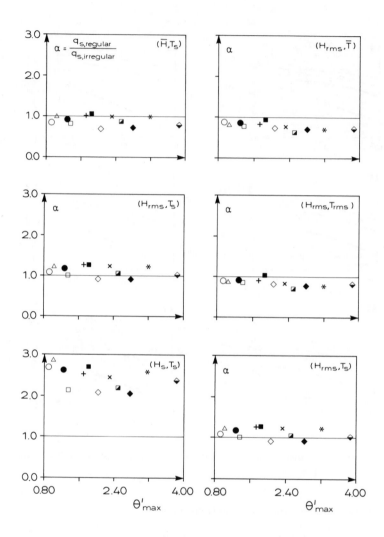

Figure 8.14 Comparison between the mean suspended sediment transport
due to regular or irregular waves, analysed by a mathematical
model. (Zyserman and Fredsøe, 1988).

where u is the resulting velocity obtained from the vectorial sum of the velocity
of the steady current U, and the wave-induced velocity U_0.

In Eq. 8.84, the current velocity can be an additional parameter in the
problem under consideration. However, for small flow velocities compared to the

wave-induced velocities, this dependence on U appears to be weak.

Zyserman and Fredsøe (1988) made simulations with the mathematical model described above, using a time series based on irregular waves represented by a Pierson-Moskowitz spectrum. Fig. 8.14 shows the results of a series of numerical tests. In all tests, the mean velocity of the current is taken equal to 10% of the root mean square value of the near-bed wave velocities. α in Fig. 8.14 is defined as the ratio between q_s calculated on the basis of the regular wave and q_s determined on the basis of the irregular wave train. In the left half of Fig. 8.14, the variation of α with the height of the representative regular wave is depicted. Three different values have been chosen, namely the mean height $\overline{H}$, the root mean square value $H_{\rm rms}$ and finally the significant wave height H_s. In the right half of Fig. 8.14, the variation in α with different selected values of the wave period is shown.

From the analysis it turns out that the best representation of the mean suspended sediment transport of the series over the whole range of values of $\overline{\theta}_{\max}$ studied is given by the combination of $H_{\rm rms}$ and T_s.

8.4 Vertical distribution of suspended sediment under broken waves

Inside the point of breaking, the surface-generated turbulence results in a much higher level of turbulent kinetic energy, especially close to the water surface, where the production of turbulence takes place around the surface roller. This leads to a significant increase in the amount of suspended sediment compared to unbroken waves of the same height and period in the same water depth.

The transition region from breaking to fully developed broken waves - the outer zone shown in Fig. 4.4 - has until now not been satisfactorily described because of the complexity in the flow close to the point of breaking as described in Section 4.1.

In the inner part of the surf zone, where the broken waves are more or less transformed to hydraulic bores, the effect of an increased turbulence level on the distribution of suspended sediment can be analysed by use of the hydrodynamic description given in Section 4.3.2. In this section the eddy viscosity was given by

$$\nu_T = l\sqrt{k} \tag{8.85}$$

in which the length scale l was prescribed by Eq. 4.74, and where k is the turbulent kinetic energy, which was calculated by a one-equation turbulence model.

Eq. 8.85 can be inserted into the diffusion equation, Eq. 8.47, which can then be solved numerically, using the same boundary conditions as given in Section 8.1.

Fig. 8.15 shows the results of such a calculation. For reasons of comparison the solution obtained for the case without wave-breaking is indicated by a dashed

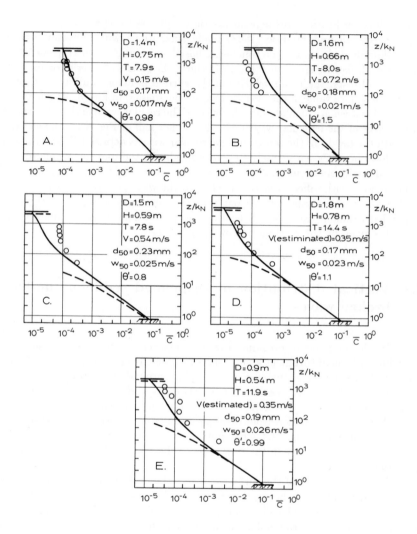

Figure 8.15 Comparison with field measurements by Nielsen et al. (1982) for spilling breaker case. (After Deigaard et al., 1986).

line. The circles represent field measurements by Nielsen et al. (1982) carried out under so heavy conditions that the bed became plane.

Fig. 8.16 shows, in dimensionless form, how the mathematical model predicts the relative vertical distribution of time-averaged suspended sediment concentration due only to the wave breaking for three different wave heights and

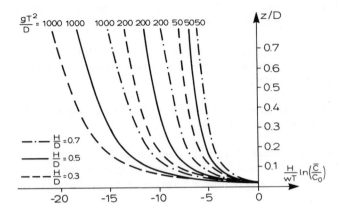

Figure 8.16 Theoretical vertical distribution of suspended sediment. (After
Deigaard et al., 1986).

periods. The concentration is normalized with c_0, the concentration 0.025 times
the water depth above the bed. The curves plotted in Fig. 8.16 are not valid
inside the wave boundary layer.

REFERENCES

Coleman, N.L. (1981): Velocity profiles with suspended sediment. J. Hydr. Res.,
19:211-229.

Deigaard, R. (1991): On the turbulent diffusion coefficient for suspended sediment. Progress Report No. 73, Inst. of Hydrodynamics and Hydraulic
Engineering, ISVA, Techn. Univ. Denmark, pp. 55-66.

Deigaard, R., Fredsøe, J. and Brøker Hedegaard, I. (1986): Suspended sediment
in the surf zone. J. Waterway, Port, Coastal and Ocean Eng. Div., ASCE,
112(1):115-128.

Einstein, H.A. and Chien, N. (1955): Effects of heavy sediment concentration near
the bed on velocity and sediment distribution. University of California,
Institute of Engineering Research, MRD, Sediment Series, Report No. 8.

Engelund, F. (1975): Steady transport of moderately graded sediment (part 2).
Progress Report No. 35, Inst. of Hydrodynamics and Hydraulic Engineering, ISVA, Techn. Univ. Denmark, pp. 31-36.

Engelund, F. and Fredsøe, J. (1976): A sediment transport model for straight alluvial channels. Nordic Hydrology, 7:293-306.

Fredsøe, J., Andersen, O.H. and Silberg, S. (1985): Distribution of suspended sediment in large waves. J. Waterway, Port, Coastal and Ocean Eng. Div., ASCE, 111(6):1041-1059.

Jonsson, I.G. (1963): Measurements in the turbulent wave boundary layer. IAHR, 10th Congr., London, 1:85-92.

Justesen, P. and Fredsøe, J. (1985): Distribution of turbulence and suspended sediment in the wave boundary layer. Progress Report No. 62, Inst. of Hydrodynamics and Hydraulic Engineering, ISVA, Techn. Univ. Denmark, pp. 61-66.

Justesen, P., Fredsøe, J. and Deigaard, R. (1986): The bottleneck problem for turbulence in relation to suspended sediment in the surf zone. 20th Coastal Engineering Conference Proceedings, CER Council, ASCE, Taipei, Taiwan, 2:1225-1239.

Kirkegaard Jensen, J. and Sørensen, T. (1972): Measurement of sediment suspension in combinations of waves and currents. Proc. 13th Int. Conf. Coast. Eng., Vancouver, Canada, 2:1097-1104.

Lundgren, H. (1972): Turbulent currents in the presence of waves. Proc. 13th Int. Conf. Coast. Eng., Vancouver, Canada, 1:623-634.

Nielsen, P. (1979): Some basic concepts of wave sediment transport. Series Paper No. 20, Inst. of Hydrodynamics and Hydraulic Engineering, ISVA, Techn. Univ. Denmark.

Nielsen, P., Green, M.O. and Coffey, F.C. (1982): Suspended sediment under waves. Technical Report No. 8216, Dept. of Geography, The Univ. of Sydney, Coastal Studies Unit..

Parker, G. (1978): Self-formed straight rivers with equilibrium banks and mobile bed. Part 1: The sand river. J. Fluid Mech., 89(1):109-126.

Smith, J.D. and McLean, S.R. (1977): Boundary layer adjustments to bottom topography and suspended sediment. In: Bottom Turbulence, Ed. by J. Nihoul, Elsevier, Amsterdam, pp. 123-151.

Soulsby, R.L. and Wainwright, B.L.S.A. (1987): A criterion for the effect of suspended sediment on near-bottom velocity profiles. J. Hydr. Res., 25(3):341-356.

Staub, C., Jonsson, I.G. and Svendsen, I.A. (1984): Variation of sediment suspension in oscillatory flow. 19th Coastal Eng. Conference, Houston, Texas, 3:2310-2321.

Sumer, B.M. and Deigaard, R. (1981): Particle motions near the bottom in turbulent flow in an open channel. Part 2. J. Fluid Mech., 109:311-338.

van de Graaff, J. (1988): Sediment concentration due to wave action. Diss. Delft University of Technology.

van Rijn, L.C. (1984): Sediment transport, Part II: Suspended load transport. J. Hydr. Eng., ASCE, 110(11):1613-1641.

Zyserman, J.A. and Fredsøe, J. (1988): Numerical simulation of concentration profiles of suspended sediment under irregular waves. Progress Report No. 68, Inst. of Hydrodynamics and Hydraulic Engineering, ISVA, Techn. Univ. Denmark, pp. 15-26.

Chapter 9. Current-generated bed waves

An erodible seabed exposed to waves and current will usually not remain stable but will form different kinds of bed waves. These bed waves have a significant influence on the flow and sediment pattern which will be evaluated in this and the coming chapter.

9.1 Bed waves in current alone

Fig. 9.1 gives a picture of the most important bed forms formed by a steady current.

Ripples

When the tractive force is increased to the point where sediment transport starts, the bed will be unstable. In the case of fine sediment, ripples are formed, while coarse sediments will usually form dunes.

Small triangular sand waves are called ripples. They are usually shorter than about 0.6 meter, and not higher than about 60 mm.

At relatively small flow velocities a viscous sublayer of thickness

$$\delta_v = \frac{11.6\nu}{U_f'} \tag{9.1}$$

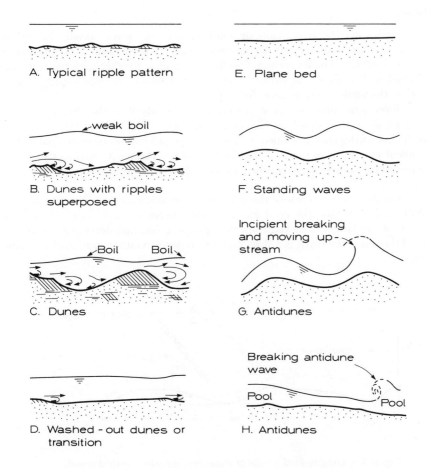

A. Typical ripple pattern

E. Plane bed

B. Dunes with ripples superposed

weak boil

F. Standing waves

C. Dunes

Boil   Boil

G. Antidunes

Incipient breaking and moving up-stream

D. Washed - out dunes or transition

H. Antidunes

Breaking antidune wave

Pool   Pool

Figure 9.1 Typical bed forms in order of increased stream power. From Simons and Richardson (1961).

is formed. This "hydraulically smooth" situation occurs when δ_v is larger than the sediment size d. It is assumed (Simons and Richardson, 1961) that ripples are formed if a viscous layer is present when the critical tractive force just is surpassed, while dunes are formed if the bed is hydraulically rough. The ripple length depends on the sediment size (and other parameters), but is essentially independent of the water depth. Three-dimensional ripple patterns have been discussed by Raudkivi (1976), see also Sleath (1984).

Dunes

Dunes are the large, more or less irregular sand waves usually formed in natural streams. This is by far the most important bed form in the field of practical river engineering. The longitudinal profile of a dune is roughly triangular, with a mild and slightly curved upstream surface and a downstream slope approximately equal to the angle of repose, see Fig. 9.2.

Flow separation occurs at the crest, reattachment in the trough, so that bottom rollers are formed on the lee side of each dune. Above this a zone of violent free turbulence is formed in which a large production (and dissipation) of turbulent energy takes place. Near the zone of reattachment, sediment particles are moved by turbulence, even when the local shear stress is below its critical value (Raudkivi, 1963).

On the upstream side of the dune the shear stress moves sediment particles uphill until they pass the crest and eventually become buried in the bed for a period. As sediment is moved from the upstream side and deposited on the lee side of the dune, the result is a slow, continuous downstream migration of the dune pattern.

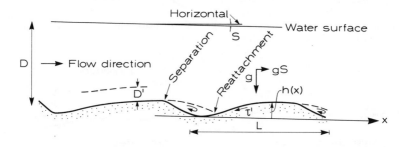

Figure 9.2 Longitudinal profile of dunes (exaggerated vertical scale).

Transition and plane bed

For increased stream power the dunes tend to wash out, i.e. they become longer and flatter and finally disappear. This happens at Froude numbers below 1, i.e. in subcritical flow, and marks the transition to the so-called upper flow regime. This change from dunes to flat bed means a rather drastic reduction of both hydraulic resistance and water depth.

Antidunes

A further increase in stream power leads from transition and plane bed to formations of the so-called antidunes and related configurations. In this case the longitudinal bed profile is nearly sinusoidal, and so is the water surface, but usually with a much larger amplitude (Fig. 9.1 F and G).

At higher Froude numbers the amplitude of the surface profiles often tends to grow until breaking occurs. After breaking, the amplitude may be small for a time, and then the process of growth and breaking is repeated.

The name "antidune" indicates the fact that the bed and surface profiles are moving upstream, particularly just before breaking. Fig. 9.1H illustrates an extreme form of antidunes occurring at high Froude numbers.

An overview of bedforms in sand-bed streams is obtained from imagining an experiment in a flume in which the discharge (or stream power) is gradually increased, as shown in Fig. 9.3. The ordinate is the total bed shear stress τ_b, and the abscissa is the mean velocity V. In the case of a fixed bed, the relation between τ_b and V would be as the dashed curve in Fig. 9.3, i.e. close to a second-order parabola, corresponding to the expression

$$\tau_b = \frac{1}{2}\rho f V^2 \tag{9.2}$$

which defines the friction factor f.

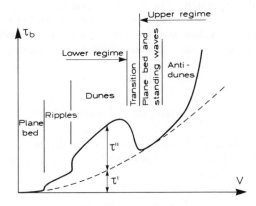

Figure 9.3 Relation between total bed shear stress τ_b and flow velocity V for different bed forms.

The occurrence of ripples and dunes obviously implies a considerable increase of the hydraulic resistance. On the other hand, plane bed, standing waves,

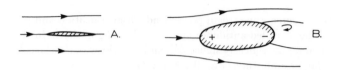

Figure 9.4 Flow around submerged bodies. A: Slender bodies give mainly
frictional drag. B: Drag on blunt bodies gives also form drag
due to large normal stresses upstream and underpressure at
the rear.

and weak antidunes bring the resistance back to skin friction only. To understand
the additional hydraulic resistance associated with the dune case, consider the
drag on a submerged slender body placed parallel to the current, Fig. 9.4A. In
this case the only drag is due to boundary layer friction (skin friction). In case
of a blunt body, however, Fig. 9.4B, separation occurs, and the pressure (normal
stress) is larger on the upstream face that on the rear. Hence not only frictional
(tangential) stresses are involved, but also the normal stresses have a resultant,
the so-called form drag. Hence the total drag consists of two components, the
friction and the form drag.

Consider a single dune with length L_D, see Fig. 9.2. The hydrostatic force
is $G = \gamma D L_D$ (per unit width), with streamwise component $G \sin \beta = GS$ which
is usually assumed to be balanced by the total shear stress $\tau_b L_D$. S is the surface
water slope which in the case of steady uniform flow is equal to the energy gradient
I. If the local normal and shear stress is called p and τ, respectively, and the local
inclination of the dune is γ, the following equilibrium equation is required

$$GS = \rho g D L_D S = \tau_0 L_D$$

$$= \int_{L_D} \tau \cos \gamma \, dx + \int_{L_D} p \sin \gamma \, dx \qquad (9.3)$$

Here the last term (form drag) is different from zero due to underpressure at the
rear side of the dune. Hence, a certain part of the total drag τ_0 is not carried by
friction. Eq. 9.3 is now written

$$\tau_b L_D = \tau' L_D + \tau'' L_D$$

or

$$\tau_b = \tau' + \tau'' \qquad (9.4)$$

where τ' is the mean skin friction (called: effective shear or tractive stress) while
τ'' is a formal contribution to τ_b actually originating from the normal stresses.
This separation of τ_b is of great importance in the theory of sediment transport,

because the bed particles are generally moved by the actual shear stress and are essentially unaffected by the normal stress.

The problem of estimating the effective shear stress τ' has not yet been solved in a completely satisfactory way. The method currently used was originally suggested by Einstein (1950).

Close to the crest the flow is converging, which means that the flow attains the character of a boundary layer with thickness D', (see Fig. 9.2). In these circumstances the mean velocity V can approximately be given as

$$\frac{V}{U'_f} = 6.2 + 2.5 \ln\left(\frac{D'}{k_N}\right) \tag{9.5}$$

where k_N is the roughness, and

$$U'_f = \sqrt{gD'I} \tag{9.6}$$

where I is the energy gradient (slope).

At first sight it would be natural to assume the roughness k_N to be equal to the mean sediment diameter d. However, the irregular surface prevailing during bed-load movements gives a somewhat larger value, so that k_N must be taken to be about 2.5 d (Engelund and Hansen, 1972). When V and d are known, D' and U'_f can be calculated from Eqs. 9.5 and 9.6.

9.2 Mechanics of dunes in a steady current

To investigate the mechanics of bed forms a little closer it is necessary to make the simplifying assumption that the bed waves are moving at a constant velocity a without any change in shape. This is of course not true in detail but it represents all the essential features of the problem with sufficient accuracy. Under these conditions the shape of the bed is described by an expression of the form

$$h = h(x - at) \tag{9.7}$$

in which h is the local height of the dune above an x-axis placed through the troughs, see Fig. 9.2.

Next, consider the sediment discharge q_T through two consecutive sections with unit spacing in the x-direction. The net outflow is $\frac{\partial q_T}{\partial x}$, which must equal the change in bed elevation when the correction for the porosity n of the bed material is taken into account, and the change in the amount of sediment stored in suspension is ignored. Hence, the equation of continuity

$$\frac{\partial q_T}{\partial x} = -(1-n)\frac{\partial h}{\partial t} \tag{9.8}$$

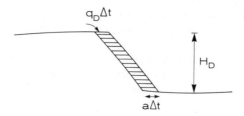

Figure 9.5 Migration of the dune front.

is obtained. If Eq. 9.7 is substituted into Eq. 9.8, it becomes obvious that the equations are satisfied by putting

$$q_T = q_T(x) = q_0 + a(1 - n)h \tag{9.9}$$

The quantity q_0 is a constant, interpreted as the value of q_T for $h = 0$, i.e. at the troughs where the bed load vanishes. For small shear stresses, q_0 becomes equal to zero because the sediment mainly moves as bed load so q_T becomes equal to q_B. This transport is related to the local shear stress, which is negligible in the trough because of flow separation. In this case the following interesting relation is obtained

$$q_B = a(1 - n)h \tag{9.10}$$

from which it is seen that the local intensity of bed load transport is proportional to the local height of the bed above the plane through the troughs. This indicates that the shear stress τ^* at the dune surface must vary from zero at the trough to a maximum at the crest, an assumption confirmed by measurements by Raudkivi (1963).

At the front of the sand waves the amount of sediment deposited is called q_D. This amount determines the migration velocity a of the sand wave by (see Fig. 9.5)

$$a = \frac{q_D}{(1 - n)H_D} \tag{9.11}$$

From Eqs. 9.10 and 9.11 it is seen that for small shear stresses q_D becomes the same as q_B. At larger shear stresses, the deposited sediment consists partly of the bed load transport at the crest and a certain fraction of the suspended sediment as sketched in Fig. 9.6. A suspended grain will contribute to q_D and to the dune migration if it by settling or diffusion moves into the separation 'bubble' before it is carried past the separation zone by the flow. Whether the grain is deposited at

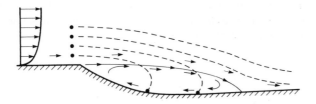

Figure 9.6 The probability for a suspended grain to settle in the trough
depends on its height.

the front or not will thus depend on its distance from the bed when it passes the
dune crest.

Now, if the sediment is transported mainly as bed load, the shape of the
bed form can be found from Eqs. 9.10 and 9.11, which when combined give

$$\frac{h}{H_D} = \frac{q_B}{q_{B,\text{top}}} \tag{9.12}$$

in which $q_{B,\text{top}}$ is the sediment transport at the crest level where $h = H_D$, H_D
being the dune height. In case of bed load only, the transport of sediment can, for
instance, be calculated by the Meyer-Peter formula, Eq. 7.60, which on a sloping
bed reads

$$q_B = 8\sqrt{(s-1)gd^3} \left(\theta^* - \frac{\theta_c}{\tan\phi}\frac{\partial h}{\partial x} - \theta_c \right)^{3/2} \tag{9.13}$$

cf. Eqs. 7.35 and 7.60, ϕ being the friction angle. In Eq. 9.13 θ^* is the local value
of the dimensionless skin friction. Inserting Eq. 9.13 into Eq. 9.12 gives

$$\frac{1}{\tan\phi}\frac{\partial h}{\partial x} + \left(\frac{\theta_{\text{top}}}{\theta_c} - 1 \right)\left(\frac{h}{H_D} \right)^{2/3} = \frac{\theta^*}{\theta_c} - 1 \tag{9.14}$$

From this equation it is realized that the variation in bed shear stress θ^*
along the sand wave must be known in order to proceed further.

The calculation of the bed shear stress along sand waves has been performed
by, among others, McLean and Smith (1986) using the law of the wake, and
Mendoza and Shen (1990), who applied a turbulent closure to calculate the flow
field.

Fredsøe (1982) used a semiempirical approach for the flow description to
calculate the dune shape. This was done by adapting the measured bed shear
stress downstream of a negative step, see also Bradshaw and Wong (1972) who
expressed the variation in the local bed shear stress τ by

$$\tau^* = \tau_{\text{top}} f\left(\frac{x}{H_s} \right) \tag{9.15}$$

in which H_s is the step height, and τ_{top} the bed shear stress at the top of the dune. The measured function f is shown in Fig. 9.7, and this function is normalized so it becomes equal to unity at the top of the dune where $x = L_D$, L_D being the dune length.

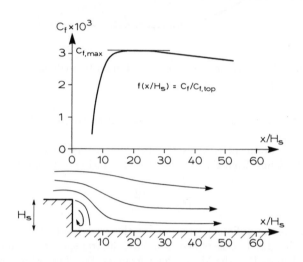

Figure 9.7 Bed shear stress variation downstream a negative step. $\tau^* = \frac{1}{2}\rho C_f V^2$.

Due to the presence of the dunes, the local water depth varies along the dune, which results in spatial changes to the depth-averaged flow velocity V^*. This velocity varies along the sand wave as

$$V^*\left(D + \frac{H_D}{2} - h\right) = q \tag{9.16}$$

neglecting undulations on the water surface due to the presence of the sand waves (small Froude numbers). In Eq. 9.16, q is the water discharge per unit width. In the case of dunes where the wave length is several times the water depth this variation in mean flow velocity can be included in the spatial variation in bed shear stress along the sand wave by writing the bed shear stress as

$$\tau^* = \tau_{top}\ f\left(\frac{x}{H_D}\right)\left(\frac{V^*}{V_{top}}\right)^2 \tag{9.17}$$

because the bed shear stress usually varies with the mean flow velocity squared. Eqs. 9.16 and 9.17 give

$$\theta^* = \theta_{top}\ f\left(\frac{x}{H_D}\right)\left[\frac{D - \frac{H_D}{2}}{D + \frac{H_D}{2} - h}\right]^2 \tag{9.18}$$

Fig. 9.8 shows two examples of calculated bed waves profiles (with exaggerated vertical scale), obtained by inserting Eq. 9.18 into Eq. 9.14.

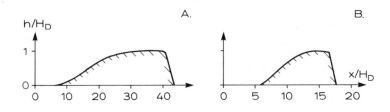

Figure 9.8 Examples of calculated bed wave profiles in current. A: $\theta_{\text{top}} = 0.057$, $H_D/D = 0.04$. B: $\theta_{\text{top}} = 0.30$, $H_D/D = 0.27$.

The dune is assumed to end just downstream from the location where the surface attains a maximum height. Downstream from this point, the flow will become divergent due to the decrease in shear stress. This will very quickly enhance the tendency towards flow separation, so a dune front will occur. Just downstream of this point, the local slope of the bed wave will be very near the angle of repose because the flow velocities in the separation zone are very small compared to the outer flow. As seen from Fig. 9.8A, the dune becomes quite long compared to its height for a Shields parameter close to the critical. This is because the effect of longitudinal slope on the bed load transport becomes significant at a low Shields number, so the maximum transport of bed load is influenced as well by the local shear stress as by the local bed slope. The combined maximum of these two effects can be located much further downstream from the maximum of the shear stress, which usually is located around 16 times the dune height downstream from the former dune front (see Fig. 9.7). At larger Shields parameters, the slope correction term on the bed load becomes insignificant, and the dune length is mainly determined by the location of maximum shear stress.

Dune height

The calculation of the bed wave shape just described does not require any knowledge on the actual height of the dune. However, this height can also be evaluated from the basic concepts derived above: from pure geometrical considerations, Eqs. 9.7 and 9.11 give

$$\frac{\partial h}{\partial t} = -a\frac{\partial h}{\partial x} = -\frac{q_D}{(1-n)H_D}\frac{\partial h}{\partial x} \tag{9.19}$$

Combining this equation together with the equation of continuity, Eq. 9.8 gives

$$\frac{\partial q_T}{\partial x} = \frac{q_D}{H_D}\frac{\partial h}{\partial x} \tag{9.20}$$

which can also be written as

$$\frac{\partial \Phi_T}{\partial x} = \frac{\Phi_D}{H_D} \frac{\partial h}{\partial x} \qquad (9.21)$$

where Φ is the dimensionless sediment transport rate, cf. Eq. 7.36.

In a steady current, Φ is a function of the Shields parameter θ^*, so Eq. 9.21 can be written as

$$\frac{\partial \Phi_T}{\partial \theta^*} \frac{\partial \theta^*}{\partial x} = \frac{\Phi_D}{H_D} \frac{\partial h}{\partial x} \qquad (9.22)$$

At the dune crest, Eq. 9.18 can be approximated by

$$\theta^* = \theta_{\text{top}} \left[\frac{D - \frac{H_D}{2}}{D + \frac{H_D}{2} - h} \right]^2 \qquad (9.23)$$

cf. Eq. 9.18. In Eq. 9.23, the weak variation in the function f is disregarded, which is a good approximation far away from the former crest. θ_{top} appearing in Eq. 9.23 is the Shields parameter due to skin friction, which can be related to the averaged skin friction θ' by

$$\theta' = \theta_{\text{top}} \frac{(D - \frac{H_D}{2})^2}{D^2} \qquad (9.24)$$

Now Eq. 9.23 gives

$$\frac{\partial \theta^*}{\partial x} = \frac{2\theta_{\text{top}}}{D - \frac{H_D}{2}} \frac{\partial h}{\partial x} \qquad (9.25)$$

which combined with Eq. 9.22 results in the following expression for the dune height

$$\frac{H_D}{D} = \frac{\Phi_D}{2\theta \frac{d\Phi_T}{d\theta}} \Big/ \left[1 + \frac{\Phi_D}{4\theta \frac{d\Phi_T}{d\theta}} \right] \qquad , \theta = \theta_{\text{top}} \qquad (9.26)$$

in which all quantities on the right-hand-side must be taken at the dune top.

Example 9.1:   Dune height at low and high Shields numbers

At low Shields numbers, both the deposited load on the dune front Φ_D and the total load at the crest Φ_T can be taken equal to the bed load transport Φ_B. Applying the Meyer-Peter formula Eq. 7.60, Eq. 9.26 now gives

$$\frac{H_D}{D} = \frac{2(\theta - \theta_c)}{7\theta - \theta_c} \qquad , \theta = \theta_{\text{top}} \qquad (9.27)$$

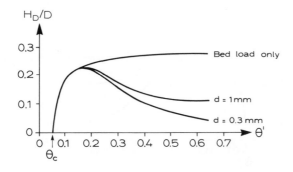

Figure 9.9 Variation in dune height with bed shear stress for different
grain sizes. $D/d = 10^3$.

At higher Shields numbers, a large part of the sediment is transported in
suspension. In this case, Φ_D as a first approximation can be taken to be Φ_B, while
Φ_T must be taken to be the sum of Φ_B and Φ_S. Now, the calculation of dune
height becomes much more complicated because Φ_S depends on several parameters
as $D/d, \theta$ and w_s/U_f. Fig. 9.9 shows the results of one numerical calculation in
which Φ_B and Φ_S are calculated separately by the method suggested by Engelund
and Fredsøe (1976), as described in Chapter 7.

It is seen that in case of fine sediment the inclusion of suspended sediment
leads to a decrease in dune height. This is the transition to plane bed as shown in
Fig. 9.1.

Dune length

In the case of dominant bed load transport, the maximum bed shear stress
is located around 16 times the dune height downstream from the former crest, see
Fig. 9.7. As the location of maximum sediment transport rate, except for very
small Shields parameters, is also the location of maximum dune height, cf. Eq.
9.12, the dune length is easily obtained from

$$L_D = 16\,H_D \qquad (9.28)$$

in which H_D is obtained from Eq. 9.27.

At higher shear stresses, where suspended sediment becomes the dominant
transport mechanism, the situation becomes a little more complex because a spatial phase lag L_s is introduced between the location of the maximum bed shear

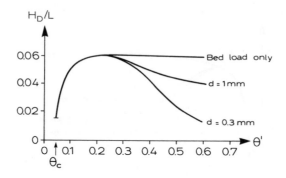

Figure 9.10 Variation in length-height ratio with bed shear stress for different grain sizes. $D/d = 10^3$. (Fredsøe, 1982).

stress and the location of the maximum suspended load transport. The maximum bed and suspended load transport can be estimated to be located around

$$\frac{L_D}{H_D} = \left[16 q_B + \left(\frac{L_s}{H_D} + 16 \right) q_S \right] \Big/ (q_B + q_S) \tag{9.29}$$

which is the weighted mean of the influence from bed load and suspended load. Eq. 9.29 suggests that the influence of suspended sediment will be that the sand waves will lengthen as described in the following example.

Example 9.2: Estimation of the phase lag of suspended sediment

The phase lag L_s is introduced because a sediment grain takes some time to settle after being picked up from the bed. L_s can be estimated from the diffusion equation for suspended sediment Eq. 8.45. In the steady case this reads

$$U \frac{\partial c}{\partial x} = w_s \frac{\partial c}{\partial z} + \frac{\partial}{\partial z} \left(\varepsilon_s \frac{\partial c}{\partial z} \right) + \frac{\partial}{\partial x} \left(\varepsilon_s \frac{\partial c}{\partial x} \right) \tag{9.30}$$

An illustrative example of the solution of Eq. 9.30 can be obtained assuming ε_s and U to be constant over the flow depth. In the uniform case, Eq. 9.30 can now be solved giving

$$c = c_{b0} \exp\left(-\frac{w_s}{\varepsilon_s} z \right) \tag{9.31}$$

Here c_{b0} is a nominal reference concentration at the bed, which differs from the real one because ε_s is over-estimated at the bottom. Now the case is considered where the bed concentration c_{b0} varies in the flow direction, for instance due to spatial changes in the bed shear stress. In this simplified example U and ε_s are assumed not to vary in the x-direction.

Because of the variation in c_{b0} the vertical distribution of the suspended sediment will deviate from the equilibrium profile given by Eq. 9.31. For simplicity the vertical distribution of the suspended sediment is still described by an exponential function, but now with a variable steepness

$$c = c_{b0} \exp\left[-\frac{w_s}{\varepsilon_s}(1+\lambda)z\right] \qquad (9.32)$$

Introducing Eq. 9.32 into the diffusion equation 9.30 and integrating over the flow depth give

$$\frac{U\varepsilon_s}{w_s}\frac{\partial}{\partial x}\left(\frac{c_{b0}}{1+\lambda}\right) = -c_{b0}w_s + c_{b0}w_s(1+\lambda) \qquad (9.33)$$

In this derivation it has been assumed that the sediment concentration vanishes towards the water surface, and the horizontal diffusion of sediment is neglected. The variation in c_{b0} is taken to be a small perturbation, and Eq. 9.33 can then be linearized to give the following differential equation for the unknown parameter λ

$$\frac{d\lambda}{dx} + \frac{w_s^2}{\varepsilon_s U}\lambda - \frac{1}{c_{b0}}\frac{\partial c_{b0}}{\partial x} = 0 \qquad (9.34)$$

This equation can be solved for a given variation in c_{b0}. As an example a periodic perturbation of c_{b0} is considered, giving a variation of c_{b0} of

$$c_{b0} = c_0 + c_1 \sin(kx) \qquad (9.35)$$

Introducing Eq. 9.35 into Eq. 9.34 and using that $c_1 \ll c_0$ give

$$\frac{d\lambda}{dx} + \frac{\lambda}{L_s} - \frac{c_1 k}{c_0}\cos(kx) = 0 \qquad (9.36)$$

where the length scale L_s has been introduced

$$L_s = \frac{\varepsilon_s U}{w_s^2} \qquad (9.37)$$

The solution to Eq. 9.36 can be written to be

$$\lambda = \frac{c_1 k}{c_0}\frac{L_s}{1+(kL_s)^2}\left(\cos(kx) + kL_s\sin(kx)\right) + c_2\exp(-x/L_s) \qquad (9.38)$$

The vanishing transient part of the solution can be ignored, i.e. $c_2 = 0$. For long wave lengths of the perturbation the dimensionless parameter kL_s is small, and Eq. 9.38 can be approximated by

$$\lambda = \frac{c_1}{c_0} kL_s \cos(kx) \qquad (9.38a)$$

Now the sediment transport rate q_s can be found as

$$q_S = \int_0^D cU\,dz = c_{b0}U\frac{\varepsilon_s}{w_s}\frac{1}{1+\lambda} \approx c_{b0}U\frac{\varepsilon_s}{w_s}(1-\lambda) =$$

$$c_0\frac{U\varepsilon_s}{w_s}\left(1 + \frac{c_1}{c_0}\sin(kx)\right)\left(1 - \frac{c_1}{c_0}kL_s\cos(kx)\right) \approx$$

$$c_0\frac{U\varepsilon_s}{w_s}\left[1 + \frac{c_1}{c_0}\left(\sin(kx) - kL_s\cos(kx)\right)\right] \approx$$

$$c_0\frac{U\varepsilon_s}{w_s}\left[1 + \frac{c_1}{c_0}\sin\left(k(x - L_s)\right)\right] \qquad (9.39)$$

If the development in the sediment concentration profile had been neglected, i.e. for quasi-uniform conditions, the suspended sediment transport rate would have been calculated as

$$q_S = \int_0^D cU\,dz = c_{b0}U\frac{\varepsilon_s}{w_s} = c_0\frac{U\varepsilon_s}{w_s}\left(1 + \frac{c_1}{c_0}\sin(kx)\right) \qquad (9.39a)$$

By comparing Eqs. 9.39 and 9.39a it is seen that the development in the concentration profile causes the sediment transport rate to have a phase lag relative to the variation in the bed concentration. The phase lag is equal to L_s. As seen from Eq. 9.37 the lag distance increases with decreasing settling velocity and with increasing flow velocity or eddy viscosity.

As seen from Eqs. 9.31 and 9.37, the length scale L_s can be written as

$$L_s = z_c\frac{U_c}{w_s} \qquad (9.40)$$

where z_c $(= \varepsilon_s/w_s)$ is the height above the bed of the centroid of the concentration profile

$$z_c = \frac{\int_0^D cz\,dz}{\int_0^D c\,dz} \qquad (9.41)$$

U_c is the mean flow velocity at the level z_c.

In a more refined analysis where U and ε_s are varying with z the formulation given by Eq. 9.40 is still valid for small spatial changes in the transport capacity.

In Fig. 9.10 the influence of phase lag for suspended sediment on the bed wave steepness is illustrated by a numerical example. For the case with no suspended sediment present at all, the wave steepness increases with θ' at small θ'-values (because of decreasing influence from the bed slope on the bed load transport), and the steepness becomes constant at high θ'-values. However, the influence of suspended sediment results in a decrease in wave steepness with θ' at high θ'-values.

In the numerical example shown in Fig. 9.10, the transport of bed load and suspended load is calculated by the Engelund-Fredsøe model described in Chapter 7.

9.3 Influence of waves on current-generated sand waves

The presence of waves together with a current will result in a significant change in the dimensions of the sand waves. Sand waves of this type are often found offshore of locations where the tidal current is sufficiently strong for sand waves to be formed.

An example, taken from Houbolt (1968), of an echo sounding showing a train of sand waves is given in Fig. 9.11. These asymmetric sand waves have a height of 5-7 m at 25-30 m water depth; the wave length is approximately 200 m.

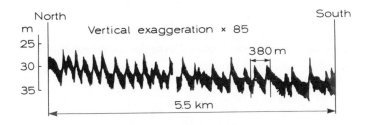

Figure 9.11 Example of echo sounding showing asymmetric sand waves, West of Imuiden, the Netherlands, after Houbolt (1968).

No theory for the behaviour of sand waves formed by a tidal current has so far been developed. However, the general behaviour of the sand waves, and the physical mechanisms behind the formation of sand waves does not seem to deviate significantly from the processes associated with dunes formed under a unidirectional current. The large sand waves formed by a tidal current can, according to Stride (1982), be considered as unidirectional current-formed waves, which have been modified to a smaller or larger extent by the reversing current.

In the following example, it is outlined how the effect of the co-presence of waves with a unidirectional current modifies the presented theoretical findings for dune height and dune length.

The sand wave height

As assumed earlier the influence from the preceding dune is weak near the dune crest, and the local bed shear stress can be calculated from the local mean velocity V by the logarithmic resistance law

$$\frac{V}{U'_f} = \frac{1}{\kappa}\left(\ln\left(\frac{30D}{k_N}\right) - 1\right) \tag{9.42}$$

in the case of pure current. This leads to the formulation given by Eq. 9.23.

In the case of combined waves and current, the variation in bed shear stress can not be described by the simple approximation presented in Eq. 9.23, because both the wave action and the current velocity vary with the water depth. However, for a given bed sediment, the spatial variation in sediment transport close to the crest is given by

$$\frac{\partial q_T}{\partial x} = \frac{dq_T}{dD}\frac{\partial D}{\partial x} = -\frac{dq_T}{dD}\frac{\partial h}{\partial x} \tag{9.43}$$

The term dq_T/dD must now be calculated by a model for sediment transport in combined waves and current.

For a given wave climate and given sediment properties, the sediment transport can, in dimensional form, be expressed by

$$q_T = q_T\left(V, D\right) \tag{9.44}$$

The variation in V with D can be found from the requirement

$$DV = \text{constant} \tag{9.45}$$

so

$$\begin{aligned}\frac{dq_T}{dD} &= \frac{\partial q_T}{\partial D} + \frac{\partial q_T}{\partial V}\frac{\partial V}{\partial D} \\ &= \frac{\partial q_T}{\partial D} - \frac{V}{D}\frac{\partial q_T}{\partial V}\end{aligned} \tag{9.46}$$

Now the equation of continuity Eq. 9.8 reads

$$\frac{\partial h}{\partial t} = \frac{-1}{(1-n)}\frac{\partial q_T}{\partial x} = \frac{-1}{(1-n)}\left(\frac{\partial q_T}{\partial D} - \frac{V}{D}\frac{\partial q_T}{\partial V}\right)\frac{\partial h}{\partial x} \tag{9.47}$$

cf. Eqs. 9.43 and 9.46. Using Eq. 9.20 the factor $\partial h/\partial x$ can be eliminated from Eq. 9.47, and the following expression for the sand wave height

$$H_D = -\frac{q_D}{\dfrac{dq_T}{dD}} = -\frac{q_D}{\dfrac{\partial q_T}{\partial D} - \dfrac{V}{D}\dfrac{\partial q_T}{\partial V}} \tag{9.48}$$

is obtained, where all sediment transport quantities must be calculated at the crest.

In case of a current without waves the first term in the denominator of Eq. 9.48 can be neglected, because the flow resistance, Eq. 9.42, is insensitive to variation in D. In case of combined waves and current this term can be significant due to the relation between the water depth and the near-bed orbital motion induced by waves of given height and period. In this case, the individual terms on the right-hand-side must be evaluated from a sediment transport model for combined wave-current motion.

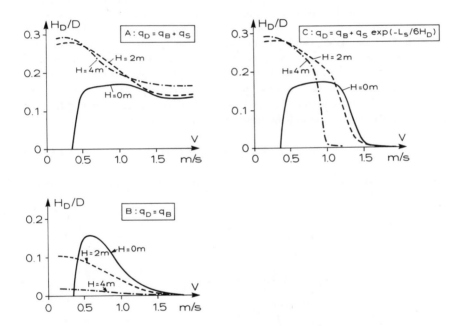

Figure 9.12 Sand wave height as a function of mean current velocity V and water wave height H; $d = 0.20$ mm, $D = 10$ m, $\gamma = 90°$. From Deigaard and Fredsøe (1986).

Fig. 9.12 shows the result of such a calculation in which the model for sediment transport by Fredsøe et al. (1985) (see Chapter 8) has been used. The calculations have been made for a water depth of 10 m and a bed material grain size of $d = 0.20$ mm. The direction of the wave motion is normal to the mean current, $\gamma = 90°$. Two wave heights are considered: $H = 2$ m and 4 m with periods of 7.2 s and 10.1 s, respectively.

In the case of these very large sand waves where the dune length is many times the water depth, a more detailed investigation should be made on how large an amount of the total sediment load q_T transported over the crest will settle on the sand wave front (q_D). Earlier in this chapter, the dune dimensions have been calculated on the basis of $q_D = q_B$.

Different assumptions for the settling of suspended sediment on the sand wave front are presented in Fig. 9.12A, B and C. In Fig.A it is assumed that all transported sediment is deposited at the sand wave front: $q_D = q_B + q_S$. In Fig.B the sand wave height is calculated, assuming that only the bed load is deposited. For pure current, the sand wave height is the same for the lower current velocities because the bed load transport is dominant. For higher current velocities the suspended load transport becomes dominant, and the sand wave height decreases if only the bed load is deposited. Under wave action, sediment will be suspended in the wave boundary layer even for small current velocities, and the calculated sand wave heights even for small current velocities are different in Fig.A and B.

In Fig. 9.12C the amount of suspended sediment which is deposited on the sand wave front is estimated on basis of the lag distance, L_s, and the length of the separation zone downstream of the front. q_D is calculated as

$$q_D = q_B + q_S \exp(-L_s/6H_D) \qquad (9.49)$$

The length scale for the lag of the suspended sediment is evaluated by the use of Eqs. 9.40 and 9.41.

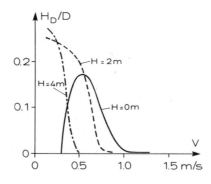

Figure 9.13 As Fig. 9.12C, but $d = 0.15$ mm.

For small current velocities L_s is small compared with H_D, and the calculated sand wave heights correspond to Fig. 9.12A. For increasing current velocities, L_s increases, and the sand wave height converges towards the results in Fig. 9.11B.

Fig. 9.12C shows that the wave conditions, especially at moderate current velocities, have a strong influence on the formation of sand waves.

The theory predicts that for a given wave climate the range of current velocities under which sand waves are formed will gradually become narrower as the water depth decreases, because the near-bed orbital motion and the suspended sediment load increase. This is in agreement with the observation that sand waves are decreasing or absent in the shallow parts of the southern North Sea, e.g. McCave and Langhorne (1982), Terwindt (1971) and Houbolt (1968).

Fig. 9.13 shows the same situation as Fig. 9.12C, but is calculated for finer sediment, namely $d = 0.15$ mm. A significant change is observed in the range of current velocities, where sand waves are formed. This range is now much narrower because of the larger amount of suspended sediment.

The sand wave length

As a first approximation the sand wave length is calculated by Eq. 9.29. The calculated steepness of the sand waves treated in Fig. 9.12 C is shown in Fig. 9.14. As the suspended load increases, the height decreases, and the length increases. The sand wave steepness therefore drops rapidly for increasing current velocity. The effect of gravity on the bed load transport in the case of pure current is indicated by the dotted line. In combined waves and current the suspended load is dominant even at small current velocities and the gravity effect is not expected to be of importance.

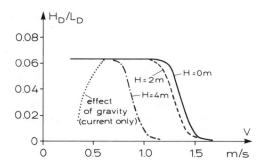

Figure 9.14 The sand wave steepness as a function of current velocity V and water wave height H, $d = 0.20$ mm, $D = 10$ m, $\gamma = 90°$.

9.4   Flow resistance due to bed waves in a current

In the presence of bed waves, the resistance to the flow consists of two parts, one originating from skin friction (or grain resistance), and another due to the expansion loss after each sand wave crest. This latter is called $\Delta H''$, and its magnitude can be estimated from the Carnot formula

$$\Delta H'' = \zeta \left(V_{\text{top}} - V_{tr} \right)^2 \Big/ 2g \qquad (9.50)$$

where V_{top} is the mean velocity before the expansion (at the crest), and V_{tr} is the smaller mean velocity downstream from the crest (at the trough). ζ is a non-dimensional coefficient depending on the flow geometry. V_{top} and V_{tr} are given by

$$V_{\text{top}} = q/(D - H_D/2)$$

$$V_{tr} = q/(D + H_D/2) \qquad (9.51)$$

in which $q = VD$ is the flow discharge per unit width.

Eq. 9.50 now becomes

$$\Delta H'' \simeq \alpha \frac{V^2}{2g} \left(\frac{H_D}{D} \right)^2 \qquad (9.52)$$

The total energy loss per unit length in the flow direction, which is equal to the energy gradient I, can be written

$$I = I' + \frac{\Delta H''}{L_D} = I' + I'' \qquad (9.53)$$

in which I' is the gradient due to friction. In uniform, steady channel flow, the total bed shear stress τ_b is related to the energy gradient I by

$$\tau_b = \rho g D I \qquad (9.54)$$

where D is the total flow depth. According to Eq. 9.53, τ_b can be divided into two parts by

$$\tau_b = \rho g D I' + \rho g D I'' = \tau' + \tau'' \qquad (9.55)$$

in which τ' is the mean value of that part of the bed shear stress which acts directly as a friction on the surface of the bed wave. The residual part τ'' corresponds to the form drag on the bed waves.

In dimensionless form, Eq. 9.47 can be written as

$$\theta = \theta' + \theta'' \tag{9.56}$$

where

$$\theta = \frac{\tau_b}{\rho g (s-1) d} \tag{9.57}$$

$$\theta' = \frac{DI'}{(s-1)d} \tag{9.58}$$

$$\theta'' = \frac{\alpha}{2} \frac{V^2}{(s-1)gd} \frac{H_D}{D} \frac{H_D}{L_D} \tag{9.59}$$

To calculate I' appearing in Eq. 9.50, an additional flow resistance formula for the skin friction is needed. For this purpose, consider the flow past a dune as shown in Fig. 9.15.

Immediately downstream the crest a wake-like flow is formed, in which a large amount of turbulent energy is produced. This is dissipated into heat further downstream, thus causing the expansion loss.

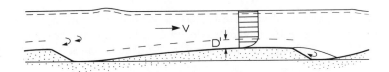

Figure 9.15 Sketch of the boundary layer developed along a dune.

At the end of the trough a boundary layer with thickness D' is formed, in which the velocity gradient is large, while the velocity distribution outside this layer is very uniform.

Engelund and Hansen (1972) used this flow picture to interpret the two quantities I' and I''. They assumed that the upper flow and the boundary layer flow are independent of each other in the sense that no significant amount of energy is exchanged between them. Hence, the energy gradient of the boundary layer flow (defined as the dissipation divided by unit weight and discharge) must be equal to that of the upper layer and that of the total flow.

$$I = \frac{\tau' V'}{\rho g V' D'} = \frac{f' \frac{1}{2} \rho V^2}{\rho g D'} = \frac{f' V^2}{2g D'} \tag{9.60}$$

in which V' is the mean velocity in the boundary layer, and f' is the skin friction coefficient defined by

$$\tau' = \frac{1}{2} \rho f' V^2 \tag{9.61}$$

As

$$I = \frac{f\,V^2}{2gD} \tag{9.62}$$

the expression

$$\frac{f'}{D'} = \frac{f}{D} \tag{9.63}$$

is obtained. This and the equation

$$\sqrt{\frac{2}{f}} = \frac{V}{\sqrt{gDI}} \tag{9.64}$$

give

$$\sqrt{\frac{2}{f'}} = \frac{V}{\sqrt{gD'I}} \tag{9.65}$$

The friction factor f' for the boundary is determined by a formula of the type

$$\sqrt{\frac{2}{f'}} = c_1 + 2.5\ln\left(\frac{D'}{k_N}\right) \tag{9.66}$$

in which k_N is the equivalent sand roughness as defined by Nikuradse, while c_1 is a constant depending on the unknown velocity distribution in the layer. As $f = f'$ for $D = D'$, Eq. 2.20 suggests that $c_1 = 6$, so Eq. 9.66 becomes

$$\frac{V}{\sqrt{gD'I}} = 6 + 2.5\ln\left(\frac{D'}{k_N}\right) \tag{9.67}$$

This equation was originally suggested by H.A. Einstein (1950), who obtained it as an analogy to his method of calculating side wall correction. The present method, however crude, has the immediate advantage of giving an interpretation of D' as the boundary layer thickness. As experimental support reference is made to a paper by Meyer-Peter and Müller (1948), who developed an expression of different appearance, but numerically very close to that of Einstein.

Combining Eqs. 9.60 and 9.62 gives the important expression

$$\tau' = \rho g D' I \tag{9.68}$$

Example 9.3:   Numerical example of calculation of θ' and θ''.

In this example, a flow resistance curve is developed for a steady uniform flow over an erodible bed. In order to clarify the procedure, a specific case has been chosen.

The sediment is assumed to be covered by sand with $d_{50} = 0.47$ mm, and $\sqrt{d_{84}/d_{16}} = 1.4$. The distribution is assumed to be log-normal as described by Eq. 7.1. The energy slope is taken to be $I = 5.1 \times 10^{-5}$.

The first step is to calculate the flow and sediment transport in the boundary layer D'. This is done in the same way as calculating the flow over a plane bed with roughness $k_N = 2.5 \, d_{50}$.

Let us assume $D' = 5$ m. Next from the flow resistance formula Eq. 9.67 the mean flow velocity V is found to be

$$V = \sqrt{gD'I} \left[6 + 2.5 \ln\left(D'/2.5 \, d_{50}\right) \right] = 1.34 \text{ m/s} \tag{9.69}$$

The skin friction velocity U_f' is given by

$$U_f' = \sqrt{gD'I} = 0.050 \text{ m/s} \tag{9.70}$$

which gives

$$\theta' = U_f'^{\,2} \Big/ (s - 1)gd_{50} = 0.33$$

Now the bed load transport and bed load transport gradient can easily be found from Eq. 7.54 and 7.58 (Engelund-Fredsøe formula) or from Eq. 7.60 (the Meyer-Peter formula). In the following example, the Engelund-Fredsøe formula has been used with $\mu_d = 0.65$. This gives

$$\Phi_B = 1.56$$

and

$$\frac{d\Phi_B}{d\theta'} = 7.1$$

for $\theta' = 0.33$.

To calculate the similar expressions for suspended sediment requires a little more work. First the mean fall velocity of sediment into suspension must be found. This can be done by using the requirement that only sediment with a fall velocity smaller than w_{cr} given by

$$w_{cr} = U_f' = 0.050 \text{ m/s}$$

will enter into suspension (cf. Eq. 7.81). This corresponds to a critical grain diameter $d_{cr} = 0.42$ mm (cf. Table 7.1). From the grain distribution curve Eq. 7.1 it is seen that 40 per cent of the sediment is finer than 0.42 mm (with $d_{85}/d_{15} = 1.4$). Hence, the mean fractile for sediment corresponds to 20 per cent or 0.32 mm which finally gives a mean fall velocity for suspended sediment equal to

$$w_s = 0.036 \text{ m/s}$$

The bed concentration of suspended sediment can be found to be $c_b = 0.104$ from the Engelund-Fredsøe formula, Eqs. 7.41, 7.58 and 7.76. By application of Einstein's (1950) diagrams the suspended load is found to be

$$\Phi_S = 1.32 \qquad \text{for } \theta' = 0.33$$

The gradient in Φ_S is most easily found by repeating this procedure for another value of the Shields parameter $\theta' + \Delta\theta'$ giving an estimated gradient of

$$\frac{d\Phi_S}{d\theta} \simeq \frac{\Phi_S(\theta' + \Delta\theta') - \Phi_S(\theta')}{\Delta\theta'} \simeq 14 \qquad \text{for } \theta' = 0.33$$

Hence, from Eq. 9.26, the relative dune height is found to be

$$\frac{H_D}{D} = 0.112$$

In this derivation, θ'_{top} has been assumed to be equal to θ'. In a refined analysis, $\frac{H_D}{D}$ can be obtained by successive iterations calculating θ'_{top} from Eq. 9.24, the first iteration giving $\theta'_{\text{top}} = 0.37$. This refinement, however, only changes the result slightly so for practical purposes it can be omitted.

In order to calculate the dune length, Eq. 9.29 is used, except for small θ'-values where Eq. 9.29 actually underestimates the length of the dunes. For θ'-values below 0.20, the solution of the differential-equation, Eq. 9.14, gives dunes with a smaller steepness as can be read from Fig. 9.10.

The phase lag for suspended sediment L_s is needed to solve Eq. 9.29. This can be done by using the expression Eq. 9.40. An easier approximation for L_s can be obtained from Eq. 9.39 by adapting the concept of a constant eddy viscosity over the entire flow depth. Engelund (1966) suggested

$$\nu_T = 0.077 \, U_f' D' \tag{9.71}$$

to be a representative value for the depth-averaged eddy viscosity. The diffusivity of sediment ε_s is taken equal to the eddy viscosity given by Eq. 9.71.

Eq. 9.31 now gives the equilibrium profile of suspended sediment. By inserting this profile into Eq. 9.41, the height of the centroid z_c of the concentration profile is found to be

$$z_c = \frac{\varepsilon_s}{w_s} \left[1 - \frac{w_s D'}{\varepsilon_s} \exp\left(-\frac{w_s D'}{\varepsilon_s}\right) \bigg/ \left(1 - \exp\left(-\frac{w_s D'}{\varepsilon_s}\right)\right) \right] \tag{9.72}$$

Using that $w_s/U_f' = 0.72$ and taking the diffusivity of sediment to be equal to the constant eddy viscosity, Eqs. 9.71 and 9.72 give

$$z_c = 0.107 D'$$

The flow velocity U_c at the level $z = z_c$ is found from

$$\frac{U_c}{U_f'} = 8.3 + 2.5 \ln\left(\frac{z_c}{2.5\, d_{50}}\right) \tag{9.73}$$

to be

$$\frac{U_c}{U_f'} = 23.6$$

From Eq. 9.39 the adaption length for suspended sediment can be written by the use of Eq. 9.71

$$\frac{L_s}{D'} = \frac{1}{13}\left(\frac{U_f'}{w_s}\right)^2 \frac{U_c}{U_f'} \tag{9.74}$$

which in the present numerical example gives

$$L_s = 3.50\ D'$$

Eq. 9.29 now gives

$$\frac{L_D}{H_D} = 16 + 14.3\ \frac{D'}{D}$$

This expression may overpredict the real phase lag because of the assumption of a constant eddy viscosity over the entire flow depth.

When the dimensions of the sand waves have been determined, *the second step* in a flow resistance analysis is to determine how much additional water can be carried because of the form drag over the dunes. This can be obtained from Eq. 9.59, which can be rewritten as

$$\theta'' = \theta' \frac{\alpha}{2} \frac{H_D}{D} \frac{H_D}{L_D}\left(\frac{V}{U_f'}\right)^2 \tag{9.75}$$

which with $\alpha = 1$ gives

$$\frac{\theta''}{\theta'} = \frac{40.5}{16 + 14.3\ \theta'/(\theta' + \theta'')}$$

From iteration this gives

$$\frac{\theta''}{\theta'} = 1.94$$

or

$$\theta'' = 0.64$$

The total dimensionless bed shear stress now becomes

$$\theta = \theta' + \theta'' = 0.33 + 0.64 = 0.97$$

The total water depth D is easily found from Eq. 9.57, which reads

$$\theta = \frac{DI}{d(s-1)} \qquad (9.76)$$

from which it is found that

$$D = 14.7 \text{ m}$$

As the starting point of this analysis was $D' = 5$ m, it is seen that the skin friction carries 34 per cent (5/14.7) of the total flow resistance, while the form drag carries the remaining part.

The same analysis can be performed for different values of D', whereby a total flow resistance curve for a current over an erodible bed can be constructed.

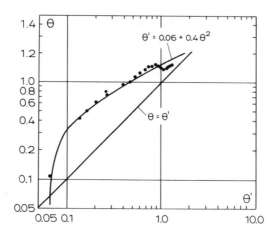

Figure 9.16 Flow resistance curve, $I = 5.1 \times 10^{-5}$, $d = 0.47$ mm, $d_{85}/d_{15} = 1.4$, s = 2.65. •: calculated values of $\theta - \theta'$.

Fig. 9.16 shows such an example for our present numerical example. One way to illustrate the flow resistance is to plot θ against θ' as done in Fig. 9.16.

Another more direct way is to plot the water discharge per unit width q against the total water depth D. This can easily be obtained from the $\theta - \theta'$

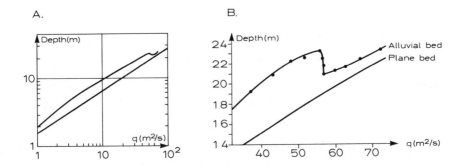

Figure 9.17 q–D curve for an alluvial stream. The data are the same as in Fig. 9.16.

relation as illustrated in the numerical example above. Fig. 9.17 shows this plot. As shown in Fig. 9.17B, it is seen that for a given range of q the water level drops when q increases. This is called transition to plane bed (see Fig. 9.1). In this region the sand waves are getting lower and longer, and the form drag decreases with a corresponding decrease in the flow depth.

In Fig. 9.16 the flow resistance formula is also plotted

$$\theta' = 0.06 + 0.4\theta^2 \tag{9.77}$$

This formula was originally proposed by Engelund (1966), who from similarity principles deduced that θ only depends on θ'

$$\theta = \theta\,(\theta') \tag{9.78}$$

The relationship given by Eq. 9.77 was then determined from the comprehensive experimental work by Guy et al. (1962). The relationship Eq. 9.78 is a valid approximation for low Shields parameters, where the influence of suspended sediment is moderate. At higher Shields parameters, Eq. 9.78 must be extended to read

$$\theta = \theta\left(\theta',\, \frac{w_{\text{susp}}}{U_f'},\, \frac{d_{85}}{d_{15}},\, \frac{D}{d},\, \mathcal{F}\right) \tag{9.79}$$

in order to explain the complicated behaviour in the transition regime. In Eq. 9.79, $\mathcal{F}$ is the Froude number $(= V/\sqrt{gD})$.

REFERENCES

Bradshaw, P. and Wong, F.Y.F. (1972): The reattachment and relaxation of a turbulent shear layer. J. Fluid Mech., 52(1):113-135.

Deigaard, R. and Fredsøe, J. (1986): Offshore sand waves. Proc. 20th Coastal Eng. Conf., CER Council, ASCE, Taipei, Taiwan, November 9-14, pp. 1047-1061.

Einstein, H.A. (1950): The bed-load function for sediment transportation in open channel flows. U.S. Dept. of Agriculture, Techn. Bulletin No. 1026, 71 p.

Engelund, F. (1966): Hydraulic resistance of alluvial streams. J. Hydr. Div., ASCE, 92(HY2):315-326.

Engelund, F. and Hansen, E. (1972): A Monograph on Sediment Transport in Alluvial Streams. 3.ed. Technical Press, Copenhagen, 62 p.

Engelund, F. and Fredsøe, J. (1976): A sediment transport model for straight alluvial channels. Nordic Hydrology, 7(5):293-306.

Fredsøe, J. (1982): Shape and dimensions of stationary dunes in rivers. J. Hydr. Div., ASCE, 108(HY8):932-947.

Fredsøe, J., Andersen, O.H. and Silberg, S. (1985): Distribution of suspended sediment in large waves. J. Waterway, Port, Coastal and Ocean Eng. Div., ASCE, 111(6):1041-1059.

Guy, H.P., Simons, D.B. and Richardson, E.V. (1966): Summary of alluvial channel data from flume experiments, 1956-1961. U.S. Geological Survey Professional Paper, 462-I.

Houbolt, J.J.H.C. (1968): Recent sediments in the southern bight of the North Sea,. Geologie en Mijnbouw, 47:245-273.

McCave, I.N. and Langhorne, D.N. (1982): Sand waves and sediment transport around the end of a tidal sand bank. Sedimentology, 29:95-110.

McLean, S.R. and Smith, J.D. (1986): A model for flow over two-dimensional bed forms. J. Hydraulic Eng., ASCE, 112(4):300-317.

Mendoza, C. and Shen, H.W. (1990): Investigation of Turbulent Flow over Dunes. J. Hydraulic. Eng., ASCE, 116(4):459-477.

Meyer-Peter, E. and Müller, R. (1948): Formulas for bed-load transport. Proc. 2nd Meet. Int. Ass. Hydr. Struct. Res., Stockholm, pp 39-64.

Raudkivi, A.J. (1963): Study of sediment ripple formation. J. Hydr. Div., ASCE, 89(HY6):15-33.

Raudkivi, A.J. (1976): Loose Boundary Hydraulics. Pergamon.

Simons, D.B., Richardson, E.V. and Albertson, M.L. (1961): Flume studies using medium sand (0.45 mm). U.S. Geol. Surv., Water-Supply Paper 1498-A.

Sleath, J.F.A. (1984): Sea Bed Mechanics. Wiley 1984.

Stride, A.H. (Ed.) (1982): Offshore Tidal Sands. Chapman and Hall, London, New York.

Terwindt, J.H.J. (1971): Sand waves in the southern bight of the North Sea. Marine Geology, 10:51-67.

Chapter 10. Wave-generated bed forms

10.1 Introduction

In an oscillatory flow, the shape of the bed forms is quite different from those found on an erodible bed exposed to unidirectional flow. Because of changes in the strength and direction of the flow the shape of the bed is also unsteady and will change during the wave period.

For short periodic waves, the volume of sediment moved back and forth during one wave period is usually small compared to the volume of sand in a bed wave. This means that the shape can be considered nearly steady, with only small fluctuations in the profile during the wave period (see Fig. 10.1A).

For long periodic waves or in the case of very fine sediment the changes in the bed wave profile can be comparable to the total volume of a bed wave as indicated in Fig. 10.1B. In this latter case, the mean bed profile becomes more elongated. The main difference between the long periodic and short periodic waves is that the flow separation in the short periodic case exhibits a very unsteady behaviour, while the separation bubble in the long periodic case is more permanent during each half wave cycle.

The ripples generated by short periodic waves (Clifton (1976) suggests that $a/d \leq 500 - 1000$) can be split up into two main groups, namely the rolling grain ripples and the vortex ripples (Bagnold, 1946).

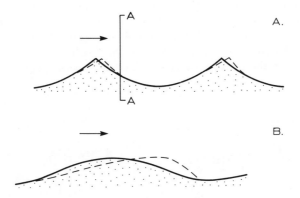

Figure 10.1 Bed forms in oscillatory flow: instantaneous (dashed) and mean (solid) profiles. A: short periodic waves. B: long periodic waves. (Partly after Sleath, 1984).

The rolling-grain ripples are formed at low Shields numbers not much larger than twice the critical Shields number. These ripples have so small a height that no real vortex formation takes place downstream from the ripple crest.

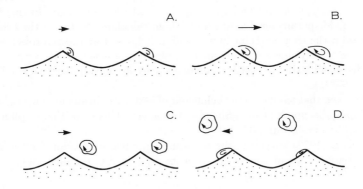

Figure 10.2 Sketch of vortices formed over a vortex ripple.

The vortex ripples are higher ripples, formed at higher Shields parameters. Fig. 10.2 sketches the unsteady behaviour of the vortices in the flow over these ripples. Just after the flow has turned to the right, a separation bubble develops downstream from the crest. This bubble expands with time (Fig. 10.2A and B) and will later be released from the bed when the pressure gradient reverses (Fig. 10.2C and D), after which the released eddy gradually loses its strength. This vortex is able to move considerable amounts of sediment away from the bed, which means that the presence of ripples increases the amount of sediment in suspension.

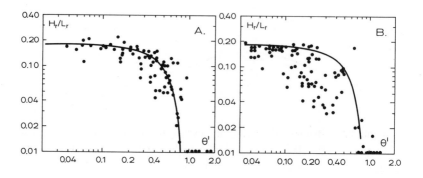

Figure 10.3 Ripple steepness versus dimensionless shear stress for laboratory data (A) and field data (B). After Nielsen (1979).

Fig. 10.3 indicates that for small values of θ', the steepness of vortex ripples H_r/L_r (H_r = ripple height and L_r = ripple length) is quite high. In this figure θ' is the Shields parameter due to skin friction, calculated by use of the ordinary expression for the friction factor in waves (Eq. 2.43) setting the bed roughness to $2.5\,d_{50}$. Fig. 10.3A shows the measured ripple steepness for regular waves, while Fig. 10.3B shows a similar case for irregular waves obtained under field conditions (Nielsen, 1979).

It is seen that similar to the behaviour of bed waves in steady flow, (Chapter 9) the ripples disappear at high Shields parameters. This transition to plane bed occurs for θ'-values around 0.8–1.0.

Nielsen (1979) suggested the following empirical expression for the ripple steepness based on regular wave experiments performed in the laboratory

$$\frac{H_r}{L_r} = 0.182 - 0.24(\theta')^{3/2} \tag{10.1}$$

This equation is the solid line in Fig. 10.3. In the case of irregular waves (field data), the steepness becomes slightly smaller as seen from Fig. 10.3B. In this case Nielsen suggested the following expression for the ripple steepness

$$\frac{H_r}{L_r} = 0.342 - 0.34(\theta')^{1/4} \tag{10.2}$$

10.2 A simple model for vortex ripples

In this section a simple description of the flow over ripples, first described by Fredsøe and Brøker (1983), is used to explain some of the most important features of the physics of vortex ripples. The description leads to an expression for the shape of the ripples and a suggestion for absolute values of H_r and L_r.

Considering an arbitrary cross section A-A (Fig. 10.1), the average net sediment transport over one wave period T must be equal to zero in order to have an equilibrium profile. The description is first limited to the case where all sediment is transported as bed load (i.e. small Shields parameters θ'). For a location of A-A just to the right-hand-side of the wave crest, a separation bubble will be formed when the flow just starts to move from left to right. (To be correct, this bubble starts to develop a little earlier. In order not to make the description too complex, this is neglected in this approximated description). The separation bubble expands with time as sketched in Fig. 10.4.

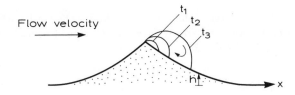

Figure 10.4 Development of separation bubble downstream of ripple crest. $t_1 < t_2 < t_3$.

This implies that a bed load particle located at A-A will be caught for a shorter or longer period of time within the separation bubble dependent on the location of A-A. In the initial stage of the separation bubble development the expansion of the bubble has not reached A-A and consequently the near-bed flow at A-A will be in the same direction as the outer flow. At a later stage, when A-A has been incorporated in the separation zone, the near-bed flow velocity reverses and the flow will be in the uphill direction towards the ripple crest.

When the flow is from right to left, the distance from the upstream crest to A-A becomes longer and the flow will consequently be directed in the same direction as the outer flow (to the left) for a longer period of time. This results in a time-averaged bed shear stress which is directed toward the crest of the ripple. In order to ensure that the resulting bed load transport is zero, the local bed slope must be so large that the effect of bed slope on the bed load transport rate counteracts the effect of the non-zero bed shear stress directed toward the crest.

The main problem in describing the mechanics of ripples is to describe the rather complex flow behaviour just above the ripples. Except for the case of very small Reynolds numbers in the laminar regime, this unsteady separated flow must be calculated by use of extensive numerical models as described later. However, some important features of the ripples can be obtained from very simple considerations.

As shown in Fig. 10.4, the growth of the separation bubble just after the outer flow velocity formed by the wave is turning to the right, can be compared with the measurements by Honji (1975) concerning the starting flow down a negative step: by suddenly moving a negative step with a certain velocity through water originally at rest, Honji measured the time variation of the location of the reattachment point downstream the step. His results are shown schematically in Fig. 10.5: at small values of the dimensionless time $t^* = Ut/h_s$ (U = outer flow velocity, h_s = step height, t = real time) the reattachment point x_r moves nearly linearly downstream with time

$$x_r = \alpha_r Ut \tag{10.3}$$

until t^* becomes 25. For larger values of t^*, x_r obtains the stationary value $x_r \sim 6\,h_s$ which is well known from steady flow (see Chapter 9). α_r is of the order 0.2–0.3 (Brøker and Fredsøe, 1983).

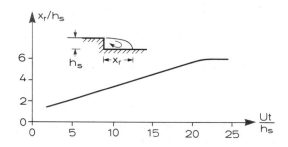

Figure 10.5 Variation in reattachment point x_r with time, after Honji (1975).

In the *steady flow* case described in Chapter 9, the shear stress distribution could be characterized as follows; cf. Fig. 9.7: inside the separation bubble the bed shear stress is directed in the opposite direction to the main flow, but is much smaller than the mean bed shear stress outside the separation bubble. After the point of reattachment, the shear stress increases rapidly to a maximum value, after which it remains almost constant (in fact it decreases slightly, due to the growth of a new boundary layer).

The picture for the unsteady case is somewhat similar to that described above (as sketched in Fig. 10.6), with the important modification that x_r is taken as the instantaneous value of the reattachment distance downstream from the step obtained by Eq. 10.3. In Fig. 10.6, an approximation to the real bed shear distribution is depicted by the dashed line: inside the separation bubble, the bed shear stress is assumed to be vanishing, while it is assumed to have a constant value downstream from the point of reattachment. An improved description could be made by taking into account the rotation of the vortex roller in unsteady flow which in fact rotates with a velocity comparable to the outer flow velocity. However, this modification does not change the following calculations significantly, so the dashed line depicted in Fig. 10.6 approximates the real distribution sufficiently accurately for the present purpose.

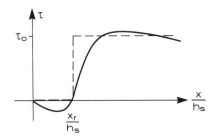

Figure 10.6 Bed shear stress distribution downstream from a negative step in steady flow (solid line, from Bradshaw and Wong, 1972). The dashed line is the analytical approximation applied in the present example.

The mean bed shear stress distribution

For simplicity, the flow velocity U away from the bed due to the wave motion as an idealization is assumed to be constant $+\ U$ (to the right) in the time $0 < t < \frac{T}{2}$ (T = wave period) and constant $-\ U$ (to the left) in the time $\frac{T}{2} < t < T$. To obtain a simple picture of the bed shear stress variation along the ripple, the flow over a hole in the bed, as sketched in Fig. 10.7A, is considered. The geometry of this hole has the main characteristics of the ripple profile, namely the rearward facing steps which are felt by the flow and which lead to the vortex formation along the bed.

For the case of flow over a hole in the bed with the length L_r as shown in Fig. 10.7A, consider the bed shear distribution in the hole at x_0 as a function of time.

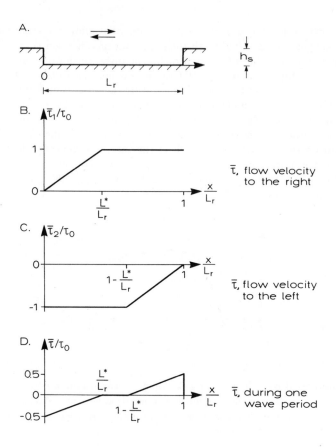

Figure 10.7 Mean bed shear distribution in a hole for the case $T/2 <$ $6h_s/(\alpha_r U)$ and $L_r > 12\ h_s$. L^* is defined in Eq. 10.5.

For $0 < t < t_1$, the bed shear stress is τ_0. At $t = t_1$ the bed shear stress becomes zero because the separation bubble attains x_0, so t_1 is determined by

$$t_1 = \frac{x_0}{\alpha_r U} \qquad (10.4)$$

cf. Eq. 10.3. Eq. 10.4 is only valid if x_0 is so small that $x_0 \leq 6\ h_s$. If $x_0 \geq 6\ h_s$, the bed shear stress is always equal to τ_0. This gives the distribution in mean value of bed shear stress $\overline{\tau_1}$ during the first half of the wave period as shown in Fig. 10.7B. When the flow turns, the distribution of $\overline{\tau_2}$ is similarly obtained and shown in Fig. 10.7C. Fig. 10.7D shows the mean distribution over one wave period $\overline{\tau}$ for the case where $6h_s/L_r \leq \frac{1}{2}$. If $6h_s/L_r \geq \frac{1}{2}$, the distribution is given by Fig. 10.8A.

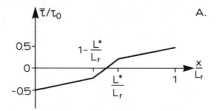

Figure 10.8A Mean bed shear stress distribution in a hole with small width, compared with Fig. 10.7:
$6h_s < L_r < 12h_s$. $T/2 < 6h_s/(\alpha_r U)$.

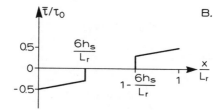

Figure 10.8B Mean bed shear stress distribution in a wide hole exposed to large wave periods, compared with Fig. 10.7:
$T/2 > 6h_s/(\alpha_r U)$. $L_r > 12\,h_s$.

The calculations shown in Fig. 10.7 are only valid if the wave period is so small that the vortex has not reached its equilibrium position. This is the same as requiring

$$\frac{T}{2} < \frac{6h_s}{\alpha_r U} \quad \text{or} \quad \frac{\alpha_r U T}{2} = L^* < 6h_s \tag{10.5}$$

In Fig. 10.8B an example is sketched where the above requirement is not fulfilled. For wave ripples, Eq. 10.5 is usually fulfilled, so the distribution in Fig. 10.8B does not occur in nature (probably because the discontinuity in bed shear distribution will form a new, smaller, wave ripple). In the following calculations, Eq. 10.5 is assumed to be fulfilled.

Bed load movement

Returning to the real ripple profile shown in Fig. 10.1 and assuming that the bed shear stress is small, all sediment moves as *bed load*. Because of the local slope, it is easier to move a particle downhill than uphill. This can be taken into account by changing the Shields parameter θ (dimensionless bed shear stress) on a plane bed to θ_γ on a bed with a longitudinal slope γ, by the relation

$$\theta_\gamma = \theta\left(1 - \frac{\tan\gamma}{\tan\phi}\right)\cos\gamma \tag{10.6}$$

where ϕ is the friction angle, cf. Eq. 7.35. Eq. 10.6 can be approximated by

$$\theta_\gamma = \theta\left(1 - \frac{1}{\tan\phi}\frac{dh}{dx}\right) \tag{10.7}$$

The requirement for a stable ripple profile is that the net sediment transport during one wave period is zero at every location on the ripple. For the case of pure bed load, the dimensionless sediment transport Φ_b is a function of θ_γ only

$$\Phi_B = \Phi_B(\theta_\gamma) \tag{10.8}$$

The continuity requirement is

$$\int_0^T \Phi_B dt = 0 \tag{10.9}$$

In the present simplified model θ is either equal to a constant value θ_0 or equal to zero, cf. the dashed line in Fig. 10.6. Hence, Eq. 10.9 is fulfilled if

$$\overline{\theta_{\gamma 1}} + \overline{\theta_{\gamma 2}} = 0 \tag{10.10}$$

where $\overline{\theta_{\gamma 1}}$ is the dimensionless version of the bed shear stress shown in Fig. 10.7B, corrected for the action of gravity by Eq. 10.7. $\overline{\theta_{\gamma 2}}$ is defined from Fig. 10.7C in the same way. Application of Eqs. 10.7 and 10.10 gives

$$\overline{\tau_1}\left(1 - \frac{1}{\tan\phi}\frac{dh}{dx}\right) + \overline{\tau_2}\left(1 + \frac{1}{\tan\phi}\frac{dh}{dx}\right) = 0 \tag{10.11}$$

or

$$\frac{dh}{dx} = \tan\phi\frac{\overline{\tau_1} + \overline{\tau_2}}{\overline{\tau_1} - \overline{\tau_2}} \tag{10.12}$$

where

$$\frac{\overline{\tau_1}}{\overline{\tau_0}} = \begin{cases} \dfrac{x}{L^*} & 0 < x < L^* \\ \\ 1 & x > L^* \end{cases} \tag{10.13a}$$

and

$$\frac{\overline{\tau_2}}{\overline{\tau_0}} = \begin{cases} -1 & 0 < x < L_r - L^* \\ -\frac{L_r - x}{L^*} & L_r - L^* < x < L_r \end{cases} \tag{10.13b}$$

By inserting Eq. 10.13 into Eq. 10.12 a differential equation for h is obtained. Three different expressions are obtained, depending on the ratio L_r/L^*:

1) $L_r/L^* \leq 1$: In this case Eq. 10.12 reads

$$\frac{dh}{dx} = \tan\phi \frac{(2x - L_r)}{L_r} \tag{10.14}$$

which has the solution

$$h = \frac{\tan\phi}{L_r}\left(x^2 - L_r x\right) + h_0 \tag{10.15}$$

This solution is depicted in Fig. 10.9. This profile has a steepness of $\tan\phi/4 \sim 0.16$ which is quite close to the measured values at low values of θ', cf. Fig. 10.3.

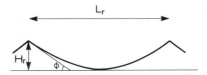

Figure 10.9 Theoretical solution for the ripple profile, $L_r < 2\,L^*$.

2) $L^* < L_r < 2L^*$: In this case, the bed shear stress distribution (depicted in Fig. 10.8A) gives the following differential equation in h

$$\frac{dh}{dx} = \begin{cases} \tan\phi \frac{(x/L^* - 1)}{(x/L^* + 1)} & 0 < x < L^* \\ \tan\phi \frac{2x - L_r}{L_r} & L_r - L^* < x < L^* \\ \tan\phi \frac{(1 + x/L^* - L_r/L^*)}{(1 - x/L^* + L_r/L^*)} & L^* < x < L_r \end{cases} \tag{10.16}$$

The solution to Eq. 10.16 gives a solution nearly identical to the one given by Eq. 10.14.

3) $L_r > 2L^*$: For this long ripple, the differential equation in h reads

$$\frac{dh}{dx} = \begin{cases} \tan\phi \dfrac{(x/L^* - 1)}{(x/L^* + 1)} & 0 < x < L^* \\[2mm] 0 & L^* < x < L_r - L^* \\[2mm] \tan\phi \dfrac{(1 + x/L^* - L_r/L^*)}{(-x/L^* + 1 + L_r/L^*)} & L_r - L^* < x < L_r \end{cases} \qquad (10.17)$$

which predicts a solution with a kind of "solitary ripple crests" on a plane bed, see Fig. 10.10. This solution will not be present in nature, because the plane part of the bed is unstable.

Figure 10.10 A "solitary ripple" train as predicted by Eq. 10.17, $L_r > 2L^*$.

Determination of ripple length

The model described above is able to calculate the ripple height and shape if the ripple length is known in advance. Fig. 10.11 shows a time series of ripple development: the ripples are normally observed to be initiated close to a small disturbance from which the ripples spread to cover the total bed. Initially, the ripple length is small but grows with time. This can be explained by the fact that the extent of the separation bubble is a function of the height of the ripple.

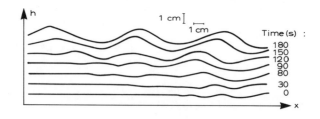

Figure 10.11 Ripple development from a plane bed. $a = 0.073$ m, $T = 1.5$ s. $\theta' = 0.105$.

When the height of the ripples is sufficiently large, the extent of the separation bubble is constrained by the limited time (half a wave period) for the expansion of the bubble which implies that no further elongation of the ripple length will take place.

An idea about the mature ripple height can be obtained by setting

$$L_r = 2L^* \qquad (10.18)$$

This corresponds to the transition between the profiles shown in the Figs. 10.9 and 10.10. As the profiles corresponding to "solitary" ripples are unlikely to occur because the plane stretch between the ripple crests will be unstable, L_r cannot be larger than $2L^*$. On the other hand, if L_r is taken smaller than $2L^*$, the ripple height will be correspondingly smaller. Due to stochastic variations, some ripples will always be higher than others and will consequently dominate the erosion/deposition pattern at the bed, eliminating the smaller bedforms. For this reason, it must be expected that the largest possible ripple height at the end will be the preferred ripple height, justifying the choice of Eq. 10.18 for determining the ripple height.

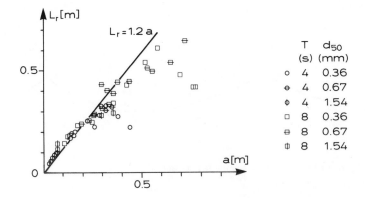

Figure 10.12 Ripple length versus near-bed water excursion amplitude. Measurements by Mogridge (1972). (From Brøker, 1985).

Now Eq. 10.18 and Eq. 10.5 give

$$L_r = 2\frac{\alpha_r}{2}UT = \alpha_r UT \qquad (10.19)$$

or

$$L_r = 4\alpha_r a \qquad (10.20)$$

in which $a = UT/4$ is the near-bed water excursion amplitude for the idealized picture, where U is constant during each half wave period. For $\alpha_r \simeq 0.3$, Eq. 10.20 gives

$$L_r = 1.2a \qquad (10.21)$$

This expression fits very well with measurements as seen from Fig. 10.12. However, at large a-values there is a systematic deviation from Eq. 10.21 which can partly be attributed to the influence of suspended sediment.

Influence of suspended sediment on ripple shape

The influence of suspended sediment on the shape of vortex ripples has been studied by Brøker (1985) by use of the model outlined above. A detailed description of suspended sediment over ripples is given in the next section, but for the present analysis a more simplified approach has been used.

When the sediment is passing the crest of the ripple, the bed load is immediately trapped in the separation bubble while a large part of the suspended sediment is transported a greater distance and will settle further downstream on the ripple (quite similar to the description of sediment transport over dunes, cf. Chapter 9).

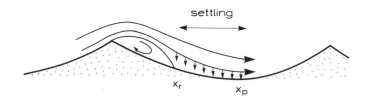

Figure 10.13 Settling of suspended sediment.

Brøker (1985) assumed that all the suspended sediment transported over the ripple crest after the flow turned to the right (see Fig. 10.13) would be deposited uniformly between the reattachment point and the point x_p, which is determined as the point where a particle, released at the crest just after flow reversal and moving outside the separation bubble (i.e. following the potential flow), has reached. The total amount of settled suspended sediment on this stretch during half a wave period must equal the amount of suspended sediment transported over the crest in the same half wave period. The latter can be calculated as the transport capacity of suspended sediment in the boundary layer formed along the ripple in front of the ripple crest and can, for instance, be calculated by the sediment transport formula developed in Chapters 7 and 8.

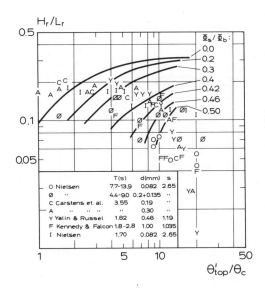

Figure 10.14 Ripple steepness versus Shields parameters. The solid lines are
calculated on the basis of a constant ratio of Φ_S/Φ_B. (After
Brøker, 1985).

By implementing this description together with the earlier description of
bed load it is possible to determine the shape of the ripples by requiring the net
transport through each vertical to be zero. The result indicates that the ripples
will become steeper, as shown in Fig. 10.14. In this figure the ripple steepnes
is calculated for different ratios of Φ_S/Φ_B and at different values of θ'_{top} (the
effective Shields parameter at the ripple crest). The parameter Φ_S/Φ_B in this
figure is assumed to be known in advance. (The curves in Fig. 10.14 for the pure
bed load case ($\Phi_S/\Phi_B = 0$) deviate slightly from the steepness predicted earlier
in this section, because in Fig. 10.14 it has also been taken into account that
bed load sediment is transported over the crest). By calculating the ratio Φ_S/Φ_B
as a function of θ' it is possible to obtain a curve for the ripple steepness as a
function of θ'. Fig. 10.15 shows such an example based on the sediment transport
descriptions in Chapters 7 and 8.

Because only a certain fraction of the suspended sediment will be exchanged
with the bottom during each wave cycle, Brøker introduced a factor ξ_s, so only
$\xi_s\Phi_S$ is assumed to be deposited on the ripple during a wave cycle. As seen
from Fig. 10.15, the ripple steepness is described quite well for ξ_s in the order of
0.15–0.20.

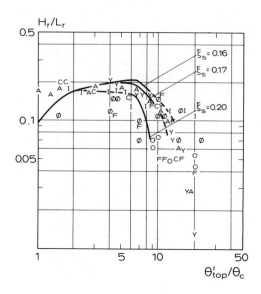

Figure 10.15 Ripple steepness versus Shields parameter. The factor ξ_s indicates the fraction of suspended sediment passing the crest and settling on the downstream ripple. The symbols are the same as in Figure 10.14. (After Brøker, 1985).

10.3 Distribution of suspended sediment over vortex ripples

The vertical distribution of suspended sediment over vortex ripples is quite different from that over a plane bed (see Chapter 8) because of the organized vertical motion connected with the vortex roller, see Fig. 10.2.

If the measured concentration is averaged over a wave period, and further averaged along the ripple, Nielsen (1979) demonstrated that the vertical distribution could be well described by

$$c = c_0 \exp(-\alpha z) \qquad (10.22)$$

in which α is a constant, and c_0 the concentration a certain distance over the bed.

Nielsen assumed that this vertical distribution could be described by introducing a diffusion coefficient ε_s for the sediment which is constant in space and

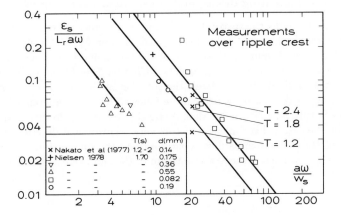

Figure 10.16 Measured diffusion coefficient for suspended sediment over ripples. After Nielsen (1979).

time. Hence, the distribution given by Eq. 10.22 becomes

$$c = c_0 \exp\left(-\frac{w_s}{\varepsilon_s}z\right) \tag{10.23}$$

Based on the plotted data by Nielsen the following empirical expression for the diffusion coefficient can be obtained

$$\frac{\varepsilon_s}{L_r a\omega} = \exp\left\{1.5 - 4500d - 1.2\ln\left(\frac{a\omega}{w_s}\right)\right\} \tag{10.24}$$

in which d must be inserted in m (SI-units). a is the amplitude of the near-bed orbital motion and ω is the cyclic frequency. From Eq. 10.23 and Fig. 10.16 it can be seen that ε_s depends rather strongly on the fall velocity of the sediment, as ε_s increases with decreasing fall velocity. This is typical for a flow where organized vertical motion of the flow is present as described in Example 8.2.

On the basis of measurements, Nielsen (1979) suggested that the reference concentration appearing in Eq. 10.23 should be given by

$$c_0 = 0.028(\theta' - \theta_c)\frac{2}{\pi}\text{Arccos}\sqrt{\frac{\theta_c}{\theta'}} \tag{10.25}$$

Mathematical modelling of suspended sediment over ripples

In order to model the suspended sediment over ripples, a detailed description of the flow is required. To describe this unsteady non-uniform flow (including flow separation) requires a numerical solution of the flow equations. Such an analysis can either be based on the $k - \varepsilon$ approach (see Chapter 2) or by use of the so-called discrete vortex method. In the latter, the rotation in the flow is discretized and represented by ideal vortices, which can be followed by a Lagrangian scheme (see for instance Sarpkaya, 1989, or Smith and Stansby, 1989). While this method does not describe the detailed turbulent structure very accurately, it describes the gross behaviour of the flow, such as flow separation and the further development and movement of the vortices formed downstream of the ripple crest, quite well.

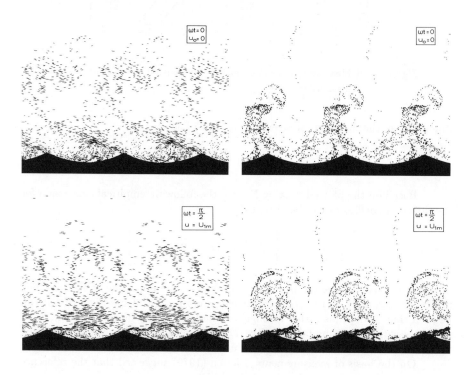

Figure 10.17 Flow pattern (at the left) and suspended sediment particles (at the right) at two phases during a wave cycle. (Hansen et al., 1991).

Fig. 10.17 shows the calculated flow field over a ripple bed at different phases by use of the discrete vortex model (Hansen et al., 1991). With such a flow description, the individual grains in suspension can be followed by a Lagrangian formulation. The change in their position during a time step Δt due to convection and settling is expressed by the flow field

$$\big(x(t+\Delta t),\ z(t+\Delta t)\big) = \big(x(t),\ z(t)\big) + \big(u(t)\Delta t,\ (w(t)-w_s)\Delta t\big) \qquad (10.26)$$

In this description the diffusion processes are not described completely correctly. Velocity fluctuations with dimensions smaller than the grid spacing, or time scales shorter than the time step cannot be simulated directly, which implies that the diffusion processes (caused by turbulent fluctuations) with length scales smaller than the grid spacing or time scales shorter than the time step cannot be directly represented. The diffusion process is chiefly important very close to the bed, where a wave boundary layer develops (Fig. 10.18). Inside this wave boundary layer, the suspended sediment can be calculated according to the plane bed case described in Chapter 8, with a bed concentration concept and an increasing sediment diffusivity ε_s away from the bed. Outside the wave boundary layer, the dominant transport term is the convective transport by the large organized eddies.

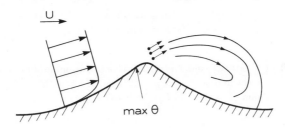

Figure 10.18 Release of suspended sediment particles close to the ripple crest.

The right-hand-side of Fig. 10.17 shows a calculation of the distribution of suspended sediment particles during one wave cycle (Hansen et al., 1991). The particle path is found from the Lagrangian scheme Eq. 10.26. The feed of particles into the outer flow is found from the transport capacity in the wave boundary layer just in front of the ripple crest, where the skin friction is largest, see Fig. 10.18. Here, the amount of sediment transported in the boundary layer during the time interval Δt is given by

$$q_s\Delta t = \int_{2d}^{\delta} c(z)u(z)dz\Delta t \qquad (10.27)$$

This amount of sediment is released in each time interval Δt a short distance away from the ripple crest as indicated in Fig. 10.18. Numerical sensitivity analyses have shown that the average concentration is only slightly affected by the exact placement of the release point. Fig. 10.19 shows a comparison between the described theoretical analysis and the measured concentration over ripples, performed by Ribberink and Al-Salem (1989) in an oscillating water tunnel. It can be seen that quite a good correlation between the magnitude of the near-bed concentrations, as well as the vertical distribution, is obtained.

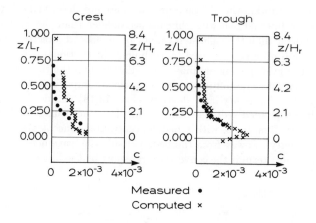

Figure 10.19 Measured and calculated concentration profiles over ripple crest and trough. $L_r = 0.17$ m, $H_r = 0.028$ m, $d_{50} = 0.21$ mm, $T = 4$ s and $U_{1m} = 0.457$ m/s.

REFERENCES

Bagnold, R.A. (1946): Motion of Waves in Shallow Water: Interaction Between Waves and Sand Bottoms. Proc. Roy. Soc., London, A 187:1-15.

Bradshaw, P. and Wong, F.Y.F. (1972): The reattachment and relaxation of a turbulent shear layer. J. Fluid Mech., 52(1):113-135.

Brøker, I.H. and Fredsøe, J. (1983): Velocity measurements in oscillatory flow along a ripple-surface. Progress Report No. 59, Inst. of Hydrodynamics and Hydraulic Engineering, ISVA, Techn. Univ. Denmark, pp. 23-32.

Brøker, I.H. (1985): Wave generated ripples and resulting sediment transport in waves. Series Paper No. 36, Inst. of Hydrodynamics and Hydraulic Engineering, ISVA, Techn. Univ. Denmark.

Clifton, H.E. (1976): Wave-formed Sedimentary Structures - A Conceptual Model. In: R.A. Davis and R.L. Ethington (Editors), Beach and Nearshore Processes. Soc. Econ. Paleontol. Mineral., Spec. Publ., 24:126-148.

Fredsøe, J. and Brøker, I.H. (1983): Shape of oscillatory sand ripples. Progress Report, No. 58, Inst. of Hydrodynamics and Hydraulic Engineering, ISVA, Techn. Univ. Denmark. pp. 19-29.

Hansen, E.A., Fredsøe, J. and Deigaard, R. (1991): Distribution of suspended sediment over wave generated ripples. Int. Symposium on the Transport of Suspended Sediments and its Mathematical Modelling, Florence, Italy, September 2-5, 1991, pp. 111-128.

Honji, H. (1975): The starting flow down a step. J. Fluid Mech., 69(2):229-240.

Mogridge, G.R. (1972): Wave Generated Bed Forms. Ph.D. Thesis, Queens University, Kingston.

Nielsen, P. (1979): Some basic concepts of wave sediment transport. Series Paper No. 20, Inst. of Hydrodynamics and Hydraulic Engineering, ISVA, Techn. Univ. Denmark.

Ribberink, J.S. and Al-Salem, A. (1989): Bed forms, near-bed sediment concentrations and sediment transport in simulated regular wave conditions. Delft Hydraulics, Report H-840, Part 3.

Sarpkaya, T. (1989): Computational Methods with Vortices - The 1988 Freeman Scholar lecture. Journal of Fluids Engineering, Vol. 111, No. 1, March 1989, pp. 5-52.

Sleath, J.F.A. (1984): Sea Bed Mechanics, Wiley 1984.

Smith, P.A. and Stansby, P.K. (1989): Postcritical Flow around a Circular Cylinder by the Vortex Method. J. of Fluids and Structures, 3:275-291.

Chapter 11. Cross-shore sediment transport and coastal profile development

In the previous chapters the hydrodynamics of waves and currents in- and outside the surf zone, and the mechanics of sediment transport under waves and currents have been considered. In this chapter these results will be integrated to determine sediment transport rates in the cross-shore direction. The interplay between the sediment transport and the coastal morphology will be described for some specific examples.

11.1 Cross-shore sediment transport

The sediment transport in the direction normal to a long uniform coast with normally incident waves has recieved considerable attention and much experimental and theoretical work has been carried out. Cross-shore sediment transport is very important because the coastal profile is formed by the erosion/deposition associated with shore-normal transport. As described in Chapter 12, a detailed model of the longshore wave-driven current and the longshore sediment transport requires as input the shape of the coastal profile. The profile of a sandy beach changes continuously and may be modified considerably during a single storm. In principle, it will therefore not be possible to make a detailed simulation of the longshore sediment transport without having a model for the cross-shore sediment transport and the development of the coastal profile. It should be noted that in reality the models of the coastal profile development have not yet reached a stage where they can be coupled to the longshore sediment transport models and the simulations are therefore normally based on profiles that have been estimated from surveys carried out during calm periods.

It is normally assumed that the situation is completely two-dimensional with no mean cross-shore flow. This situation represents the experimental conditions encountered in an ordinary wave flume, however, in nature the assumption of zero mean cross-shore flow will, in many cases, not be valid. Small deviations from a uniform coastal profile may generate horizontal circulation currents and even a completely uniform situation may be unstable, ending up with a series of horizontal circulation currents and rip currents.

Nevertheless, the strictly two-dimensional situation is of considerable interest as it is well documented and because all the mechanisms are also active in the more complex three-dimensional flow situation. The situation with zero cross-shore flux is actually more complex than that with a strong mean current, because many different mechanisms will contribute to the resulting sediment transport without the possibility of excluding any in advance. With a strong current the effect of the mean shear stress will be dominant for the mean current and the sediment transport.

In the following, the mechanisms of the cross-shore sediment transport are described, for conditions outside and within the surf zone. The formulation of the sediment transport model is divided in two parts: a) the hydrodynamic description the main purpose of which is to model the mean current velocity distribution in addition to the variation of the eddy viscosity and the bed shear stress, and b) the description of the sediment concentrations and the resulting sediment load.

11.1.1 Hydrodynamics outside the surf zone

Outside the surf zone the energy dissipation and the turbulence are mainly confined to the near-bed wave boundary layer and the main effort is therefore concentrated on describing the conditions in the boundary layer and its influence on the mean flow and sediment transport. The hydrodynamics and sediment transport are described by the methods given in Chapters 1-4 and in Chapter 8.

First a flow situation which can be described by potential flow theory outside the wave boundary layer is considered. The criterion for potential theory to be valid is that the wave-averaged shear stress must be zero outside the boundary layer.

Streaming

The phenomenon of streaming was treated in Section 2.4, where a description of how the non-uniformity of the boundary layer under real waves causes a change in the mean shear stress over the boundary layer was given. The displacement in the non-uniform wave boundary layer causes small vertical velocities that grow from zero at the bed to v_∞ outside the boundary layer. The detailed variation of this velocity can be determined from the velocity distribution in time and space of the wave boundary layer. The streaming gives a jump in the mean shear stress of $\Delta\tau$ over the wave boundary layer.

As described in Example 6.1 a perfect force balance with zero mean shear stress over the entire water column (outside the wave boundary layer) can be obtained by an adjustment in the slope of the mean water surface. This situation is then consistent with a potential flow outside the boundary layer. Because the mean shear stress is different from zero in the wave boundary layer, the mean flow velocity is also non-zero. The mean flow velocity increases through the wave boundary layer to become constant outside the boundary layer.

Non-linear waves

In shallow water where the waves are close to breaking they become very non-linear, and the variation of the near-bed orbital motion can deviate significantly from the sinusoidal prediction of first order wave theory. If the wave boundary layer is laminar, the oscillatory boundary layer will have zero mean shear stress for zero mean horizontal flow near the bed. This is because the solution to the laminar boundary layer is linear and a complete solution can be obtained by making a Fourier decomposition of the wave-induced motion. Due to the linearity, the complete laminar solution can be obtained by adding the solutions corresponding to each Fourier component. As the mean shear stress is zero for each harmonic component, it will also be zero for the total solution.

For a turbulent oscillatory boundary layer the mean shear stress is not necessarily zero for zero mean flow. This can be illustrated by considering the simplified example of a constant friction factor f_w

$$\tau_b = \frac{1}{2} f_w u_0 |u_0| \tag{11.1}$$

where τ_b is the instantaneous bed shear stress and u_0 is the near-bed wave-induced orbital velocity. The velocity u_0 is composed of two harmonics, as found from the second order wave theory

$$u_0 = U_{1m} \cos(\omega t) + U_{2m} \cos(2\omega t) \tag{11.2}$$

where ω is the wave angular frequency. It is assumed that the second harmonic is small, i.e.

$$a_2 = \left| \frac{U_{2m}}{U_{1m}} \right| << 1 \tag{11.3}$$

When the mean shear stress is calculated from Eqs. 11.1 and 11.2, it is found to be

$$\bar{\tau} = \frac{4}{3\pi} a_2 \tau_{\max} \tag{11.4}$$

where $\tau_{\max}$ is the maximum bed shear stress during a wave period. If a mean shear stress of zero is required, it is necessary to add a constant current velocity at the top of the wave boundary layer U_δ to the orbital motion. The magnitude of U_δ can be determined by the same principles that were applied in Section 3.1, cf. Eq. 3.8

$$-\bar{\tau} = \frac{2}{\pi} f_w U_{1m} U_\delta = \frac{4}{\pi} \tau_{\max} \frac{U_\delta}{U_{1m}} \tag{11.5}$$

or

$$U_\delta = -\frac{1}{3} a_2 U_{1m} = -\frac{1}{3} U_{2m} \tag{11.6}$$

A wave motion with this mean velocity, constant over the vertical, is thus seen to fulfill the requirements for a potential wave motion without any mean shear stress, provided that Eq. 11.1 determines the bed shear stress. When a more detailed model for the wave boundary layer is used, the mean flow giving zero mean shear stress can be determined by trial and error. The mean current obtained from these considerations must be combined with the mean flow associated with the streaming described above.

The wave drift

The wave motion itself implies a net flux of water, the wave drift q_{drift}. As described in Chapter 1, it can be calculated by time-averaging the instantaneous discharge through a fixed cross section. For linear shallow water waves this gives

$$q_{\mathrm{drift}} = \frac{1}{T} \int_0^T u_0 (D + \eta) dt = \overline{u_0 \eta} = \frac{D}{c} \overline{u_0^2} \tag{11.7}$$

The wave drift can alternatively be calculated by Lagrangian considerations, following the path of the individual water particles. For shallow water waves the particles will have a slow forward drift, because as they move forward they move with the wave which is itself propagating with celerity c, while their backward motion is against the wave. The forward motion of the particle is therefore of slightly longer duration than the backward motion. The mean Lagrangian drift velocity u_ℓ is found to be

$$u_\ell = \frac{1}{c} \overline{u^2} \tag{11.8}$$

As seen from Eqs. 11.7 and 11.8 the two alternative approaches give the same magnitude for q_{drift}, but while the Eulerian calculation gives a discharge concentrated between the wave trough and crest, the Lagrangian gives (for linear shallow water waves) an even distribution over the mean water depth.

Example 11.1:   Wave drift of suspended sediment

The Lagrangian drift velocity must be included in the flow field when calculating suspended load transport under waves, as long as the vertical orbital velocity is not taken into account when solving the diffusion equation for suspended sediment. In the plane situation this equation reads (cf. Chapter 8)

$$\frac{dc}{dt} = \frac{\partial c}{\partial t} + u\frac{\partial c}{\partial x} + w\frac{\partial c}{\partial z} = \frac{\partial}{\partial z}\left(\varepsilon_s\frac{\partial c}{\partial z}\right) + w_s\frac{\partial c}{\partial z} \tag{11.9}$$

If the convection terms are included in the description and the vertical and horizontal flow velocities are determined correctly, then the wave drift of the suspended sediment will be determined correctly by an Eulerian analysis, and the Lagrangian drift velocity must not be included in the description. Consider the following simplified example:

In the lower half of the water column there is suspended fine sediment with the constant concentration c_0. In the upper half the concentration is zero. Settling and diffusion of sediment are neglected. The water waves can be described by linear shallow water wave theory and all effects of the boundary layer are neglected. From the velocity field the upper boundary z_{top} of the concentration field is found to vary as

$$z_{\text{top}} = \frac{1}{2}(D + \eta) \tag{11.10}$$

The horizontal flow velocity is the orbital velocity u_0. The wave drift of the sediment can, by Eulerian analysis, be determined as

$$q_{sd} = \frac{1}{T}\int_0^T c_0\frac{1}{2}(D + \eta)u_0\,dt = \frac{1}{2}c_0 q_{\text{drift}} \tag{11.11}$$

In most sediment transport models for waves and current, however (e.g. the one described in Chapter 8), the vertical flow velocities in- and outside the oscillatory boundary layer are not calculated and the convective terms in Eq. 11.9 are neglected. In the present example this corresponds to taking z_{top} to be a constant and equal to $D/2$. With this simplification the wave sediment drift determined by an Eulerian analysis becomes zero, which is obviously incorrect. A good first approximation is obtained by using the Lagrangian drift velocity u_ℓ. When u_ℓ is calculated from Eq. 11.8, the velocity distribution in the wave boundary layer must be taken into account. In the present example the Lagrangian wave sediment drift is found to be

$$q_{sd\ell} = \int_0^D u_\ell\bar{c}\,dz = \frac{D}{2}\frac{1}{c}\,\overline{u_0^2}\,c_0 = \frac{1}{2}q_{\text{drift}}c_0 \tag{11.12}$$

which is identical to the correctly determined Eulerian wave sediment drift in Eq. 11.11.

The mean turbulent flow

Until now, the requirement of zero shear stress has been maintained for the flow outside the wave boundary layer to be described by potential theory. Three contributions to the net flow have been identified: streaming in the wave boundary layer, effect of non-linear waves on the turbulent wave boundary layer and the wave drift. The sum of the three net discharges is different from zero. In order to satisfy the continuity equation which in the present case states that the net cross-shore flux is zero, the mean water surface must have a slope S giving a mean shear stress which drives a current to compensate for the net flux of the potential wave motion. The mean water surface slope gives the triangular shear stress distribution

$$\bar{\tau} = \rho g S (D - z) \tag{11.13}$$

This mean shear stress must then be included in the hydrodynamic model describing the turbulent wave-current boundary layer which includes the effect of the three wave phenomena described above and fulfills the continuity equation.

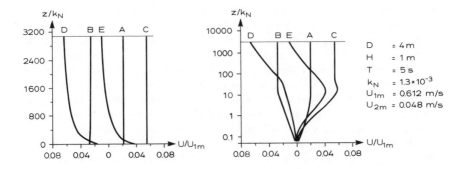

Figure 11.1 The contribution to the mean flow in the case of non-breaking waves, linear scale (left) and logarithmic scale (right). A: effect of streaming. B: effect of asymmetric wave motion. C: Lagrangian drift velocity. D: combined mean flow. E: combined mean flow plus Lagrangian drift velocity.

An example is given in Fig. 11.1, where each of the contributions to the mean flow are shown. The wave motion is calculated by second order Stokes theory urbulence is modelled by a mixing length model. The velocity profile E an velocity profile plus the Lagrangian drift velocity and gives a zero depth-integrated discharge. The logarithmic scale shows the detail in the wave boundary layer, while the linear scale shows the velocity distribution outside the wave boundary layer. According to the mixing length model the non-linearity of the wave motion gives a mean velocity outside the wave boundary layer of -0.016 m/s, which is in agreement with the simple estimate in Eq. 11.6.

11.1.2 Cross-shore sediment transport outside the surf zone

When a satisfactory hydrodynamic description has been obtained, the sediment transport is calculated as bed load and suspended load transport. In addition to the mean current velocity determined by the hydrodynamic model, there is a contribution to the net sediment transport due to the non-linearities in the sediment transport model. If the near-bed wave-orbital velocity (and the bed shear stress) is larger in the onshore motion than in the offshore motion, a net sediment transport in the onshore direction is induced because the bed load transport and the bed concentration are non-linear functions of the friction velocity.

Another effect is due to the time variation of the eddy viscosity and the sediment concentration at different levels. It is important to have the correct phases of the concentrations and the flow velocities in order to calculate the net sediment transport. In extreme cases the phase variation may cause the suspended transport to go against the direction of the maximum shear stress during a wave period, but normally it only modifies the magnitude of the net transport.

The presence of wave ripples can be of significance in this respect, because, as described in Chapter 10, the sediment brought into suspension during one half wave period is carried in the vortex in the lee of each ripple. Only in the next half wave period are the vortex and sediment ejected into the flow over the ripples (at the same time as a new vortex is formed with suspended sediment).

If the bed has a slope β, this will also give a contribution to the net transport. The increased density due to the suspended sediment carried by the water gives a mean shear stress contribution of

$$\tau = \int_z^D \tan\beta(s-1)\rho g \bar{c} dz'$$ (11.14)

where s is the relative density of the sediment. This contribution should then be included in the hydrodynamic model in order to calculate the mean flow with this gravity effect included. The inclusion of this term will require an iterative procedure, first a hydrodynamic calculation should be made neglecting the effect of the bed slope, followed by calculating the sediment concentration field and determining the slope-induced shear stress from Eq. 11.14. Finally, a new hydrodynamic simulation including the gravity effect term is made and the net sediment transport is found from this second calculation of mean flow.

There are other effects of a sloping bed which can be included in cross-shore sediment transport models. The sloping bed causes the wave-orbital motion to be divergent/convergent when water moves off- or onshore. This has an effect on the turbulence and the bed shear stress in the wave boundary layer and will create a mean flow near the bed going uphill (Justesen, 1988). Another effect is that the shoaling process will cause the waves to be skew with a steeper shore-facing surface slope. This also affects the turbulent wave boundary layer and gives a mean shear stress. The mean shear stress in the boundary layer under very skew waves was analyzed in Example 2.6. These two mechanisms can, however, generally be expected to be less important.

In this section a number of mechanisms responsible for the net sediment transport under non-breaking waves have been described. At present no theoretical transport model describing this situation has been thoroughly verified by comparison with field or laboratory data. The main difficulty lies in the many different contributions that are all small compared to the gross sediment motion due to the wave-orbital motion. As will be described in the following section the cross-shore sediment transport is an order of magnitude larger in the surf zone than outside. Still the cross-shore transport under non-breaking waves is important for the development of the coastal profile, because at a given point the waves will be non-breaking during calm conditions for much longer time than the relative short periods of storms-induced wave-breaking.

11.1.3   Cross-shore sediment transport in the surf zone

The conditions in the surf zone are characterized by the strong energy dissipation and production of turbulence caused by the wave-breaking. All the mechanisms that are found to give contributions to the cross-shore sediment transport outside the surf zone are also relevant in the surf zone, but their significance is much smaller because the energy dissipation in the wave boundary layer is small compared to the energy loss due to wave-breaking.

As described in Chapter 6, it is necessary to introduce shear stresses when the energy loss takes place near the surface in order to obtain a force balance and these shear stresses are important for the velocity distribution. There is also an extra contribution to the continuity equation in the water carried shorewards with the surface roller following each wave front of the broken waves in the surf zone.

The distribution of the shear stresses together with the continuity equation gives the mean velocity profile, described in Chapter 6, with the strong offshore-directed undertow near the bed and an onshore mean flow near the surface. The suspended sediment concentration profiles are influenced by the high turbulence level due to the wave-breaking, as described in Section 8.4. The concentration distribution is much more even over the vertical than outside the surf zone. Nevertheless, the concentrations near the bed are the largest, and the resulting sediment transport goes offshore with the undertow.

The magnitude of the sediment transport in the surf zone relative to the one outside is illustrated by an example. The mean water depth is $D = 2$ m, the wave height is half the water depth $H = 1$ m and the period is $T = 5$ s. The simulation has been made under the assumption of breaking as well as non-breaking waves. With an offshore bed slope $\tan \beta = 1/50$ and grain size $d = 0.2$ mm the non-breaking wave gives a net transport of 0.82 m^3/m day, while the breaking waves give a transport of -11.2 m^3/m day, where the negative sign shows an offshore transport.

The model outlined here is valid for spilling breakers and for the inner surf zone where the broken waves are rather similar to bores. The complex flow pattern near the plunge point of plunging breakers is not described. The plunging wave can generate a jet that penetrates down to the bed and large coherent vortices may be formed that can have a large effect on the local sediment distribution and flux. No quantitative model has yet been formulated which can take this into account.

11.2   Development of the coastal profile

The coastal profile can vary considerably during a year or even a single storm event. Longshore bars can be formed and existing bars can be shifted or destroyed. The cross-shore sediment transport plays an important role in the development of the coastal profile and with the assumptions made for the cross-shore sediment transport, a model which describes the morphological development can .be formulated. The main assumption is that the situation is strictly two-dimensional, i.e. that the net discharge in the direction parallel to the coastline is zero.

The morphological model consists of an onshore/offshore sediment transport model which calculates the detailed variation of the transport across the profile. From the sediment transport field, the development of the coastal profile is calculated by the continuity equation for the sediment:

$$\frac{\partial h}{\partial t} = -\frac{1}{1-n}\frac{\partial q_{sx}}{\partial x} \tag{11.15}$$

where h is the bed level, n is the porosity of the bed and q_{sx} is the sediment transport rate in the x-direction. In practice, the sediment transport model and the continuity equation will have to be solved numerically. Normally a finite difference scheme is used, so that the hydrodynamic conditions and the sediment transport rate are calculated at each grid point at time t. By use of the continuity equation the bed topography after a morphological time step (time $t + \Delta t$) is determined. It is not trivial to select the numerical scheme for solving the continuity equation. The aspects of numerical analysis will not be treated in detail, but it can be mentioned that a modified Lax-Wendroff scheme (see Abbott, 1985) has proven to behave satisfactorily (Deigaard et al., 1988).

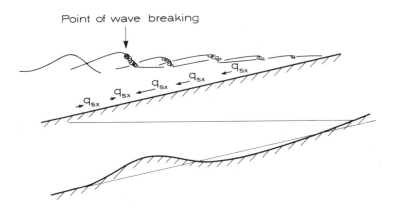

Figure 11.2 The cross-shore sediment transport, and the formation of a breaker bar.

From the section on cross-shore transport it is seen that the undertow gives an offshore directed transport in the surf zone and that the transport outside the surf zone is weak, with a tendency to be in the onshore direction. This means that a longshore bar will tend to be formed on a constant slope profile as a result of the cross-shore transport (Fig. 11.2). The cross-shore sediment transport is considered to be an important factor for the formation of breaker bars, but other mechanisms can also be of significance, such as the large vortices generated by plunging breakers and the low frequency oscillations generated by wave groups.

Several different morphological models have been developed along the principles outlined above, Dally and Dean (1984), Nairn (1988), Deigaard et al. (1988), Hedegaard et al. (1991) and Roelvink (1991). All these models predict the initial formation of a bar on an originally plane beach profile due to the offshore-directed transport in the surf zone, but deviate in the later stages of the development due to differences in the formulation in the wave conditions inshore of the bar and the details of sediment transport modelling. An important aspect in the cross-shore sediment transport model is the smoothing of the calculated transport. In case of regular waves the calculated sediment transport may have a discontinuity at the point of wave-breaking (Fig. 11.3). This is unrealistic, as the undertow profile needs some distance before it becomes fully developed. In order to describe

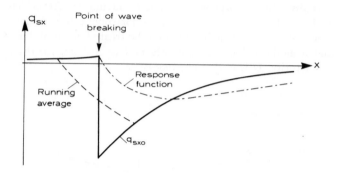

Figure 11.3 Smoothing of the calculated transport distribution.

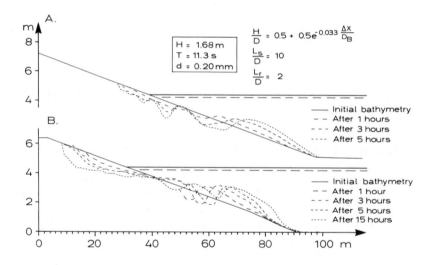

Figure 11.4 A. Simulated evolution (only the first 5 hours). B. Measured
evolution, Saville (1957). After Hedegaard et al. (1991).

this effect the calculated sediment transport field is smoothed at each time step
in the morphological model before it is used to calculate the bed level change.
The smoothing has for example been made as a running average or by applying a
response function in the form

$$\frac{dq_{sx}}{dx} = \frac{q_{sxo} - q_{sx}}{L_r} \tag{11.16}$$

where q_{sxo} is the transport rate obtained directly from the cross-shore sediment transport model and q_{sx} is the sediment transport rate used for calculating the morphological development. In Eq. 11.16 the x-axis is directed onshore. L_r is the length scale for adaptation of the sediment transport. The effect of the two types of smoothing is shown in Fig. 11.3. It can be seen that both types of smoothing cause the maximum of the offshore-directed transport (negative) to be shifted shoreward of the breaking point. This effect is significant in the formation of a trough inshore of the bar. Further, the running average causes the offshore-directed transport to start offshore of the breaking point which smooths out the front face of the bar.

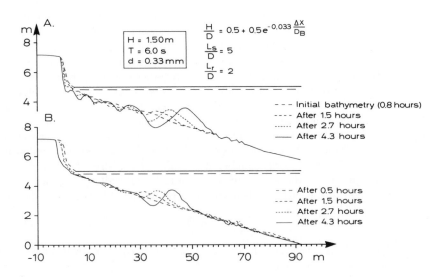

Figure 11.5 A. Simulated evolution. B. Measured evolution, Dette and Uliczka (1986). After Hedegaard et al. (1991).

Hedegaard et al. (1991) have made simulations using different combinations of the two types of smoothing. Figs. 11.4 and 11.5 show comparisons between measurements and simulated coastal profile developments. In both cases the smoothing has been made by first taking a running average and subsequently using the response function. In the comparison with measurements by Saville (1957) the running average is made over a length L_s equal to ten times the local water depth D, and the length scale L_r for the response is taken to be $2D$. In the simulation of the experiments by Dette and Uliczka (1986) the running average is made over a length L_s of $5D$, and L_r is $2D$.

It is seen how a coastal profile model can represent the formation of the first bar and its gradual offshore migration. Further, the formation of the second and third bar inshore of the first is reproduced in the simulation of Saville's (1957) measurements. However, the model does not satisfactorily reproduce the retreat of the coastline and in general, coastal profile models cannot yet be expected to give reliable long term predictions.

Example 11.2:   Dean's model for equilibrium beach profiles

A model for the equilibrium beach profile expected under a given wave condition was developed by Dean (1977). The model considers the high level of turbulence in the surf zone, due to the energy dissipation of the breaking and broken waves. Chapter 4 described how the turbulence level is related to the wave energy loss which acts as a source for the production of turbulence. Similar arguments are used by Dean and lead to the hypothesis that the bed on a beach profile can only withstand a certain rate of energy dissipation (and thus production of turbulence). If this level is exceeded, it will lead to a reshaping of the beach profile to give a wider surf zone with a lower intensity of the wave energy loss. Dean considers the energy loss per unit bed area, as well as per unit water volume. The latter has shown to be most successful and is therefore treated in the following.

Dean (1977) introduces a number of simplifications in order to obtain simple analytical expressions: the waves are described as linear shallow water waves and the wave height in the surf zone is taken to be proportional to the water depth

$$H = KD \qquad (11.17)$$

The analysis is based on the energy conservation equation, which for steady conditions reads (cf. Section 4.2.2)

$$\frac{dE_f}{dx} = \overline{\mathcal{D}} \qquad (11.18)$$

where E_f is the wave energy flux and $\overline{\mathcal{D}}$ is the depth-integrated energy dissipation. x in this example is positive in the offshore direction.

The wave energy flux is given as (cf. Section 1.2)

$$E_f = \frac{1}{8}\rho g H^2 \sqrt{gD} \qquad (11.19)$$

Introducing Eqs. 11.17 and 11.19 into Eq. 11.18 gives the following expression for the mean energy dissipation per unit volume

$$\overline{\varepsilon} = \frac{\overline{D}}{D} = \frac{1}{D}\frac{d}{dx}\left(\frac{\rho g K^2}{8}\sqrt{g}D^{5/2}\right) =$$

$$\frac{5}{24}\rho g \, K^2 \sqrt{g} \, \frac{d(D^{3/2})}{dx} \tag{11.20}$$

Following Dean's hypothesis, that $\overline{\varepsilon}$ is constant for a given bed material on an equilibrium beach profile, leads to the following shape of the beach profile

$$D^{3/2} = \frac{24}{5}\frac{\varepsilon(d)}{\rho g K^2 \sqrt{g}} \, x \tag{11.21}$$

or

$$D \propto x^{2/3} \tag{11.22}$$

where $\varepsilon(d)$ signifies that the equilibrium energy dissipation is a function of the grain size d of the bed material.

The relation in Eq. 11.22 has been supported by many field measurements. A similar relation was proposed by Bruun (1954) as an empirical result based on analyses of beach profiles under very different conditions.

It should be noted that Eq. 11.21 describes a monotonic profile, with the water depth always increasing with the distance from the shore. In the case of a barred profile it can, at most, only be expected to describe a part of the profile.

REFERENCES

Abbott, M.B. (1985): Computational Hydraulics. Pitman Publishing Limited, London, 326 pp.

Bruun, P. (1954): Coast erosion and the development of beach profiles. Beach Erosion Board, Tech. Memo No. 44.

Dally, W.R. and Dean, R.G. (1984): Suspended sediment transport and beach profile evolution. J. Waterway, Port, Coastal and Ocean Eng., ASCE, 110(1):15-33.

Dean, R.G. (1977): Equilibrium beach profiles: U.S. Atlantic and Gulf coasts. Dept. of Civil Engineering, Ocean Engineering Report No. 12, Univ. of Delaware, 45 pp.

Deigaard, R., Hedegaard, I.B. and Fredsøe, J. (1988): A model for off/onshore sediment transport. Proc. IAHR Symposium on Mathematical Modelling of Sediment Transport in the Coastal Zone. Copenhagen, pp. 182-192.

Dette, H. and Uliczka, K. (1986): Seegangsserzeugte Wechselwirkung zwischen Vorland und Vorstrand sowie Küstenschutzbauwerk. Technischer Bericht N. 3 - SBF 205/TP A6, Universität Hannover.

Hedegaard, I.B., Deigaard, R. and Fredsøe, J. (1991): Onshore/offshore sediment transport and morphological modelling of coastal profiles. Proc. Coastal Sediments'91, ASCE, Seattle, pp. 643-657.

Justesen, P. (1988): Turbulent wave boundary layers. Series paper No. 43, Inst. of Hydrodynamics and Hydraulic Engineering, ISVA, Techn. Univ. Denmark, 226 pp.

Nairn, R.B. (1988): Prediction of wave height and mean return flow in cross-shore sediment transport modelling. Proc. IAHR Symposium on Mathematical Modelling of Sediment Transport in the Coastal Zone, Copenhagen, pp. 193-202.

Roelvink, J.A. (1991): Modelling of cross-shore flow and morphology. Proc. Coastal Sediments'91, ASCE, Seattle, pp. 603-617.

Saville, T. (1957): Scale effects in two-dimensional beach studies. Trans. 7th Meeting, IAHR, Lisbon, Vol. 1, A3.

Chapter 12. Longshore sediment transport and coastline development

When waves approach the coast at an oblique angle, a longshore current will be generated as described in Chapter 5 and on a sandy coast the waves and the current may transport considerable amounts of sediment along the coast. The resulting annual longshore transport at a given site will often be a dominant factor in the sediment budget and significant beach erosion or accretion can be caused by interference with the longshore sediment transport.

12.1 Longshore sediment transport

The longshore sediment transport will often manifest itself through the coastal erosion or accretion around structures. Examples are given in Fig. 12.1, where structures are blocking the longshore transport or littoral drift partially or completely. In Fig. 12.1A, a harbour is constructed on a coast with a net littoral drift from left to right. Initially, this harbour completely blocks the longshore transport and the result is rapid accretion on the up-drift side and erosion on the down-drift side. If the coast is very long and straight, the accretion and erosion will continue and the coastline will move offshore on the up-drift side, until the littoral drift starts to pass the left-hand breakwater. In this situation the harbour entrance becomes shallower and it will be necessary to carry out maintenance dredging or to extend the breakwater in order to increase the volume of sediment that can be deposited.

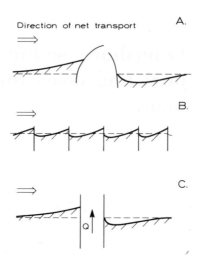

Figure 12.1 Erosion and deposition caused by structures interfering with the longshore transport. A: Harbour. B: Groyne field. C: Jetties at river mouth.

In Fig. 12.1B a field of groynes has been constructed on an eroding coastline. The groynes have locally modified the longshore transport and the coastline along the groyne field has been stabilized. Fig. 12.1C shows a river mouth which has been stabilized by jetties. Before the construction of the jetties, the littoral drift may have caused a gradual migration of the river mouth in the down-drift direction until the river course became too extended at which time the river would break through the longshore spit to form a new mouth, up-drift of the former exit, with a more direct flow to the sea. In this way the river mouth can shift continuously, which can be very inconvenient if the river is used for navigation. The jetties fix the position of the river mouth, but also cause erosion on the down-drift side and accretion on the up-drift side.

12.1.1 The CERC-formula

One of the oldest and still most successful methods for calculating the long-shore sediment transport is known as the CERC-formula, or the SPM-method after the Shore Protection Manual (Coastal Engineering Research Center, 1984). The idea that the longshore sediment transport is mainly driven by the incoming waves rather than by tides and ocean currents became generally accepted early in the 20th century and a formula relating the longshore sediment transport rate to the height and direction of the deep water waves was established by Munch-Petersen (Svendsen, 1938). Munch-Petersen's formula can be considered as a forerunner of the CERC-formula which is based on an empirical correlation between the transport rate and the quantity $P_{\ell s}$ defined as

$$P_{\ell s} = E_{fb} \cos(\alpha_b) \sin(\alpha_b) \qquad (12.1)$$

where α_b is the angle between the waves and the coast at the point of breaking, and E_{fb} is the wave energy flux at the point of wave-breaking. There has been considerable discussion on the physical interpretation of the parameter $P_{\ell s}$, which has been termed 'The longshore energy flux factor'. Longuet-Higgins (1972) made an analysis where the shear component of the radiation stress was used in the derivation of $P_{\ell s}$ and in explaining the connection between this parameter and the longshore sediment transport.

Komar and Inman (1970) made a dimensionally correct formula that was based on a large number of field and laboratory data

$$I_\ell = K_c P_{\ell s} \qquad (12.2)$$

where K_c is a constant equal to 0.77, and I_ℓ is the submerged weight of the transported sediment

$$I_\ell = \rho(s - 1)gQ_\ell \qquad (12.3)$$

where ρ is the density of the water, s is the relative density of the sediment, g is the acceleration of gravity, and Q_ℓ is the longshore sediment transport rate measured as solid volume. In case of irregular waves $P_{\ell s}$ is calculated on the basis of H_{rms}, which gives a good estimate of the wave energy flux. Fig. 12.2 shows the CERC-formula as given by Eq. 12.2 plotted together with the field data originally considered by Komar and Inman (1970).

The field measurements that are required for establishing a single point in Fig. 12.2 are very extensive. The field measurements have been reviewed critically by Greer and Madsen (1978), and it was found that in fact only few of the original data points could be considered to be totally reliable. A more recent addition to the data set has been established by Mangor et al. (1984) who analyzed the

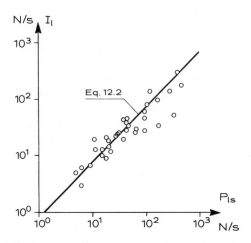

Figure 12.2 Measured longshore sediment transport and calculated values of $P_{\ell s}$, plotted together with the CERC-formula, after Komar and Inman (1970).

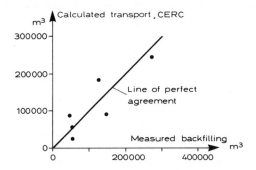

Figure 12.3 Comparison between measured backfilling in a large trench, and longshore transport calculated by the CERC-formula, after Mangor et al. (1984).

sedimentation in a trench dredged for the shore approach of two marine pipelines. The volume of the trench was 600,000 m³. The trench was surveyed, and the backfilling could be determined for six periods during which the wave conditions were also monitored. In Fig. 12.3 the measured backfilling of the trench is shown together with the corresponding calculation of the longshore sediment transport, using the CERC-formula.

Example 12.1:   The CERC-formula in dimensional form

The CERC-formula is frequently met in versions with non-homogeneous units, for example as given by the Shore Protection Manual. One of the traditionally most common versions reads

$$Q'_\ell = 7500 P_{\ell s} \qquad (12.4)$$

where Q'_ℓ is the longshore transport rate given in cubic yards per year deposited volume, and $P_{\ell s}$ is calculated by use of U.S. Customary Units entering the significant wave height H_s as the wave height. A dimensionally inhomogeneous formula gives of course as good results as a dimensionally correct one, as long as the parameters are handled correctly, and Eq. 12.4 can easily be shown to be equivalent to Eq. 12.2.

Eq. 12.4 can be written in dimensional homogeneous form by including the units in the constant

$$Q'_\ell = 7500 \left(\frac{\text{yd}^3}{\text{year}} \right) \left(\frac{\text{s}^3}{\text{slug} \times \text{ft}} \right) \frac{1}{16} \rho g H_{bs}^2 c_{gb} \sin(2\alpha_b) \qquad (12.5)$$

where

$$1 \text{ ft} = 0.305 \text{ m}$$

$$1 \text{ yd} = 3 \text{ ft}$$

$$1 \text{ slug} = 14.6 \text{ kg}$$

1 slug is the mass which is given an acceleration of 1 ft/s^2 by a force of one pound. Eq. 12.5 can be made equivalent to Eq. 12.2 by multiplying by $(1 - n)(s - 1)\rho g$ at both sides and noting that

$$\rho = 1.99 \text{ slug/ft}^3$$

$$g = 32.1 \text{ ft/s}^2$$

The porosity is taken to be $n = 0.4$, and Eq. 12.5 can now be written as

$$Q'_\ell \, (1-n) \, (s-1) \, \rho g = Q_\ell \, (s-1) \, \rho g =$$

$$7500 \left(\frac{\text{yd}^3}{\text{year}} \right) \left(\frac{\text{s}^3}{\text{slug ft}} \right) (1-n)(s-1)\rho g \frac{1}{16}\rho g H_{bs}^2 c_{gb} \sin(2\alpha_b) =$$

$$7500 \left(\frac{27\text{ft}^3}{3.15 \cdot 10^7 \text{s}} \right) \left(\frac{\text{s}^3}{\text{slug ft}} \right) 0.6 \times 1.65 \times 1.99 \left(\frac{\text{slug}}{\text{ft}^3} \right) 32.1 \left(\frac{\text{ft}}{\text{s}^2} \right) P_{\ell s} =$$

$$0.41 P_{\ell s} \qquad\qquad (12.6)$$

or

$$\rho g(s-1)Q_\ell = 0.41 P_{\ell s} \qquad\qquad (12.7)$$

which is similar to Eq. 12.2. The difference of about a factor two between the two coefficients is mainly due to the use of H_{rms} as input to Eq. 12.2 and H_s to Eq. 12.7.

The CERC-formula has only the characteristics of the incoming waves as input. This is not realistic, as the sediment transport must be expected to depend both on the sediment and on the coastal profile. Fig. 12.4 shows a plot made by Dean et al. (1982) where the coefficient K_c in the CERC formula (Eq. 12.2) is plotted as a function of the grain size. It can be seen how the transport becomes smaller for the coarse sediment and larger for the finer sediment.

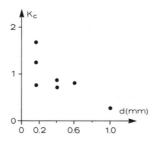

Figure 12.4 Variation in K_c-factor with grain size according to Dean et al. (1982).

It can be argued that neither the grain size nor the coastal profile are completely independent parameters because the sediment sizes may be sorted on the coast and the profile is continuously reshaped by the waves, as described in Section 11.2. The coefficient in the CERC-formula should thus represent an average where the sorting of the sediment and the beach profile are in equilibrium with the waves.

Kamphuis et al. (1986) have made an extensive analysis of field and laboratory data and propose a longshore sediment transport formula, where Q_ℓ is proportional to the beach slope and inversely proportional to the sediment size

$$Q_\ell = 1.28 \, \frac{\tan \beta H_{bs}^{3.5}}{d} \sin(2\alpha_{bs}) \qquad (12.8)$$

where $\tan \beta$ is the beach slope, H_{bs} is significant wave height of the breaking waves, and α_{bs} is the corresponding wave angle at breaking. Equation 12.8 is not dimensionally homogeneous; the coefficient 1.28 requires that the wave height and the particle size are entered in metres, and Q_ℓ is then calculated as kg/s.

12.1.2 Longshore sediment transport models

The longshore sediment transport is closely associated with the wave-driven currents and several longshore sediment transport formulae have been based on arguments that the sediment is stirred up by the waves and then transported along the coast by the wave-driven longshore current (Longuet-Higgins, 1972; and Inman and Bagnold, 1963).

Bijker (1971) made the first detailed longshore sediment transport model, using the longshore current model of Longuet-Higgins (1970) for a beach with constant slope in combination with a sediment transport model for waves and current.

Deigaard et al. (1986) developed a model which includes a longshore current model for arbitrary coastal profiles. The flow resistance in the longshore current model is calculated from the combined wave-current boundary layer and is consistent with the sediment transport description. The sediment load is calculated as bed load and suspended load and the concentration profile of the suspended sediment is calculated by the turbulent diffusion equation with a time-varying diffusivity. The turbulent exchange coefficient is composed of contributions from the oscillatory boundary layer, the current boundary layer and from the wave-breaking in the surf zone, cf. Chapter 8. Fig. 12.5 gives an example simulation showing the littoral current profile and the distribution of littoral sediment transport rate across a coastal profile with one longshore bar.

A detailed analysis was made of the calculated sediment transport on a coast with a constant slope $\tan \beta$. The dependence of the sediment transport on the bed slope was investigated and it was found that the transport is approximately proportional to the square root of the bed slope. The dimensionless longshore sediment transport rate was therefore defined as

$$\Phi_\ell = \frac{Q_\ell}{H_0 \sqrt{\tan \beta} \sqrt{(s-1)gd^3}} \qquad (12.9)$$

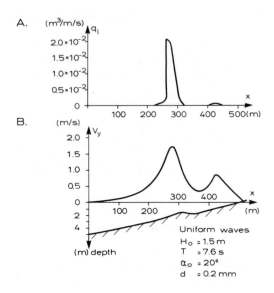

Figure 12.5 Example of model results. A: Littoral sediment transport rates.
B: Littoral current and coastal profile.

where Q_ℓ is the longshore sediment transport rate, and H_0 is the deep water wave height.

Φ_ℓ will be a function of the following dimensionless parameters: deep water wave height H_0/d, deep water wave steepness H_0/L_0, where $L_0 =$ deep water wave length, the angle α_0 between the coast and the deep water wave crests and finally the settling dimensionless velocity w^* of the sediment defined as

$$w^* = \frac{w_s}{\sqrt{gd}} \tag{12.10}$$

As noted above Φ_ℓ is only weakly dependent on the beach slope β.

The variation in the different parameters with the deep water wave direction is shown in Fig. 12.6 for a single combination of wave height, wave steepness, and settling velocity. Besides the sediment transport, the maximum longshore current velocity and the $P_{\ell s}$-factor are shown. The latter indicates the variation in the longshore sediment transport according to the CERC-formula (Eq. 12.2). The variation in Q_ℓ with the deep water wave angle, shown in Fig. 12.6A, is typical for a large range of H_0/d, w^* and H_0/L_0. A curve fitted to this variation is

$$\frac{Q_\ell}{Q_{\ell\,\text{max}}} = \left(\sin\left(2\alpha_0 \left[1 - 0.4 \frac{\alpha_0}{90°}\left(1 - \frac{\alpha_0}{90°}\right)\right]\right) \right)^{5/2} \tag{12.11}$$

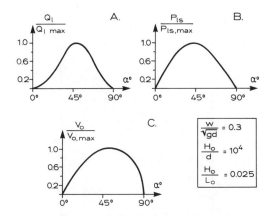

Figure 12.6 A: Variation in littoral drift; B: "Longshore energy flux factor";
C: Maximum littoral current velocity with α_0.

which may be used as an approximation for the variation of the sediment transport
with the wave direction.

The dimensionless sediment transport Φ_0 at $\alpha_0 = 45°$ is very close to the
maximum. Φ_0 is shown in Fig. 12.7 as a function of H_0/d, w^* and H_0/L_0.
The sediment transport model is only valid for a flat bed without ripples, which
corresponds to a minimum wave height H_0 of approximately 2000 d. For $H_0/d <$
3×10^4 the curves in Fig. 12.7 can be roughly approximated by the expression

$$\Phi_0 = 0.1 \left(\frac{H_0}{d}\right)^{2.3} \left(\frac{H_0}{L_0}\right)^{0.5} \exp(-6.1w^*) \tag{12.12}$$

which reproduces the curves of Fig. 12.7 with an accuracy of about 50%. Eq.
12.12, together with Eq. 12.11, can be used to obtain analytical estimates for
a plane coastal profile. For a real profile, e.g. with bars, a complete simulation
must be performed, calculating the longshore current velocity distribution and the
sediment transport at a number of points.

In Fig. 12.8 a comparison has been made between the longshore sediment
transport rate calculated by the mathematical model and the prediction made
by the CERC-formula. The comparison is made for a deep water wave direction
of $\alpha_0 = 45°$ and a beach slope of $\tan\beta = 0.01$. The comparison is made for
different sediment sizes and for different wave heights and periods. The line of
perfect agreement is shown as a solid line in the region where the CERC-formula
is supported by empirical data. The two models can be seen to predict littoral
drifts of the same order of magnitude for a grain size of 0.2 mm, which is a very

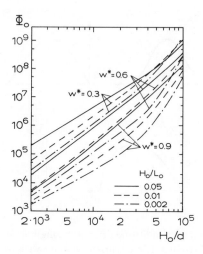

Figure 12.7 Variation of longshore transport with grain size, wave steepness, and wave height, $\alpha_0 = 45°$.

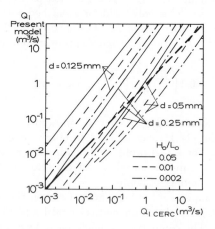

Figure 12.8 Comparison between the longshore transport model and CERC-formula for $\alpha_0 = 45°$.

common beach material. Changes in α_0 or $\tan \beta$ would result in a displacement of the curves, but no change in their shape.

The variation with the wave direction has been illustrated in Fig. 12.6 from which it can be seen that, according to the mathematical model, the transport is relatively smaller for small and large values of α_0 than predicted by the CERC-formula. The other most important parameter in the CERC-formula is the wave height and the calculated transport is approximately proportional to the wave height to a power 2.5. The dependency on the wave height is slightly stronger in the mathematical model where the power on the wave height is about 3–3.5.

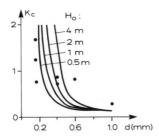

Figure 12.9 Variation in K_c-factor with grain size. Theoretical prediction is based on $H_0/L_0 = 0.025$, $\beta = 0.01$, and $\alpha_0 = 45°$. The measured values (filled circles) are from Dean et al. (1982).

The mathematical model is very sensitive to variation in the grain size, which does not even enter the basic formulation of the CERC-formula. As shown in Fig. 12.9 the dependency on the grain size is slightly larger than found by Dean et al. (1982). This difference may partly be explained by the presence of wave ripples. As described in Chapter 10 the ripples can increase the amount of suspended load relative to a situation with a plane bed. The wave ripples are formed at comparatively small values of Shields parameter, i.e. under given hydrodynamic conditions, ripples will be present for coarse sediment and not for fine sediment.

The described mathematical model for littoral drift was further developed by Deigaard et al. (1988) to include several effects present in the natural environment. The description includes irregular waves, coastal currents and the effect of a wind shear stress.

Effect of irregular waves

The effect of irregular waves on the littoral current has been included as described in Chapter 5. The heights of the deep water waves are assumed to follow a Rayleigh distribution, and the distribution of the driving force is determined accordingly. Furthermore, the total driving force is reduced due to the effect of the directional spreading of the deep water waves. The directional spreading is characterized by the cosine spreading function, cf. Eq. 5.40. Fig. 12.10 shows velocity profiles of the longshore current on a beach profile with a constant slope $\tan \beta = 0.01$. The longshore current is calculated assuming regular waves (uniform wave height) with and without directional spreading, and assuming irregular waves with Rayleigh distributed wave heights.

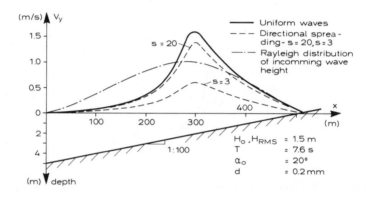

Figure 12.10 Longshore velocity profiles for regular and irregular waves, and for waves with directional spreading.

The local sediment transport rate is strongly influenced by the irregularity of the waves. The two most important effects are: A) The near-bed orbital motion is irregular which has to be taken into account when calculating the near-bed wave boundary layer and the suspended sediment concentrations. B) In the surf zone only a fraction of the waves at a given point will be breaking/broken, as some of the smaller waves will be breaking further inshore. These two effects are treated independently.

By analyzing a series of simulations, each with an irregular time series of the near-bed orbital motion, it was found that a satisfactory estimate of the time-averaged sediment transport could be obtained by a simulation with regular waves using the wave height $H = H_{rms}$ and the significant wave period $T = T_s$.

The effect of breaking and broken waves on the sediment transport is modelled according to the turbulence model described in Chapters 4 and 8, describing the production, vertical spreading and decay of turbulence generated by the passage of each front of a spilling breaker or broken wave. The input parameters to this model, wave height and period, are modified to reflect the time between each passage of a wave front and the rate energy dissipation.

Fig. 12.11 shows an example of the distribution of the longshore transport in the coastal profile, calculated for regular and irregular waves. The main effect of including irregular waves is a much smoother distribution of the transport and a reduction of the total transport by a factor of 0.3 in this particular case. The directional spreading of the waves $s = 20$ and $s = 3$ causes a reduction of the sediment transport over the regular unidirectional rate by a factor 0.7 and 0.1, respectively.

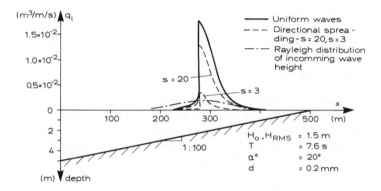

Figure 12.11 Distributions of the longshore sediment transport across a plane coast.

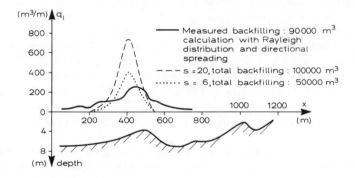

Figure 12.12 Comparison between measured backfilling, cf. Mangor et al. (1984), and calculated backfilling.

A comparison has been made between the calculated longshore sediment transport and the backfilling of a 1,600 m long trench (volume: 200,000 m³) dredged through a three-bar coastal profile at the Danish North Sea Coast, cf. Mangor et al. (1984). The backfilling during a storm with measured wave heights up to $H_s = 4.75$ m in the spring of 1982 is considered. At this period the trench (width 90 m, depth 10 m) was only dredged through the outer bar. The total backfilling of the trench in this period was measured at 90,000 m³ (solid) grain volume. Fig. 12.12 shows the distribution of the backfilling and the calculated distribution of the longshore transport on the outer bar during the storm. The wave heights have been assumed to follow the Rayleigh distribution. The calculations have been made for two directional spreadings, s, of which the smallest one is the most likely at that specific location.

Effect of wind and current

In addition to radiation stress gradients, wind and coastal currents can contribute to the forces driving the longshore current. The effect of wind shear stress and a coast-parallel current has been included in the model for the longshore current as described in Chapter 5. The wind shear stress is expressed by a friction factor (given the value: $f_w = 0.005$) and the wind speed, U_{10} at 10 m elevation

$$\tau_w = \frac{\rho_a}{2} f_w U_{10}^2 \qquad (12.13)$$

ρ_a is the density of air. τ_w is resolved into a shore-normal and a shore-parallel component according to the angle α_w between the wind direction and the shore-normal direction. The shore-normal component is included in the calculation of the set-up and the shore-parallel component is added to the driving forces of the littoral current. The coastal current is assumed to be driven by a shore-parallel gradient of the water surface, S, and the contribution to the driving force for the littoral current is $\rho g D S$, where D is the local water depth.

The effect of wind and current has been analyzed for the basic conditions: $H_0 = 1.5$ m, $T = 7.6$ s, $d = 0.2$ mm, and a coastal slope of 1:100. Three wave directions are considered: $\alpha_0 = 10°, 20°$, and $45°$. Three wind speeds have been considered: $U_{10} = 5$ m/s, 10 m/s, and 20 m/s, the wind is assumed to be parallel to the direction of wave propagation at deep water. Three coastal current velocities, measured at 5 m water depth, are analyzed: $V = -0.25$ m/s, 0.25 m/s, and 0.50 m/s.

The results are summarised in Fig. 12.13 showing the transport normalized by the situation without wind and coastal current, plotted against the total driving force normalized by the driving force due to waves alone. The driving force due to wind or water surface slope has been integrated over the area between the breaker line and the coastline, while the driving force due to the waves is the shear radiation stress S_{xy} at the breaker line. The driving forces due to waves, wind and water surface slope can easily be calculated and generalized plots similar to

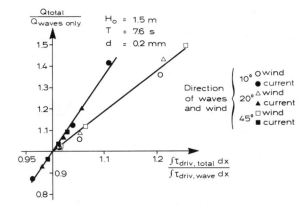

Figure 12.13 Sediment transport including the effect of wind or current, normalized by the transport due to waves only versus the total
driving forces, normalized by driving forces due to waves only.

Fig. 12.13 could then be used to estimate the relative importance of the different
contributions. The difference between the effect of wind and current, revealed in
Fig. 12.13, is due to the difference in distribution across the coastal profile; wind
shear stress is constant, while current-induced bed shear stress decreases with
decreasing depth. For the same integrated force, the coastal current will thus give
a larger contribution near the breaker line where the sediment concentration and
transport are highest.

Effect of rip channels

On a coast with longshore bars the bar will often be interrupted by rip
channels. In the rip channels there will be offshore-directed rip currents. The
mechanism behind the rip currents is discussed in detail in Chapter 5: Incoming
waves break on the foreslope of the bar, but as they pass the crest of the bar and
enter the deeper waters of the trough, they cease to break and are reformed as
non-breaking waves and are shoaled and refracted once again until breaking re-
occurs on the inner beach. At the holes, the larger depths will allow the waves to
continue propagating towards the shoreline without breaking. As a result of this
the wave set-up, produced by the cross-shore component of the radiation stress
S_{xx}, will be smaller at the holes and larger behind the bar.

This results in a net pressure gradient accelerating the water in the trough
towards the holes. Afterwards this mass of water flows through the holes in the
seaward direction in the form of rip-currents. The amount of water carried by
the rips is compensated for by the water transported over the crest of the bar.

Under these conditions, a flow will exist along the trough behind the bar even for normally indicent waves. The effect of this circulation current on the longshore sediment transport has been studied by Zyserman and Fredsøe (1988), using the depth-integrated flow model described in Example 5.7 to calculate the wave-driven currents.

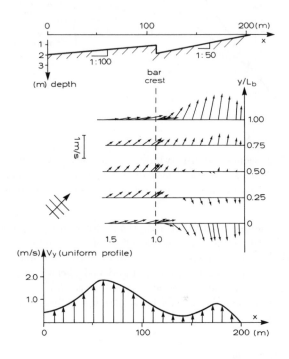

Figure 12.14 Wave-generated currents on a barred coast with and without rip channels.

Fig. 12.14 shows the flow field obtained by this model for a case with bar length $L_b = 180$ m, the width of the holes was $Y_b = 50$ m and the coastal profile is shown in Fig. 12.14. The parameters for the incoming uniform, unidirectional waves are: $H_0 = 1.2$ m, $T = 7.5$ s, $\alpha_0 = 45°$. For comparison, the corresponding longshore current velocity for a uniform coast is also shown.

Calculated current fields have been combined with the sediment transport model to describe the cross-shore variation of the longshore transport q_ℓ, as well as the total longshore transport Q_ℓ for several transversal sections along the bar. Due to the variability of the longshore current profile along the bar, the total transport varies from one section to another. In order to compare the sediment transport with that corresponding to the uniform situation, the average value of Q_ℓ along the bar was calculated. Fig. 12.15 shows the relation between the average longshore sediment transport when rip-currents are present Q_ℓ and the transport for the uniform situation $Q_{\ell 0}$ as a function of the ratio between Y_b and L_b, where Y_b has been kept constant at 50 m. The characteristics of the shore and of waves are the same as those shown in Fig. 12.14.

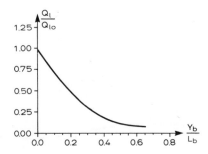

Figure 12.15 Influence of rip currents on the longshore transport of sediment.
Coastal profile and wave conditions as for Fig. 12.14.

It can be seen that the longshore sediment transport decreases drastically for decreasing distance between the rip channels, i.e. for shorter bars. This is explained by the reduction in the magnitude of the longshore current velocity due to an increased influence of the rip currents. This example shows that in addition to the local shape of the coastal profile, the along-shore variation in the profile can be very important for the magnitude of the longshore sediment transport under given wave conditions.

12.2   Modelling of coastline development

The change of the coastal profile that was treated in Section 11.2 can mainly to be considered a short term process which is related to single storm events or seasonal variation in the wave climate. If the long-term sediment budget for a coast is considered, it will often be found that the dominant term is due to variation in the longshore sediment transport along the coast and the development of the coastline can therefore be modelled by calculating the coastal erosion or accretion from the longshore sediment transport. Note that in this section the x-axis is aligned with the coastline for convenience.

A.

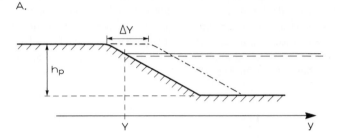

B.

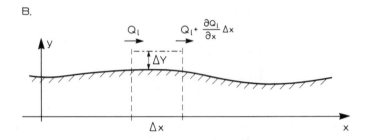

Figure 12.16 A: The coastal profile in a morphological coastline model. B: The longshore sediment transport and coastal accretion/erosion.

The simplest coastline model, the one-line model, assumes that the coastal profile is maintained constant but is shifted in the onshore or offshore direction as the result of erosion or accretion. The active height of the profile is called h_p, Fig. 12.16A. The relation between the accretion/erosion of the coast and the longshore sediment transport can be formulated by the continuity equation for sediment. The x-axis is roughly parallel to the coastline, Fig. 12.16B. The distance of the coastline from the x-axis is Y. If, during the short time interval Δt there is a change in coastline position (accretion) of ΔY, the amount of accumulated sediment (solid volume) over the distance Δx is

$$\Delta x \Delta Y h_p (1 - n) \tag{12.14}$$

The net inflow of sediment during the time Δt is

$$\left(Q_\ell - \left(Q_\ell + \frac{\partial}{\partial x} \frac{dQ_\ell}{dx} \Delta x \right) \right) \Delta t = -\frac{\partial Q_\ell}{\partial x} \Delta x \Delta t \tag{12.15}$$

If it is assumed that the accretion is only due to the longshore sediment

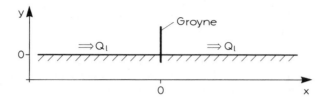

Figure 12.17 The coastline just after construction of the groyne.

transport, Eqs. 12.14 and 12.15 give the continuity equation

$$\frac{\partial Y}{\partial t}(1-n)h_p = -\frac{\partial Q_\ell}{\partial x} \tag{12.16}$$

On a long straight coast the longshore sediment transport can be determined from the wave climate, i.e. statistics for wave height and direction. If the coast is given a different orientation, the entire calculation can be carried out once more, and in this way the longshore sediment transport can be determined as a function of the coast orientation

$$Q_\ell = Q_\ell\left(\frac{\partial Y}{\partial x}\right) \tag{12.17}$$

Inserting this expression in the continuity equation gives

$$\frac{\partial Y}{\partial t} = \frac{-1}{(1-n)h_p}\frac{dQ_\ell}{d(\partial Y/\partial x)}\frac{\partial(\partial Y/\partial t)}{\partial x} = \frac{-1}{(1-n)h_p}\frac{dQ_\ell}{d(\partial Y/\partial x)}\frac{\partial^2 Y}{\partial x^2} \tag{12.18}$$

This is the equation which is solved in order to make a coastline development model. It is a parabolic partial equation which must normally be solved numerically, for example by the finite difference Crank-Nicholson method. The establishment of the coast orientation and the longshore sediment transport will require a large number of individual calculations of the longshore transport rate and it will normally not be possible to establish any analytical relationship; instead a table can be established with corresponding values of Q_ℓ and $\partial Y/\partial x$. This table may be the same for the entire stretch of coast under consideration or it may vary, e.g. due to local sheltering by islands or shoals lying off the coast.

Example 12.2: Analytical solution to coastline model

The first calculations with the coastline model were made by Pelnard-Considère (1956) who used an analytical solution to Eq. 12.18. In many cases this technique can be used to obtain a first estimate of the morphological development, e.g. in connection with the establishment of a structure. To illustrate the solution a simple example is considered, Fig. 12.17.

A groyne is constructed on a long straight coastline. It is perpendicular to the coast and blocks the longshore transport completely. The longshore sediment transport is calculated by use of the approximation formulae given in Eqs. 12.9 -12.12.

The deep water wave conditions are given in Table 12.1; they consist of only two wave situations with a combined occurrence of 5 per cent. In this example the water is therefore calm for 95 per cent of the time.

The height of the active part of the profile is assumed to be 7.5 m with a slope of 1:100. The mean grain size of the sediment is 0.2 mm with a settling velocity of 0.027 m/s. This gives a net longshore sediment transport rate of 340,000 m^3/yr.

Table 12.1 Offshore wave climate, determining the longshore sediment
transport.

Wave height	Wave period	Wave direction	Occurrence
2.5 m	7.5 s	0°	2.5 per cent
2.0 m	7.0 s	20°	2.5 per cent

The longshore sediment transport is calculated for different orientations of the coast. Fig. 12.18 shows the longshore transport as function of $\partial Y/\partial x$. The calculated transport can, with good agreement, be approximated by a straight line

$$Q_\ell = 2.27 \times 10^6 \left(0.15 - \frac{\partial Y}{\partial x} \right) m^3/yr \qquad (12.19)$$

Inserting Eq. 12.19 into Eq. 12.18 gives a description of the coastline development

$$\frac{\partial Y}{\partial t} = -\frac{1}{(1-n)h_p} \left(-2.27 \times 10^6 m^3/yr \right) \frac{\partial^2 Y}{\partial x^2} =$$

$$5.04 \cdot 10^5 m^2/yr \; \frac{\partial^2 Y}{\partial x^2} = K_1 \frac{\partial^2 Y}{\partial x^2} \qquad (12.20)$$

which is the equation for heat conduction with constant coefficients. The initial condition for Eq. 12.20 is

$$t = 0: \quad Y = 0 \;\; \text{for all } x \qquad (12.21)$$

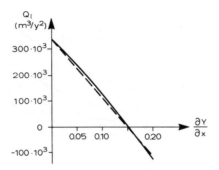

Figure 12.18 The calculated relation between $\partial Y/\partial x$ and Q_ℓ and (dotted line) approximation by a straight line.

The boundary condition at the groyne is that the longshore transport is zero. From the calculated transport relation in Fig. 12.17 this is found to give an orientation of the coastline of : $\partial Y/\partial x = 0.15$. The boundary conditions are therefore

$$\frac{\partial Y}{\partial x} = 0.15 = Y_0' \text{ for } x = \pm\, 0$$

and

$$Y \to 0 \text{ for } x \to \pm\, \infty \qquad (12.22)$$

With these boundary conditions the solution to Eq. 12.20 can be written

$$Y = Y_0' \frac{1}{\sqrt{\pi}} \left[\sqrt{4K_1 t}\, \exp\left(-\frac{x^2}{4K_1 t}\right) - x\sqrt{\pi}\left(1 - F\left(\frac{x}{\sqrt{4K_1 t}}\right)\right) \right] \qquad (12.23)$$

where F is the error function defined by

$$F(u) = \frac{2}{\sqrt{\pi}} \int_0^u \exp\left(-s^2\right) ds \qquad (12.24)$$

The coastline development calculated by Eq. 12.23 is shown in Fig. 12.19 for $x < 0$. Due to the symmetry of the problem the coastlines at $x > 0$ can be found by rotating the coastlines at $x < 0$ 180° around $(x, Y) = (0, 0)$.

Just at the updrift side of the groyne the accretion of the coastline can be written

$$Y(x = 0^-) = Y_0' \sqrt{\frac{4K_1 t}{\pi}} \qquad (12.25)$$

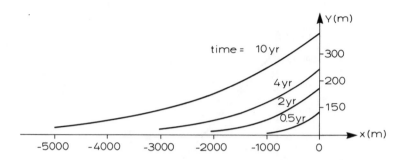

Figure 12.19 The coastline development at the updrift side of the groyne.

This very simple analytical solution can only be used as a first approxima-
tion, because the detailed variation in the relation between Q_ℓ and $\partial Y/\partial x$ cannot
be represented as it can in a numerical model. The along-shore variation of the
transport relation can be particularly important. Close to a groyne the sheltering
effect will cause the coastline to deviate from the solution given by Eq. 12.23. The
sheltering will cause the beach sediment to accumulate around the groyne, Fig.
12.20. Immediately after construction of a groyne this effect may cause a certain
erosion even on the updrift side of the groyne.

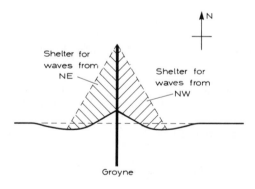

Figure 12.20 The sheltering effect around a groyne immediately after construction.

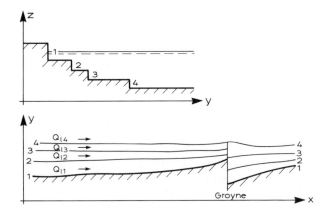

Figure 12.21 The coastal profile and plan form in a multi-line model.

The one-line model for coastline development is the simplest approach and requires many simplifying assumptions to be introduced before the calculations can be carried out. Several attempts have been made to refine the coastline modelling, for example the description of the coastal profile has been made more detailed by introducing more than one calculation line. The profile is then schematized as a stepped curve (Fig. 12.21) and the longshore transport is then estimated for each step, depending on the shape of the profile. In addition, the distribution of the cross-shore sediment transport over the profile must be estimated. The continuity equation is then solved for each line giving the accretion or erosion of the lines. In this way the development of the coastal profile is modelled in addition to the general coastal development.

In spite of the additional detail given by the multi-line models they have not been very successful, mainly because it has been difficult to specify realistic relations for the cross-shore sediment transport and for the distribution of the longshore transport. The result is a model that is more detailed than the one-line model, but it also requires much more calibration and in the end does not provide significantly more new information than it requires for calibration.

REFERENCES

Bijker, E.A. (1971): Longshore transport computations. J. Waterways, Harbors and Coastal Eng. Div., ASCE, 97(WW4):687-701.

Coastal Engineering Research Center (1984): Shore protection manual. Fourth ed., U.S. Army Corps of Engineers, Washington.

Dean, R.G., Berek, E.P., Gable, C.G. and Seymour, J. (1982): Longshore transport determined by an efficient trap. Proc. 18th Coast. Eng. Conf., Cape Town, pp. 954-968.

Deigaard, R., Fredsøe, J. and Hedegaard, I.B. (1986): Mathematical model for littoral drift. J. Waterway, Port, Coastal and Ocean Eng., ASCE, 112(3):351-369.

Deigaard, R., Fredsøe, J., Hedegaard, I.B., Zyserman, J. and Andersen, O.H. (1988): Littoral drift model for natural environments. Proc. 21st Coastal Eng. Conf., ASCE, Malaga, pp. 1603-1617.

Greer, M.N. and Madsen, O.S. (1978): Longshore sediment transport data: A review. Proc. 16th Coastal Eng. Conf., Hamburg, pp. 1563-1576.

Inman, O.L. and Bagnold, R.A. (1963): Littoral processes. In: "The Sea". Ed. Hill, M.N., Interscience, New York, 3:529-533.

Kamphuis, J.W., Davies, M.H., Nairn, R.B. and Sayao, O.J. (1986): Calculation of littoral sand transport rate. Coastal Engineering, 10(1):1-22.

Komar, P.D. and Inman, D.L. (1970): Longshore sand transport on beaches. J. Geophys. Res., 75(30):5914-5927.

Longuet-Higgins, M.S. (1970): Longshore currents generated by obliquely incident waves. J. Geophys. Res., 75(33):6778-6801.

Longuet-Higgins, M.S. (1972): Recent progress in the study of longshore currents. In: "Waves on Beaches". Ed. Meyer, R.E., Academic Press, New York, pp. 203-248.

Mangor, K., Sørensen, T. and Navntoft, E. (1984): Shore approach at the Danish North Sea coast, monitoring of sedimentation in a dredged trench. Proc. 19th Coastal Engineering Conf., Houston, pp. 1816-1829.

Pelnard-Considère, R. (1956): Essai de théorie de l'évolution des formes de rivage en plages de sable et de galets. Soc. Hydrotechnique de France, Quatrièmes Journées de l'Hydraulique. Les Énergies de la Mer, Tome I, Question III, pp. 289-298.

Svendsen, S. (1938): Munch-Petersen's formel for materialvandring. Stads- og Havneingeniøren, No. 12.

Zyserman, J. and Fredsøe, J. (1988): The effect of rip-currents on longshore sediment transport. Proc. 2nd Int. Symposium on Wave Research and Coastal Engineering, Hannover, pp. 127-138.

Appendix I. Wave table: Sinusoidal waves

Symbols: H_o and L_o are the wave height and wave length at infinite water depth

General relations: $\omega = \frac{2\pi}{T}, \; k = \frac{2\pi}{L}, \; c = L/T, \; L_o = \frac{gT^2}{2\pi}$
$= 1.56T^2$ (SI-units)

Abbreviations: $G = 2kD/\sinh(2kD)$

Surface elevation profile: $\eta = \frac{H}{2}\sin(\omega t - kx)$

Wave celerity: $\qquad c = \sqrt{\frac{g}{k}\tanh(kD)}$

Group velocity: $\qquad c_g = \frac{c}{2}(1 + G)$

Horizontal orbital velocity: $\qquad u = \frac{\pi H}{T}\frac{\cosh(kz)}{\sinh(kD)}\sin(\omega t - kx)$

Vertical orbital velocity: $\qquad v = -\frac{\pi H}{T}\frac{\sinh(kz)}{\sinh(kD)}\sin(\omega t - kx)$

Horizontal particle movement: $\qquad x = x_o + \frac{H}{2}\frac{\cosh(kz)}{\sinh(kD)}\cos(\omega t - kx)$

Vertical particle movement: $\qquad z = z_o + \frac{H}{2}\frac{\sinh(kz)}{\sinh(kD)}\sin(\omega t - kx)$

Pressure in excess of the hydrostatic: $\qquad p^+ = \rho g\,\frac{H}{2}\frac{\cosh(kz)}{\cosh(kD)}\sin(\omega t - kx)$

Radiation stress, pressure: $\qquad \overline{F}_p = \rho g H^2 G/16$

Radiation stress, momentum: $\qquad \overline{F}_m = \rho g H^2 G\,(1 + G)/16$

Energy flux: $\qquad E_f = \rho g H^2(1 + G)\,c/16$

$\frac{D}{L_0}$	$\tanh(kD)$	$\frac{D}{L}$	kD	$\sinh(kD)$	$\cosh(kD)$	G	$\frac{H}{H_0}$
0.000	0.000	0.0000	0.000	0.000	1.00	1.000	∞
002	112	0179	112	113	01	992	2.12
004	158	0253	159	160	01	983	1.79
006	193	0311	195	197	02	975	62
008	222	0360	226	228	03	967	51
0.010	0.248	0.0403	0.253	0.256	1.03	0.958	1.43
015	302	0496	312	317	05	938	31
020	347	0576	362	370	07	918	23
025	386	0648	407	418	08	898	17
0.030	0.420	0.0713	0.448	0.463	1.10	0.878	1.13
035	452	0775	487	506	12	858	09
040	480	0833	523	548	14	838	06
045	507	0888	558	588	16	819	04
0.050	0.531	0.0942	0.592	0.627	1.18	0.800	1.02
055	554	0993	624	665	20	781	1.01
060	575	104	655	703	22	762	0.993
065	595	109	686	741	24	744	981
070	614	114	716	779	27	725	971
0.075	-0.632	0.119	0.745	0.816	1.29	0.707	0.962
080	649	123	774	854	31	690	955
085	665	128	803	892	34	672	948
090	681	132	831	929	37	655	942
095	695	137	858	968	39	637	937
0.10	0.709	0.141	0.886	1.01	1.42	0.620	0.933
11	735	150	940	08	48	587	926
12	759	158	0.994	17	54	555	920
13	780	167	1.05	25	60	524	917
14	800	175	1.10	33	67	494	915
0.15	0.818	0.183	1.15	1.42	1.74	0.465	0.913
16	835	192	20	52	82	437	913
17	850	200	26	61	90	410	913
18	864	208	31	72	1.99	384	914
19	877	217	36	82	2.08	359	916
0.20	0.888	0.225	1.41	1.94	2.18	0.335	0.918

$\frac{D}{L_0}$	$\tanh(kD)$	$\frac{D}{L}$	kD	$\sinh(kD)$	$\cosh(kD)$	G	$\frac{H}{H_0}$
0.20	0.888	0.225	1.41	1.94	2.18	0.335	0.918
21	899	234	47	2.05	28	313	920
22	909	242	52	18	40	291	923
23	918	251	57	31	52	271	926
24	926	259	63	45	65	251	929
0.25	0.933	0.268	1.68	2.60	2.78	0.233	0.932
26	940	277	74	75	2.93	215	936
27	946	285	79	2.92	3.09	199	939
28	952	294	85	3.10	25	183	942
29	957	303	90	28	43	169	946
0.30	0.961	0.312	1.96	3.48	3.62	0.155	0.949
31	965	321	2.02	69	3.83	143	952
32	969	330	08	3.92	4.05	131	955
33	972	339	13	4.16	28	120	958
34	975	349	19	41	53	110	961
0.35	0.978	0.358	2.25	4.68	4.79	0.100	0.964
36	980	367	31	4.97	5.07	091	967
37	983	377	37	5.28	37	083	969
38	984	386	43	61	5.70	076	972
39	986	395	48	5.96	6.04	069	974
0.40	0.988	0.405	2.54	6.33	6.41	0.063	0.976
41	989	415	60	6.72	6.80	057	978
42	990	424	66	7.15	7.22	052	980
43	991	434	73	7.60	7.66	047	982
44	992	443	79	8.07	8.14	042	983
0.45	0.993	0.453	2.85	8.59	8.64	0.038	0.985
46	994	463	91	9.13	9.18	035	986
47	995	472	2.97	9.71	9.76	031	987
48	995	482	3.03	10.3	10.4	028	988
49	996	492	09	11.0	11.0	026	990
0.50	0.996	0.502	3.15	11.7	11.7	0.123	0.990
∞	1.000	∞	∞	∞	∞	0.000	1.000

Appendix II. Derivation of the k-equation

The k-equation given by Eq. 2.76 can be derived from the Navier-Stokes equation, which in tensor-notation reads

$$\frac{dU_i}{dt} = -\frac{1}{\rho}\frac{\partial \rho}{\partial x} + \nu \frac{\partial^2 U_i}{\partial x_j \partial x_j} + \rho g_i \qquad (\text{II.1})$$

The instantaneous velocity is written as

$$U_i = u_i + u_i' \qquad (\text{II.2})$$

where u_i' is the turbulent fluctuation. Equation II.1 is multiplied by u_i' and time-averaged.

$$\overline{\rho u_i' \frac{d(u_i + u_i')}{dt}} = \overline{u_i' \frac{\partial p}{\partial x_i}} + \overline{u_i' \rho g_i} + \overline{u_i' \mu \frac{\partial^2 (u_i + u_i)}{\partial x_j \partial x_j}} \qquad (\text{II.3})$$

The left-hand-side of Eq. II.3 gives

$$\overline{u_i' \frac{d(u_i + u_i')}{dt}} = \overline{u_i' \frac{\partial(u_i + u_i')}{\partial t}} + \overline{u_i'(u_j + u_j')\frac{\partial(u_i + u_i')}{\partial x_j}}$$

$$= \overline{u_i' \frac{\partial u_i}{\partial t}} + \overline{u_i \frac{\partial u_i'}{\partial t}} + \overline{u_i' u_j \frac{\partial u_i}{\partial x_j}} + \overline{u_i' u_j' \frac{\partial u_i}{\partial x_j}} + \overline{u_j u_i' \frac{\partial u_i'}{\partial x_j}}$$

$$+ \overline{u_i' u_j' \frac{\partial u_i'}{\partial x_j}} = 0 + \frac{\partial k}{\partial t} + 0 + \overline{u_i' u_j' \frac{\partial u_i}{\partial x_j}} + u_j \frac{\partial k}{\partial x_j} + \overline{u_j' \frac{\partial k}{\partial x_j}} \qquad (\text{II.4})$$

In evaluation of the last term, the continuity equation

$$u_{i,i}' = 0 \qquad (\text{II.5})$$

has been used. The time-averaged equation II.4 now becomes

$$\rho\frac{\partial k}{\partial t} + \rho u_j\frac{\partial k}{\partial x_j} + \rho\frac{\partial}{\partial x_j}\overline{(u_j'k)} + \overline{\rho u_i'u_j'}\frac{\partial u_i}{\partial x_j}$$

$$= -\overline{u_i'\frac{\partial p}{\partial x_i}} + \overline{\mu u_i'\frac{\partial^2 u_i'}{\partial x_j\partial x_j}} \tag{II.6}$$

or

$$\rho\frac{dk}{dt} = -\frac{\partial}{\partial x_j}(\overline{u_j'p} + \rho\overline{u_j'k}) - \overline{\rho u_i'u_j'}\frac{\partial u_i}{\partial x_j} + \overline{\mu u_i'\frac{\partial^2 u_i'}{\partial x_j\partial x_j}} \tag{II.7}$$

where

$$k = \frac{1}{2}\overline{u_i'u_i'} \tag{II.8}$$

For a boundary layer, Eq. II.7 can be reduced to

$$\rho\frac{dk}{dt} = -\frac{\partial}{\partial z}(\overline{v'p'} + \rho\overline{v'k'}) - \rho\overline{u'v'}\frac{\partial u}{\partial z} + \overline{\mu u_i'\frac{\partial^2}{\partial x_j\partial x_j}(u_i')} \tag{II.9}$$

$$\text{I} \qquad \text{II} \qquad \text{III} \qquad \text{IV} \qquad \text{V}$$

in which u' and v' are the longitudinal and vertical velocity fluctuation, and p' and k' are the fluctuating parts of pressure and turbulent energy.

The terms in Eq. A.9 are identified as follows:

I is the total change in k, which consists of two parts, the local time derivative $\partial k/\partial t$ and the convective transport $u_j\partial k/\partial x_j$.

II and III represent a net diffusion of turbulent energy. II is the increase in turbulent energy due to the work performed on a fluid particle, because of fluctuations in pressure. III is the similar increase in k, because of the fluctuations in velocities. That the terms represent a diffusion process can be seen as follows: if the terms II and III are integrated across the boundary layer, both integrals become zero (the fluctuations are zero at the bed and outside the boundary layer). Hereby, II and III do not change the overall level of k, but redistribute the energy in the boundary layer. In turbulent models, these two terms are usually modelled similar to an ordinary diffusion process by

$$-(\overline{v'p'} + \rho\overline{v'k'}) = \rho\frac{\nu_T}{\sigma_k}\frac{\partial k}{\partial z} \tag{II.10}$$

in which ν_T is the eddy viscosity.

IV describes the production of turbulent kinetic energy. Because the term $-\rho\overline{u'v'}$ represents the ordinary Reynolds stresses, this can be written as

$$-\rho\overline{u'v'}\frac{\partial u}{\partial z} = \tau\frac{\partial u}{\partial z} \tag{II.11}$$

Here, τ can be modelled by the use of the eddy viscosity concept

$$\tau = \rho \nu_T \frac{\partial u}{\partial z} \tag{II.12}$$

Alternatively, some models instead relate τ directly to k, like

$$\tau = 0.3\rho k \tag{II.13}$$

By application of Eq. II.13, τ does not necessarily need to be zero where $\partial u/\partial z$ is zero.

V represents the energy dissipation. The conversion of turbulent energy to heat is know to occur through a cascade process, where the energy in the large eddies is transferred to energy into eddies of smaller size. The dissipation occurs mainly in the very small eddies and is usually given by

$$\overline{\mu u_i \frac{\partial^2 u_i}{\partial x_j \partial x_j}} = -c_2 \rho \frac{k^{3/2}}{\ell_d} \tag{II.14}$$

in which c_2 is a dimensionless number and ℓ_d the length scale of turbulence.

If we apply the eddy viscosity concept for τ, Eq. II.12, the energy equation for boundary layers Eq. II.9 can now be written as

$$\rho \frac{dk}{dt} = \rho \frac{\partial}{\partial y} \left(\frac{\nu_T}{\sigma_k} \frac{\partial k}{\partial z} \right) + \rho \nu_T \left(\frac{\partial u}{\partial z} \right)^2 - c_2 \rho \frac{k^{3/2}}{\ell_d} \tag{II.15}$$

In Eq. II.15, k, ν_T, and u are unknown quantities. For this reason, two additional equations in the same quantities are required in order to calculate these unknown quantities. One equation is the flow equation, which for the boundary layer flow reads

$$\rho \frac{du}{dt} = -\frac{1}{\rho} \frac{\partial p}{\partial x} + \frac{\partial}{\partial x_j} \left(\nu_T \frac{\partial u}{\partial x_j} \right) \tag{II.16}$$

The other equation relates the eddy viscosity to the turbulent kinetic energy k. The eddy viscosity has the dimension $L^2 T^{-1}$ and can be formed as a product of a length scale and a velocity scale. A measure for the velocity scale is $\sqrt{k}$, which is a measure of the velocities in the eddies with most kinetic energy. A measure for the size of these eddies is the length scale ℓ_d, see Eq. 2.77. Hereby, the eddy viscosity can be written as

$$\nu_T = \ell_d \sqrt{k} \tag{II.17}$$

Eqs. II.16 and II.17 form together with Eq. II.15 a closed system from which u, k and ν_T can be found.

Example A.1:   k-equation in the equilibrium layer in uniform steady flow

In example 2.2, the logarithmic shape of the velocity profile was deduced in the equilibrium layer by some simple considerations about the turbulence.

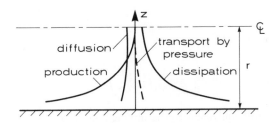

Figure II.1 Measurements of the energy balance in pipe flow (after Laufer, 1954)

Fig. II.1 represents some of Laufer's measurements in a pipe flow. It is seen that close to the wall the production and the dissipation of energy dominate other terms. In the equilibrium layer, which covers the lower 10-20% of the flow from the wall, the production and the dissipation are nearly equal to each other, so

$$\nu_T \left(\frac{\partial u}{\partial z}\right)^2 = c_2 \frac{k^{3/2}}{\ell_d} \tag{II.18}$$

see also Example 2.2, Eqs. 2.28 and 2.29.

By application of the usual eddy viscosity concept Eq. II.12, Eq. II.18 gives

$$\left(\frac{\tau}{\rho}\right)^2 = \nu_T c_2 \frac{k^{3/2}}{\ell_d} \tag{II.19}$$

or, by introduction of Eq. II.17 on the right-hand-side

$$\left(\frac{\tau}{\rho}\right)^2 = c_2 k^2 \tag{II.20}$$

This formula is identical with Eq. II.13, so in the equilibrium layer there is no conflict between the two different formulations Eqs. II.12 and II.13 of modelling the shear stress. τ can be related to ℓ_d and the velocity gradient in the following way:

From the eddy viscosity concept we have

$$\left(\frac{\tau}{\rho}\right)^2 = \nu_T^2 \left(\frac{\partial u}{\partial z}\right)^2 = \ell_d^2 k \left(\frac{\partial u}{\partial z}\right)^2 \tag{II.21}$$

cf. Eqs. II.12 and II.17. From Eq. II.20, k is given by

$$k = \frac{\tau}{\rho} \frac{1}{\sqrt{c_2}} \tag{II.22}$$

which inserted into Eq. II.21 gives

$$\frac{\tau}{\rho} = \frac{1}{\sqrt{c_2}} \ell_d^2 \left(\frac{\partial u}{\partial z}\right)^2 \tag{II.23}$$

This expression is total equivalent to the mixing length expression

$$\frac{\tau}{\rho} = \ell^2 \frac{\partial u}{\partial z} \left| \frac{\partial u}{\partial z} \right| \tag{II.24}$$

cf. Eq. 2.68 and Example 2.8. The mixing length theory can therefore be regarded as a turbulence model for the region in which we have local equilibrium. Furthermore, the mixing length ℓ and the length scale ℓ_d are closely connected, as seen from Eqs. II.23 and II.24

$$\ell_d = \sqrt[4]{c_2}\ell \sim 0.53\ell \tag{II.25}$$

where $c_2 = 0.08$.

REFERENCE

Laufer, J. (1954): The structure of turbulence in fully developed pipe flow. Rep. 1174, U.S. National Advisory Committee for Aeronautics, National Bureau of Standards, 18 p.

Appendix III. Additional references

Chapter 1.

Dalrymple, R.A. (1976): Wave-induced mass transport in water waves. J. Waterways, Harbors and Coastal Engineering, ASCE, 102(3):225-264.

Dean, R.G. (1970): Relative validities of water wave theories. J. Waterways, Harbors Coastal Engineering, Div., ASCE, 96(1):105-119.

Dean, R.G. (1974): Evaluation and development of water wave theories for engineering application. Coastal Engineering Research Center, Spec. Report 1.

Horikawa, K. (1978): Coastal Engineering. Univ. of Tokyo Press, Tokyo, 402 p.

Hsu, J.R.C., Silvester, R. and Tsuchiya, Y. (1980): Boundary-layer velocities and mass transport in short-crested waves. J. Fluid Mech., 99:321-342.

Ippen, A.T. (1966): Estuary and Coastline Hydrodynamics. McGraw-Hill Book Company, Inc., New York, St. Louis, Toronto, London, San Francisco, Sydney, 744 p.

Isaacson, M. (1976): The second approximation to mass transport in cnoidal waves. J. Fluid Mech., 78:445-457.

Isaacson, M. (1978): Mass transport in shallow water waves. J. Waterway, Port, Coastal and Ocean Eng., ASCE, 104(2):215-225.

Johns, B. (1970): On the mass transport induced by oscillatory flow in a turbulent boundary layer. J. Fluid Mech., 43:177-185.

Le Mehaute, B. (1976): An Introduction to Hydrodynamics and Water Waves. Springer Verlag, Düsseldorf.

Silvester, R. (1974): Coastal Engineering I. Elsevier Scientific Publishing Company, Amsterdam, London, New York, 457 p.

Soulsby, R.L. (1987): Calculating bottom orbital velocity beneath waves. Coastal Engineering, 11:371-380.

Chapter 2 and 3.

Asano, T. and Iwagaki, Y. (1984): Bottom turbulent boundary layer in wave-current co-existing systems. Proc. 19th Coastal Eng. Conf., ASCE, pp 1419-1438.

Brevik, I. and Aas, B. (1980): Flume experiment on waves and currents. 1. Rippled bed. Coastal Engineering, 3:149-177.

Brevik, I. (1980): Flume experiment on waves and currents. 2. Smooth bed. Coastal Engineering, 4:89-110.

Davies, A.G. (1990): A model of the vertical structure of the wave and current bottom boundary layer. Chapter 10 in Modeling Marine Systems, Ed. Alan M. Davies, CRC Press, Boca Raton, Florida, 2:263-297.

Jonsson, I.G. (1990): Wave-Current Interactions. The Sea: Ocean Engineering Science, Vol. 9, Part A, pp 65-120.

Kemp, P.H. and Simons, R.R. (1982): The interaction between waves and a turbulent current. Waves propagating with the current. J. Fluid. Mech., 116:227-250.

Myrhaug, D. (1986): Modeling rough turbulent wave boundary layers. In Encyclopedia of Fluid Mechanics. Edited by N.P. Cheremisinoff, Gulf Publishing, Houston, Texas, 6:1299-1335.

O'Connor, B.A. and Yoo, D.H. (1988): Mean bed friction of combined wave/current flow. Coastal Engineering, 12:1-21.

Peregrine, D.H. and Jonsson, I.G. (1983): Interaction of waves and currents. Misc. Rep. No. 83-6, U.S. Army, Corps of Eng., Fort Belvoir, Va.

Thomas, G.P. (1981): Wave-current interactions: an experimental and numerical study. Part 1. Linear waves. J. Fluid Mech., 110:457-474.

Chapter 4.

Basco, D.R. (1985): A qualitative description of wave breaking. J. Waterway, Port, Coastal and Ocean Eng., ASCE, 111:171-188.

Battjes, J.A. (1974): Surf similarity. Proc. 14th Coastal Eng. Conf., ASCE, pp 466-480.

Battjes, J.A. (1988): Surf-Zone Dynamics. Ann. Rev. Fluid Mech., 20:257-293.

Battjes, J.A. and Sakai, T. (1981): Velocity field in a steady breaker. J. Fluid Mech., 111:427-437.

Buhr Hansen, J. (1990): Periodic Waves in the Surf Zone: Analysis of Experimental Data. Coastal Engineering, 14:19-41.

Buhr Hansen, J. (1991): Air Entrainment in Surf Zone Waves. 3rd Int. Conf. on Coastal and Port Eng. in Developing Countries, Mombasa, 2:1357-1371.

Hattori, M. and Aono, T. (1985): Experimental study on turbulence structures under spilling breakers. In: The Ocean Surface, ed. Y. Toba, H. Mitsuyasu, Dordrecht: Reidel, pp 419-424.

Jansen, P.C.M. (1986): Laboratory observations of the kinematics in the aerated region of breaking waves. Coastal Engineering, 9:453-477.

Johns, B. (1979): The modelling of the approach of bores to a shoreline. Coastal Engineering, 3:207-219.

Meyer, R.E. (1972): Waves on Beaches and Resulting Sediment Transport. Academic Press, New York, London, p 462.

Miller, R.L. (1976): Role of vortices in surf zone prediction: sedimentation and wave forces. In: Beach and Nearshore Sedimentation, Soc. Econ. Paleontol. Mineral, Spec. Publ. 24, ed. R.A. Davis, R.L. Ethington, pp 92-114.

Nadaoka, K., Hino, M. and Koyano, Y. (1989): Structure of the turbulent flow field under breaking waves in the surf zone. J. Fluid Mech., 204:359-387.

O'Connor, B.A. and Yoo, D.H. (1987): Turbulence modelling of surf zone mixing processes. Proc. Spec. Conf. on Coastal Hydrodynamics, ASCE, Newark, Delaware, pp 371-383.

Oelerich, J. and Dette, H.H. (1988): About the Role of Trough Fill-up on Wave Energy Dissipation on Barred Beaches. IAHR Symp. on Mathematical Modelling of Sediment Transport in the Coastal Zone, Copenhagen, Denmark, pp 203-214.

Okayasu, A., Shibayama, T. and Mimura, N. (1987): Velocity field under plunging waves. Proc. 20th Coastal Eng. Conf., ASCE, pp 660-674.

Peregrine, D.H. and Svendsen, I.A. (1978): Spilling breakers, bores and hydraulic jumps. Proc. 16th Coastal Eng. Conf., ASCE, pp 540-550.

Svendsen, I.A. and Madsen, P.A. (1984): A turbulent bore on a beach. J. Fluid Mech., 148:73-96.

Tada, Y., Sakai, T. and Obana, E. (1990): Variation of Surf Zone Turbulence in a Wave Period. Proc. 22nd Coastal Eng. Conf., ASCE, pp 716-728.

Thornton, E.B. and Guza, R.T. (1983): Transformation of wave height distribution. J. Geophys. Res., 88:5925-5938.

Chapter 5.

Aagaard, T. (1991). Multiple-Bar Morphodynamics and Its Relation to Low-Frequency Edge Waves. Journal of Coastal Research, Fort Lauderdale, Florida, 7(3):801-813.

Bakker, W.T. (1971): The Influence of the Longshore Variation of the Wave Height on the Littoral Current. Study Report WWK 71-19: Ministry of Public Works (Rijkswaterstaat), The Hague, The Netherlands.

Basco, D.R. (1983): Surf zone currents. Coastal Engineering, 7:331-355.

Battjes, A.J., Sobey, R.J. and Stive, M.J.F. (1990): Nearshore Circulation. The Sea: Ocean Engineering Science, Vol. 9, Part A, pp 467-493.

Bowen, A.J. (1969): Rip currents, 1: Theoretical investigations. J. Geophys. Res., 74:5467-5478.

Bowen, A.J. and Inman, D.L. (1971): Edge waves and crescentic bars. J. Geophys. Res., 76:8662-8671.

de Vriend, H.J. and Stive, M.J.F. (1987): Quasi-3D modelling of nearshore currents. Coastal Engineering, 11:565-601.

Galvin, C. (1991): Longshore currents in two laboratory studies: Relevance to theory. J. Waterway, Port, Coastal and Ocean Eng., ASCE, 91: 44-59.

Guza, R.T. and Thornton, E.B. (1982): Swash oscillations on a natural beach. J. Geophys. Res., 87:483-491.

Hazen, D.G., Greenwood, B. and Bowen, A.J. (1990): Nearshore Current Patterns on Barred Beaches. Proc. 22nd Coastal Eng. Conf., ASCE, pp 2061-2072.

Hino, M. (1974): Theory on formation of rip-current and cuspidal coast. Proc. 14th Coastal Eng. Conf., ASCE, pp 901-919.

Inman, D.L. (1957): Wave-generated ripples in nearshore sands. U.S. Army Corps of Engineers, Beach Erosion Board Tech. Memo 100, 67 pp.

Sonu, C.J. (1972): Field observation of near-shore circulation and meandering currents. J. Geophys. Res., 77:3232-3247.

Thornton, E.B. and Guza, R.T. (1986): Surf zone longshore currents and random waves: field data and models. J. Phys. Oceanogr., 16:1165-1178.

Wind, H.G. and Vreugdenhill, C.B. (1986): Rip-current generation near structures. J. Fluid Mech., 171:459-476.

Yoo, D.H. and O'Conner, B.A. (1988): Numerical Modelling of Waves and Wave-induced Currents of Groyned Beach. IAHR Symp. on Mathematical Modelling of Sediment Transport in the Coastal Zone, Denmark, pp 127-136.

Chapter 6.

de Vriend, H. and Kitou, N. (1990): Incorporation of Wave Effects in a 3D Hydrostatic Mean Current Model. Proc. 22nd Coastal Eng. Conf., ASCE, pp 1005-1018.

Longuet-Higgins, M.S (1983): Wave Set-up, Percolation and Undertow in the Surfzone. Proc. of the Royal Society of London. A390, pp 283-291.

Nadoaka, K. and Kondoh, T. (1982): Laboratory Measurements of Velocity Field Structure in the Surf Zone by LDV. Coastal Eng. in Japan, 25:125-145.

Okayasu, A., Watanabe, A. and Isobe, M. (1990): Modeling of Energy Transfer and Undertow in the Surf Zone. Proc. 22nd Coastal Eng. Conf., ASCE, pp 123-135.

Osborne, P.D. and Greenwood, B. (1990): Set-Up Driven Undertows on a Barred Beach. Proc. 22nd Coastal Eng. Conf., ASCE, pp 227-240.

Sanchez-Arcilla, A., Collado, F., Lemos, M. and Rivero, F. (1990): Another Quasi-3D Model for Surf-Zone Flows. Proc. 22nd Coastal Eng. Conf., ASCE, pp 316-329.

Stive, M.J.F. (1988): Cross-shore Flow in Waves Breaking on a Beach. Ph.D. Thesis, Techn. Univ. Delft 1988, also published as Delft Hydraulics Communication No. 395, 1988, 39 p.

Stive, M.J.F. and de Vriend, H.J. (1987): Quasi 3-D nearshore current modelling: wave-induced secondary currents. Proc. of Special Conf. on Coastal Hydrodynamics, ASCE, New York, pp 356-370.

Svendsen, I.A. and Buhr Hansen, J. (1988): Cross-Shore Currents in Surf-Zone Modelling. Coastal Engineering, 12:23-42.

Chapter 7.

Bailard, J.A. and Inman, D.L. (1980): An energetics bedload model for a plane sloping beach, local transport. J. Geophys. Res., 86(C3):2035-2043.

Bakker, W.T. and Kesteren, W.G.M. (1986): The dynamics of oscillatory sheet flow. Proc. 20th Coastal Eng. Conf., ASCE, pp 940-954.

Lamberti, A., Montefusco, L. and Valiani, A. (1991): A granular-fluid model of the stress transfer from the fluid to the bed. In: Sand transport in rivers, estuaries and the sea. Eds.: Soulsby, R.L. and Bettess, R., Balkema, Rotterdam.

Wiberg, P.L. and Smith, J.D. (1985): A Theoretical Model for Saltating Grains in Water. J. of Geophys. Res., 90(C4):7341-7354.

Wilson, K.C. (1989): Friction of wave-induced sheet flow. Coastal Engineering, 12:371-379.

Chapter 8.

Asano, T. (1990): Two-plane flow model on oscillatory sheet-flow. Proc. 22nd Coastal Eng. Conf., pp 2372-2384.

Bailard, J.A. (1981): An energetics total load sediment transport model for a plane sloping beach. J. of Geophys. Res., 86(C11):10938-10954.

Beach, R.A. and Sternberg, R.W. (1988): Wave-Current Interactions in the Inner Surf Zone and Their Influence on Suspended Sediment Transport. IAHR Symp. on Mathematical Modelling of Sediment Transport in the Coastal Zone, Copenhagen, Denmark, 1988, pp 156-165.

Celik, I. and Rodi, W. (1991): Suspended sediment-transport capacity for open channel flow. J. Hyd. Eng., ASCE, 117(2):191-204.

Galappatti, G. and Vreugdenhil, C.B. (1985): A depth-integrated model for suspended sediment transport. J. Hydraul. Res., 23(4):359-377.

Hagatun, K. and Eidsvik, K.J. (1986): Oscillating turbulent boundary layers with suspended sediment. J. Geophys. Res., 91(C11):13045-13055.

Hallermeier, R.J. (1980): Sand Motion Initiation by Water Waves: Two Asymptotes. J. Waterway, Port, Coastal and Ocean Eng., ASCE, 106(3):299-318.

Hanes, D.M. (1989): The Structure of Intermittent Sand Suspension Events Under Mild Wave Conditions. Proc. IAHR Symp. on Sediment Transport Modeling, ASCE, New Orleans, Louisiana.

Horikawa, K. and Watanabe, A. (1970): Turbulence and sediment concentration due to wave. Coastal Eng. in Japan, 13:15-24.

Horikawa, K., Watanabe, A. and Katori, S. (1982): Sediment transport under sheet flow condition. Proc. 18th Coastal Eng. Conf., ASCE, 2:1335-1352.

Komar, P.D. and Miller, M.C. (1974): Sediment threshold under oscillatory waves. Proc. 14th Coastal Eng. Conf., ASCE, pp 756-775.

Losada, M.A., Desiré, J.M. and Merino, J. (1987): Coastal Engineering, 11:159-173.

Madsen, O.S. and Grant, W.D. (1976): Sediment transport in the coastal environment, M.I.T. Ralph M. Parsons Lab. Report 209.

Nielsen, P. (1986): Suspended sediment concentrations under waves. Coastal Engineering, 10:23-31.

O'Connor, B.A. and Nicholson, J. (1988): A three-dimensional model of suspended particulate sediment transport. Coastal Engineering, 12:157-174.

Soulsby, R.L. (1991): Aspects of sediment transport by combined waves and currents. IAHR Symp. on The Transport of Suspended Sediments and its Mathematical Modelling, Florence, Italy, pp 709-722.

Van Rijn, L.C. (1991): Sediment transport in combined waves and currents. In: Sand transport in rivers, estuaries and the sea. Eds.: Soulsby, R.L. and Bettess, R., Balkema, Rotterdam.

Vongvisessomjai, S. (1986): Profile of suspended sediment due to wave action. J. Waterway, Port, Coastal and Ocean Eng., ASCE, 112(1):35-53.

Chapter 9.

Allen, J.R.L. (1968): Current Ripples. North-Holland, Amsterdam.

Engelund, F. and Fredsøe, J. (1982): Sediment ripples and dunes. Annual Review of Fluid Mechanics, 14:13-37.

Engelund, F. (1974): The development of oblique dunes. Progress Report No. 31, Inst. of Hydrodynamics and Hydraulic Engineering, ISVA, Techn. Univ. Denmark, pp 25-30.

Engelund, F. (1974): The development of oblique dunes, part 2. Progress Report No. 32, Inst. of Hydrodynamics and Hydraulic Engineering, ISVA, Techn. Univ. Denmark, pp 37-40.

Fredsøe, J. (1974): The development of oblique dunes, part 3. Progress Report No. 33, Inst. of Hydrodynamics and Hydraulic Engineering, ISVA, Techn. Univ. Denmark, pp 15-22.

Fredsøe, J. (1974): The development of oblique dunes, part 4. Progress Report No. 34, Inst. of Hydrodynamics and Hydraulic Engineering, ISVA, Tecn. Univ. Denmark, pp 25-29.

Fredsøe, J. (1979): Unsteady flow in straight alluvial streams: modification of individual dunes. J. Fluid Mech., Vol. 91, part 3, pp 497-512.

Fredsøe, J. (1981): Unsteady flow in straight alluvial streams. Part 2. Transition from dunes to plane bed. J. Fluid Mech., 102:431-453.

Chapter 10.

Blondeaux, P. (1990): Sand ripples under sea waves. Part 1. Ripple formation. J. Fluid Mech., 218:1-18.

Blondeaux, P. and Vittori, G. (1990): Oscillatory flow and sediment motion over a rippled bed. Proc. 22nd Coastal Eng. Conf., ASCE, pp 2186-2199.

Dingler, J.R. and Inman, D.L. (1976): Wave-formed ripples in nearshore sands. Proc. 15th Conf. Coastal Eng., ASCE, pp 2109-2126.

Horikawa, K. and Ikeda, S. (1990): Characteristics of oscillating flow over ripple models. Proc. 22nd Coastal Eng. Conf., ASCE, pp 662-674.

Inman, D.L. and Bowen, A.J. (1962): Flume experiments on sand transport by waves and currents. Proc. 8th Coastal Eng. Conf., ASCE, pp 137-150.

Kennedy, J.F. and Falcon, M. (1965): Wave-generated sediment ripples. M.I.T. Hydrodyn. Lab., Rep. 86.

Lofquist, K.E.B. (1980): Measurements of oscillatory drag on sand ripples. Proc. 17th Coastal Eng. Conf., ASCE, pp 3087-3106.

Longuet-Higgins, M.S. (1981): Oscillating flow over steep sand ripples. J. Fluid Mech., 107:1-35.

Nakato, T., Locher, F.A., Glover, J. R. and Kennedy, J.F. (1977): Wave entrainment of sediment from rippled beds. J. Waterway, Port, Coastal and Ocean Eng., 103(1):83-99.

Nielsen, P. (1988): Three simple models of wave sediment transport. Coastal Engineering, 12:43-62.

Sleath, J.F.A. (1985): Energy dissipation in oscillatory flow over rippled beds. Coastal Engineering, 9:159-170.

Tunstall, E.B. and Inman, D.L. (1975): Vortex Generation by Oscillatory Flow Over Rippled Surfaces. J. of Geophys. Res., 80(24):3475-3484.

Vittori, G. and Blondeaux, P. (1990): Sand ripples under sea waves. Part 2. Finite-amplitude development. J. Fluid Mech., 218:19-40.

Vongvisessomjai, S. (1984): Oscillatory ripple geometry. J. Hyd. Eng., ASCE, 110(3):247-266.

Chapter 11.

Birkemeier, W.A. (1985): Field Data on Seaward Limit of Profile Change. J. Waterway, Port, Coastal and Ocean Eng., ASCE, 111:598-602.

Boczar-Karakiewicz, B., Bona, J.L. and Chapalain, G. (1989): Mathematical Modeling of Sand Bars in Coastal Environments. Proc. Int. Symp., Sediment Transport Modeling, ASCE, New Orleans, Louisiana.

Bruun, P. (1983): Review of conditions for uses of the Bruun rule of erosion. Coastal Engineering, 7:77-89.

Dalrymple, R.A. and Thompson, W.W. (1976): Study of equilibrium beach profiles. Proc. 15th Coastal Eng. Conf., ASCE, pp 1277-1296.

DeVries, J.W. and Bailard, J.A. (1988): A Simple Beach Profile Response Model. IAHR Symp. on Mathematical Modelling of Sediment Transport in the Coastal Zone, Copenhagen, Denmark, pp 215-224.

Hallermeier, R.J. (1981a): Seaward Limit of Significant Sand Transport by Waves: An Annual Zonation for Seasonal Profiles. Coastal Engineering Technical Aid, No. 81-2, U.S. Army Engineer Waterways Experiment Station, Vicksburg, MS.

Hallermeier, R.J. (1981b): A Profile Zonation for Seasonal Sand Beaches from Wave Climate. Coastal Engineering, 4:253-277.

Houston, S.H. and Dean, R.G. (1990): Method of Prediction of Bar Formation and Migration. Proc. 22nd Coastal Eng. Conf., ASCE, The Netherlands, pp 2145-2158.

Larson, M. and Kraus, N.C. (1989): SBEACH: Numerical Model for Simulating Storm-Induced Beach Change; Report 1, Empirical Foundation and Model Development. Technical Report CERC-89-9, U.S. Army Engineer Waterways Experiment Station, Vicksburg, MS.

Larson, M., Kraus, N.C. and Byrnes, M.R. (1990): SBEACH: Numerical Model for Simulating Storm-Induced Beach Change; Report 2, Numerical Formulation and Model Tests. Technical Report CERC-89-9, U.S. Army Engineer Waterways Experiment Station, Vicksburg, MS.

Nairn, R.B. (1990): Prediction of Cross-Shore Sediment Transport and Beach Profile Evolution. Ph.D. Thesis, Imperial College, University of London, 391 pp.

Nairn, R.B., Roelvink, J.A. and Southgate, H.N. (1990): Transition Zone Width and Implications for Modelling Surfzone Hydrodynamics. Proc. 22nd Coastal Eng. Conf., ASCE, pp 68:81.

Quick, M.C. (1991): Onshore-offshore sediment transport on beaches. Coastal Engineering, 15:313-332.

Roelvink, J.A. and Stive, M.J.F. (1988): Large Scale Tests of Cross-Shore Sediment Transport of the Upper Shoreface. IAHR Symp. on Mathematical Modelling of Sediment Transport in the Coastal Zone, Copenhagen, Denmark, pp 137-147.

Roelvink, J.A. and Stive, M.J.F. (1989): Bar-generating cross-shore flow mechanisms on a beach. J. Geophys. Res., 94:4785-4800.

Sallenger Jr., A.H. and Howd, P.A. (1989): Nearshore bars and the break-point hypothesis. Coastal Engineering, 12:301-313.

Steetzel, J.H. (1987): A model for beach and dune profile changes near dune revetments. Proc. Coastal Sediments' 87, ASCE, pp 87-97.

Swart, D.H. (1974): Offshore Sediment Transport and Equilibrium Beach Profiles. Doctorate Dissertation, Dept. of Civ. Engineering, Delft Univ. of Technology. Also: Publication No. 131: Delft Hydraulics Laboratory. And also: Report M918, part 2: Delft Hydraulics Laboratory, Delft, The Netherlands.

Uliczka, K. and Dette, H.H. (1988): Application of beach profile hindcasting on prototype and field data. Proc. 2nd Int. Symp. on Wave Research and Coastal Engineering, Hannover, pp 103-124.

Vellinga, P. (1986): Beach and Dune Erosion during Storm Surges. Ph.D. Thesis, Techn. Univ. Delft 1986, also published as Delft Hydraulics Communications No. 372, 1986, 169 p.

Watanabe, A. and Dibajnia, M. (1988): Numerical Modelling of Nearshore Waves, Cross-shore Sediment Transport and Beach Profile Change. IAHR Symp. on Mathematical Modelling of Sediment Transport in the Coastal Zone, Copenhagen, Denmark, pp 166-174.

Chapter 12.

Bakker, W.T. (1968): The Dynamics of A Coast with A Groyne System. Proc. 11th Coastal Eng. Conf., ASCE, pp 992-517.

Bijker, E.W. (1967): Some Considerations about Scales for Coastal Models with Movable Bed. Doctorate Dissertation, Delft Univ. of Technology. Also appeared as: Publication No. 50, Delft Hydraulics Laboratory, Delft, The Netherlands.

Bruun, P., Mehta, A.J. and Jonsson, I.G. (1978): Stability of Tidal Inlets. Elsevier Scientific Publishing Company, Amsterdam, Oxford, New York, 506 p.

Deigaard, R., Hedegaard, I.B., Holst Andersen, O. and Fredsøe, J. (1989): Engineering Models for Coastal Sediment Transport. Proc. Int. Symp., Sediment Transport Modeling, ASCE, New Orleans, Louisiana.

de Vriend, H.J. (1987): 2DH mathematical modelling of morphological evolutions in shallow water. Coastal Engineering, 11:1-27.

Hanson, H. (1985): Seawall Constraint in the Shoreline Numerical Model. J. Waterway, Port, Coastal and Ocean Engineering, ASCE, 111(6):1079-1083.

Hanson, H. (1989): GENESIS: a Generalized Shoreline Change Numerical Model. J. Coastal Res., 5(1):1-27.

Hanson, H. and Kraus, N.C. (1990): Shoreline Response to a Single Transmissive Detached Breakwater. Proc. 22nd Coastal Eng. Conf., ASCE, pp 2034-2046.

Horikawa, K. (1981): Coastal sediment processes. Ann. Rev. Fluid Mech., 13:9-32.

Johnson, H.K. and Kamphuis, J.W. (1988): N-line Morphology Model for a Large Initially Conical Sand Island. IAHR Symp. on Mathematical Modelling of Sediment Transport in the Coastal Zone, Copenhagen, pp 275-289.

Kamphuis, J.W. (1990): Littoral Transport Rate. Proc. 22nd Coastal Eng. Conf., ASCE, pp 2402-2415.

Kamphuis, J.W. (1991): Alongshore Sediment Transport Rate. J. Waterway, Port, Coastal and Ocean Eng., ASCE, 117(6):624-641.

Komar, P.D. (1976): Beach Processes and Sedimentation. Prentice-Hall, Inc., Englewood Cliffs, New Jersey, 429 p.

Kraus, N.C. (1983): Applications of a Shoreline Prediction Model. Proceedings Coastal Zone '89, ASCE, pp 632-645.

Kraus, N.C., Hanson, H. and Larson, M. (1988): Threshold for Longshore Sand Transport and Application to a Shoreline Change Simulation Model. IAHR Symp. on Mathematical Modelling of Sediment Transport in the Coastal Zone, Copenhagen, Denmark, pp 117-126.

Kraus, N.C., Gingerich, K.J. and Rosati, J.D. (1989): DUCK85 Surf Zone Sand Transport Experiment. Technical Report CERC-89-5, U.S. Army Engineer Waterways Experiment Station, Vicksburg, MS., 48 pp.

Kraus, N.C. and Soichi Harikai (1983): Numerical model of the shoreline change at Oarai Beach. Coastal Engineering, 7:1-28.

Lippmann, T.C. and Holman, R.A. (1990): The Spatial and Temporal Variability of Sand Bar Morphology. J. of Geophys. Res., 95(C7):11,575-11,590.

McDougal, W.G. and Hudspeth, R.T. (1989): Longshore current and sediment transport on composite beach profiles. Coastal Engineering, 12:315-338.

O'Connor, B.A. and Nicholson, J. (1989): Modeling Changes in Coastal Morphology. Proc. Int. Symp., Sediment Transport Modeling, ASCE, New Orleans, Louisiana.

Silvester, R. (1974): Coastal Eng. II. Elsevier Scientific Publishing Company, Amsterdam, London, New York, 338 p.

Van de Graaff, J. and Van Overeem, J. (1979): Evaluation of sediment transport formulae in Coastal Engineering practice. Coastal Engineering, 3:1-32.

Watanabe, A. (1982): Numerical models of nearshore currents and beach deformation. Coastal Engineering in Japan, 25:147-161.

Zenkovich, V.P. (1969): Processes of Coastal Development. Oliver and Boyd Ltd., Edinburgh and London, 738 p.

Subject Index